# UNITEXT for Physics

More information about this series at http://www.springer.com/series/13351

Claudio Chiuderi · Marco Velli

# Basics of Plasma Astrophysics

Springer

Claudio Chiuderi
Department of Physics and Astronomy
University of Firenze
Florence
Italy

Marco Velli
Earth Planetary and Space Sciences
University of California
Los Angeles, CA
USA

ISSN 2198-7882
ISSN 2198-7890 (electronic)
ISBN 978-88-470-5870-5
ISBN 978-88-470-5280-2 (eBook)
DOI 10.1007/978-88-470-5280-2

Springer Milan Heidelberg New York Dordrecht London


Softcover reprint of the hardcover 1st edition 2015


Printed on acid-free paper

Springer is part of Springer Science+Business Media (www.springer.com)

# Preface

Plasma physics is a discipline that has evolved in an almost explosive way in the past 80 years. Born as a rather marginal subject, the study of the physics of ionized gases, it had a first important application to the theory of the propagation of electromagnetic waves in the Earth's ionosphere. Its importance for astrophysics had an early recognition thanks to the pioneering work of Hannes Alfvén, that prepared the ground for many of the successive developments. A second fundamental field of application of plasma physics opened up when the controlled thermonuclear fusion became one of the leading project in "big science." The interactions between the scientific communities engaged in the main two fields of application, laboratory plasma physics and plasma astrophysics, very limited at beginning, have constantly increased in time with a reciprocal benefit.

Plasmas are the main constituents of the Universe. Therefore, the interpretation of the large amount of information provided by modern observations, both from the ground and from space, requires a working knowledge of plasma physics. The acquisition of such a working knowledge, however, is a slow process since the subject is far from our everyday experience and has the disturbing tendency of becoming complex, even in apparently simple situations. We are convinced that the study of plasma physics must be started relatively early in the university career, as soon as the student has mastered the basic concepts of fluid dynamics and electromagnetic theory.

This book is an outgrowth of the lecture notes of the course taught by the authors for a number of years at the University of Florence and an expanded version of a book previously published in Italian. Given the large number of excellent textbook in plasma physics already available, the obvious question is: why another one? Our purpose in writing this book has been to address a particular audience, namely students with a reasonable good background in mathematics and physics at the undergraduate level who intend to orient their future activity in astrophysics. Our ambition has been to bring the student from the very basic concepts of plasma physics up to some of the more active research fields. We have attempted to maintain a certain balance between mathematical rigor and physical intuition. The possibility of following explicitly the technical details of calculations allows the student to appreciate the power of mathematics and to acquire the tools necessary to attack new problems on his own. Mathematics can be boring, but is reassuring. We have also tried, whenever possible, to provide physical explanations of the phenomena under discussion, in order to develop the physical insight that allows to understand in a deep and intuitive way the inner workings of nature.

The book is divided in two unequal parts. The first seven chapters cover the fundamental aspects of plasma physics, presented in a traditional way. Although the main emphasis of the book is on fluid models and, in particular, on magnetohydrodynamics, we have attempted to show the logical path that brings us to those models. Therefore, we have included a short outline of kinetic theory, also to prepare the ground for more advanced topics such as the Landau damping and the collisionless shocks. The advantages and the limits of each model are clearly spelled out. The long chapter on instabilities contains, apart from a discussion of the more common cases, such as the Rayleigh–Taylor, Kruskal–Shafranov, and Kelvin–Helmholtz instabilities, the discussion of certain types of instabilities not normally present in introductory textbooks.

The last four chapters are more demanding for the reader as they cover the most advanced material: the properties of collisional and collisionless shocks, magnetic reconnection, the main features of turbulence and the build-up and maintenance of cosmical magnetic fields. All these areas are the subject of an intense research effort worldwide and the students have some trouble finding appropriate description of these subjects at an introductory level. We have tried to provide such an introduction, hoping to bring the reader to the point where he/she will be able to approach and understand the more specialized literature. A Section on Problems and Questions is placed at the end of each Chapter. It contains a small number of problems that should help the student to verify the degree of understanding acquired. Solutions are given for all problems.

When writing this book, we have greatly benefited from numerous discussions and suggestions from our friends and colleagues. We wish to thank them collectively here.

Florence, December 2013
Pasadena

Claudio Chiuderi
Marco Velli

# Contents

# Chapter 1
# An Introduction to Plasma Physics

**Abstract** A brief introduction to Plasma Physics is presented, focusing on ionized gases, the plasma constituting the greatest part of the visible (baryonic) Universe. Basic plasma features such as ionization degree and the condition of quasi-neutrality are discussed. A heuristic derivation of fundamental plasma parameters, such as plasma frequency and Debye length are given. Finally, the different possible approaches to the physical description of plasma dynamics are introduced.

Plasma physics studies the equilibrium and dynamics of globally neutral collections of charged particles, where the interactions between particles and the self-consistent long-range electromagnetic fields dominate over the Coulomb force between nearest neighbors. Plasmas may be thought of as the **fourth state** of matter, in the sense that they display a distinguishing set of characteristics clearly separating them from the traditionally considered physical states of solid, liquid and gaseous matter. Just as the solid to liquid and liquid to gas phase-transitions generally occur in materials as heat is added, so a transition to the plasma state requires additional energy: once the temperature becomes sufficient to ionize atoms, a gas made up of positively charged ions and electrons is formed. Such charged particles may move about freely, generating electric currents and charge densities which modify the electric and magnetic fields present in the system. However, though plasmas are dubbed the fourth state of matter, the formation of the plasma state does not occur as a classical phase-transition, rather the properties of the gas change gradually with increasing ionization. An ionized gas is just an example of a plasma, and is the most frequent **natural plasma** present in the universe. Examples range from the Heliosphere, which is the plasma cavity created by the supersonic expansion of the solar corona into the interplanetary medium, to the magnetospheres of planets, to the magnetospheres of neutron stars, to the jets in Active Galactic Nuclei (AGN), to the ionized gases of the interstellar and intergalactic mediums. The heliospheric plasma, together with the planetary magnetospheres embedded within, is the only natural plasma for which accurate measurements of both macroscopic properties, such as velocity, temperature and densities, as well as microscopic ones, such as the shape of the distribution functions, have been carried out in situ. Approximately 95 % of matter in the Universe

 
C. Chiuderi and M. Velli, *Basics of Plasma Astrophysics*,
UNITEXT for Physics, DOI 10.1007/978-88-470-5280-2_1

is found in the plasma state. From this point of view, our environment on Earth is a fortunate exception.

Even though one often thinks of a plasma as just a collection of charged particles, it is really the particles and self-consistent electromagnetic fields together that define the plasma state. A quantum mechanical point of view, where particles, fields, as well as their excitations may be thought of as (quasi) particles, helps in understanding this concept. In the case of plasmas these include real particles (electrons, ions, atoms), the electromagnetic fields (photons) and the plasma waves (plasmons, phonons). The interactions between the various plasma components, whether they be wave-particle or wave-wave interactions or particle-particle collisions, may then all be described in terms of generalized particle interactions. This framework will be helpful in clarifying important plasma concepts such as collisionless damping, discussed in a subsequent chapter, even though the description adopted in this text will be classical rather than quantum mechanical.

According to our first rough definition of a plasma, even the conducting electrons in a metal constitute a plasma, since the lattice ions form a fixed background and the dynamics of the electron "gas" is governed by electromagnetic interactions. We will not discuss such a plasma specifically in this text, which is devoted to the more common case of ionized gases as produced in the laboratory and in nature.

What degree of ionization is required for a gas to be considered a plasma? Answering this question requires calculating the conductivity of a partially ionized gas, made up of electrons, positive ions and neutral atoms. Given the large mass ratio of ions and neutrals to electrons, the motions of the former may be neglected in the equation of motion for the electrons (of charge $-e$ and mass $m_e$),[1] which in the presence of a constant electric field $\boldsymbol{E}$ is:

$$m_e \frac{d\boldsymbol{v}}{dt} = -e\boldsymbol{E} - m_e \nu_c \boldsymbol{v}$$

or

$$\frac{d\boldsymbol{v}}{dt} + \nu_c \boldsymbol{v} = -\frac{e}{m_e}\boldsymbol{E}$$

with $\boldsymbol{v}$ the electron velocity. The term proportional to the *collision frequency* $\nu_c$ is the drag force coming from collisions between electrons and both ions and neutrals. A stationary state is achieved when

$$\boldsymbol{v} = -\frac{e}{m_e \nu_c}\boldsymbol{E}.$$

[1] Throughout the text $e$ will denote the proton charge, while a generic charge will be written as $e_0$.

Electrons carry a current density $\boldsymbol{J} = -en_e\boldsymbol{v}$, and, defining the electric conductivity $\sigma$ through $\boldsymbol{J} = \sigma\boldsymbol{E}$, we find:

$$\sigma = \frac{e^2 n_e}{m_e \nu_c}, \tag{1.1}$$

where $n_e$ is the electron number density. One may rather generally define the collision frequency as

$$\nu_c = \frac{\bar{v}}{\lambda} = n\sigma_c\bar{v} = \alpha n; \qquad \alpha = \sigma_c\bar{v},$$

where $\bar{v}$ is the root mean square (rms) electron speed (typically the electron thermal speed), $\lambda$ the mean free path, $\sigma_c$ the collisional cross section for the interaction considered (not to be confused with the conductivity, $\sigma$) and $n$ the number density of scattering centers. In our case the collision frequency $\nu_c$ is the sum of a collision frequency due to electron-ion interactions $\nu_c^i$ and one due to electron-neutral interactions $\nu_c^a$, $\nu_c = \nu_c^i + \nu_c^a$, with corresponding collision cross sections $\sigma_c(i)$, $\sigma_c(a)$ and $\alpha^i$, $\alpha^a$, so that

$$\nu_c = \alpha^i n_i + \alpha^a n_a = \alpha^i n_i(1 + \frac{\alpha^a}{\alpha^i}\frac{n_a}{n_i}).$$

In terms of the degree of ionization, $\chi$,

$$\chi = \frac{n_i}{n_i + n_a} = \frac{1}{1 + (n_a/n_i)},$$

$\nu_c$ becomes

$$\nu_c = \alpha^i n_i(1 + \frac{\alpha^a}{\alpha^i}\frac{1-\chi}{\chi}),$$

and recalling that $n_e = Zn_i$, where $Ze$ is the ion charge,

$$\sigma = \frac{Ze^2}{m_e\alpha^i}\frac{1}{1 + (\alpha^a/\alpha^i)(1/\chi - 1)} = \frac{\sigma_{\max}}{1 + (\alpha^a/\alpha^i)(1/\chi - 1)}.$$

Here $\sigma_{\max}$ is the maximum conductivity at complete ionization of the plasma, $\chi = 1$. An estimate of how the ionization degree affects conductivity can be obtained by asking for what ionization $\chi_0$ the conductivity has half its maximum value $\sigma = \sigma_{\max}/2$:

$$\frac{\alpha^a}{\alpha^i}(1/\chi_0 - 1) = 1.$$

Since

$$\frac{\alpha^a}{\alpha^i} = \frac{\sigma_c(a)}{\sigma_c(i)},$$

and the cross section for ion-neutral interactions is much smaller than that for ion-electron scattering (this value is found experimentally to be about $10^{-2}$) we find that an ionization degree of only about 1 % is sufficient for the plasma to have a conductivity half of that of the completely ionized gas. With an ionization degree of 8 % the conductivity reaches a level of about $0.9\,\sigma_{\max}$.

## 1.1 Saha's Equation

The degree of ionization $\chi$ depends on the physical parameters, namely density and temperature, which characterize the thermodynamical equilibrium of the plasma. Consider an ensemble of particles of energy $E_m$ at a temperature $T$. Their number density follows from Boltzmann's formula

$$n_m = g_m \exp(-E_m/kT),$$

$g_m$ giving the number of states with energy $E_m$ and $k$ is Boltzmann's constant. The ratio $\mathcal{R}_{lm}$ between the density of states with energies $E_l$ and $E_m$ is therefore given by :

$$\mathcal{R}_{lm} = \frac{n_l}{n_m} = \frac{g_l}{g_m} \exp[-(E_l - E_m)/kT].$$

Using $i$ to denote the ionized state and 0 the fundamental state of a neutral atom, it follows that

$$\frac{n_i}{n_0} = \frac{g_i}{g_0} \exp[-I/kT],$$

where $I$ is the ionization energy. A good approximation for the number of states $g$ is found from quantum mechanics to be given by:

$$\frac{g_i}{g_0} \simeq \left(\frac{m_e kT}{2\pi\hbar^2}\right)^{3/2} \frac{1}{n_i} \simeq 2.4 \times 10^{15} \frac{T^{3/2}}{n_i},$$

and *Saha's equation* follows

$$n_i/n_0 \simeq 2.4 \times 10^{15} \frac{T^{3/2}}{n_i} \exp(-I/kT). \tag{1.2}$$

For hydrogen the ionization energy is $13.6\,eV$, and the ratio of ions to neutrals is given by

$$n_i/n_0 \simeq 2.4 \times 10^{15} \frac{T^{3/2}}{n_i} \exp(-1.58 \times 10^3/T).$$

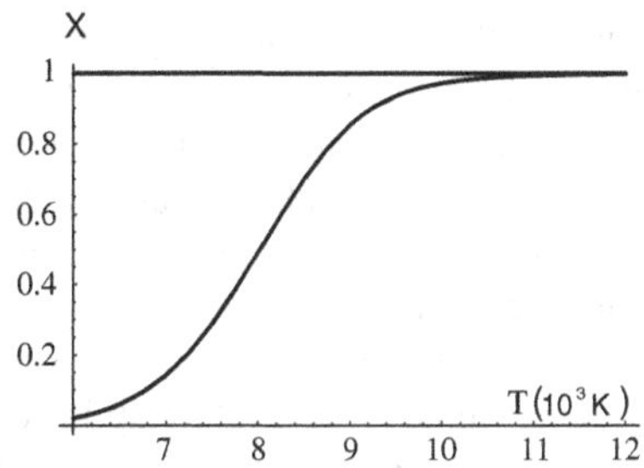


**Fig. 1.1** $\chi$ as a function of T for $n_{tot} = 10^{13}\,\text{cm}^{-3}$

Introducing the degree of ionization:

$$\chi = \frac{n_i}{n_0 + n_i} = \frac{n_i}{n_{tot}}$$

one finds from Saha's equation that

$$\frac{1-\chi}{\chi^2} \simeq 4.14 \times 10^{-16}\, n_{tot} T^{-3/2} \exp(1.58 \times 10^3/T).$$

**In the preceding formulae and throughout the text the numerical values for constants refer to the *Gaussian, or mixed cgs*, system of units**, where all electromagnetic quantities are calculated using the *electromagnetic cgs* system, except the electric charge for which the *electrostatic cgs* system is used. As a result, the speed of light $c$ and its powers often appears in formulae, but this inconvenience is outweighed by the presence of only three fundamental units, length (cm), mass (g) and time (s), simplifying dimensional analysis considerably. Temperatures are given in degrees Kelvin (K) except where explicitely stated.

A plot of the degree of ionization for hydrogen as a function of temperature at a density of $n_{tot} = 10^{13}\,\text{cm}^{-3}$ is shown in Fig. 1.1. Remarkably, the ionization of a hydrogen gas is essentially complete ($\chi = 1$) at temperatures of order $10^4$ K, much less than the temperature corresponding to the ionization potential $T \simeq I/k \simeq 1.58 \times 10^5$ K.

## 1.2 The Debye Length

A minimal degree of ionization is not the only property required for an ionized gas to be considered a plasma: another important characteristic is the property of *quasi-neutrality*.

A charged particle in a plasma interacts with all other particles directly via the Coulomb force, an interaction with a long distance range, as the electrostatic potential falls off only slowly with distance—as $1/r$. However, in a plasma charged particles move freely, so one expects positively charged particles to attract negative charges and to repel positive ones. As a result a positively charged particle in a plasma will be surrounded by a volume with prevalently negative charge. This leads to a shielding of the charge and to an electrostatic potential that nearly vanishes beyond some distance $r_s$ from the charge. We therefore expect the total charge over volumes with a linear scale greater than $r_s$ to vanish. In other words if $Q(r)$ is the total charge within a sphere of radius $r$, quasi-neutrality implies that $Q(r_s) \simeq 0$.

To estimate $r_s$, let's start by introducing an extra charge $e_0 > 0$ into a purely hydrogen plasma in some point that we take as the origin, $\boldsymbol{r} = 0$. An electric field will appear which by Gauss' theorem may be written as:

$$\nabla \cdot \boldsymbol{E} = -\nabla^2 \Phi = 4\pi q = 4\pi e(n_i - n_e) + 4\pi e_0 \delta(\boldsymbol{r}),$$

where $q$ is the charge density, $e$, as stated previously, is the proton charge and $\Phi$ is the electrostatic potential defined by

$$\boldsymbol{E} = -\nabla \Phi$$

while $\delta(\boldsymbol{r})$ denotes the Dirac $\delta$-function.

If the plasma is in equilibrium at a temperature $T$, the electron density will be given by the Boltzmann distribution:

$$n_e = n_0 \exp[-(-e\Phi)/kT], \tag{1.3}$$

where $n_0$ is the average density in the absence of the extra charge. Similarly, the protons will be distributed according to

$$n_i = n_0 \exp[-(e\Phi/kT)]. \tag{1.4}$$

The equation for the electrostatic potential $\Phi$ therefore becomes:

$$\begin{aligned} \nabla^2 \Phi &= 4\pi n_0 e \left[\exp(e\Phi/kT) - \exp(-e\Phi/kT)\right] - 4\pi e_0 \delta(\boldsymbol{r}) \\ &= 8\pi n_0 e \sinh(e\Phi/kT) - 4\pi e_0 \delta(\boldsymbol{r}), \end{aligned} \tag{1.5}$$

which, assuming $e\Phi/kT \ll 1$ (an approximation we will justify later), can be written to first order as

$$\nabla^2 \Phi = 8\pi n_0 e \left(e\Phi/kT\right) - 4\pi e_0 \delta(\boldsymbol{r}) \tag{1.6}$$

Expressing the laplacian operator explicitly, assuming the space charge to be distributed isotropically, so that $\Phi$ is only a function of the distance from the origin $r$, leads to

$$\frac{1}{r^2}\frac{\partial}{\partial r}(r^2\frac{\partial \Phi}{\partial r}) = 8\pi n_0 e(e\Phi/kT) - 4\pi e_0\delta(\boldsymbol{r}) \\ = \frac{\Phi}{\lambda^2} - 4\pi e_0\delta(\boldsymbol{r}), \tag{1.7}$$

where we have defined

$$\lambda = \sqrt{\frac{kT}{8\pi e^2 n_0}}.$$

To integrate, observe that the preceding equation may be rewritten as:

$$\frac{d^2}{dr^2}(r\Phi) = \frac{r\Phi}{\lambda^2} - 4\pi e_0 r\delta(\boldsymbol{r})$$

and, since the $\delta$-function term on the right hand side (rhs) now vanishes identically,

$$\Phi = e_0\frac{\exp(-r/\lambda)}{r}, \tag{1.8}$$

where we have imposed that the potential become the usual Coulomb potential in a vacuum, i.e. ($n_0 \to 0, \lambda \to \infty$). The *Debye length*, $\lambda_D$, is defined as

$$\lambda_D = \sqrt{\frac{kT}{4\pi e^2 n_0}} = \sqrt{2}\lambda \simeq 6.9\,T^{1/2}n_0^{-1/2}, \tag{1.9}$$

(notice the extra $\sqrt{2}$ factor with respect $\lambda$).

Figure 1.2 shows the standard vacuum Coulomb potential (dashed line), together with the shielded potential, Eq. (1.8) (continuous line), for a case where $\lambda_D = 0.07$ in (arbitrary) units of length.

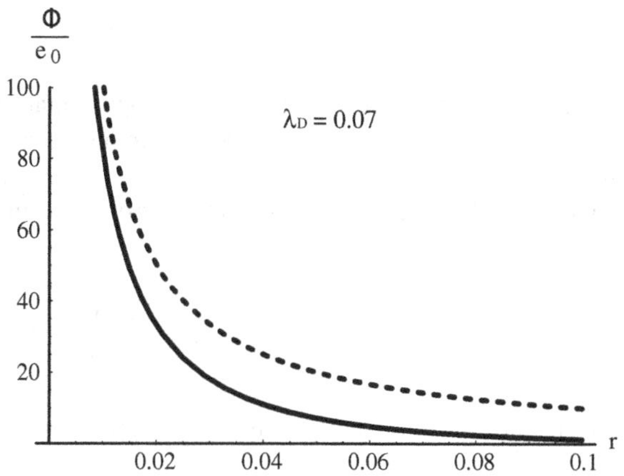


**Fig. 1.2** The Coulomb (*dashed*) and shielded potentials (*continuous*) as a function of r, shown for $\lambda_D = 0.07$ in units of r

Fig. 1.3 The total charge $Q(r)$ as a function of $r$

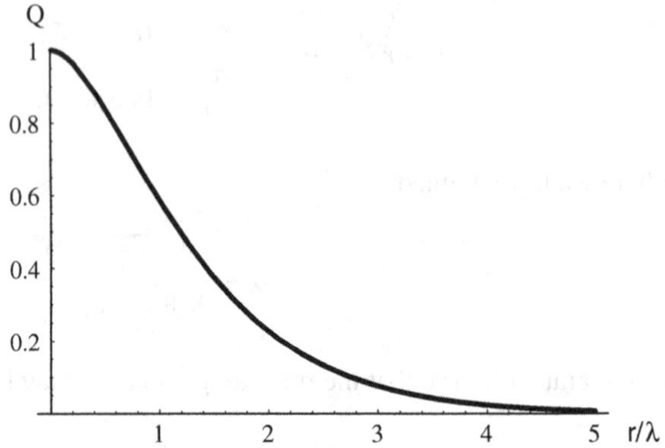


The screening effect becomes evident calculating the total charge inside a sphere of radius $r$,

$$\begin{aligned}
Q(r) &= \int_V q dV = \int_0^r 4\pi q r^2 \mathrm{d}r = \int_0^r (-\nabla^2 \Phi) r^2 \mathrm{d}r \\
&= -\int_0^r \left(\frac{\Phi}{\lambda^2}\right) r^2 \mathrm{d}r + e_0 \int_V \delta(\boldsymbol{r}) \mathrm{d}V \\
&= -\frac{e_0}{\lambda^2} \int_0^r r \exp(-r/\lambda) \mathrm{d}r + e_0 \\
&= e_0 (1 + r/\lambda) \exp(-r/\lambda).
\end{aligned} \tag{1.10}$$

We find that $Q(r)$ also falls of exponentially in units of $\lambda$ and, as shown in Fig. 1.3, it nearly vanishes beyond a distance of a few $\lambda_s$ from the bare charge.

The use of the Boltzmann distribution to derive the densities of particles over regions with a linear dimension of order of the Debye length is only meaningful if the mean inter-particle distance $\bar{d}$ is much smaller than that, $\bar{d} << \lambda_D$, which also implies, because $\bar{d} \simeq n^{-1/3}$, that the number of particles inside a volume of order $\lambda_D^3$ must be very large:

$$n\lambda_D^3 \gg 1. \tag{1.11}$$

This inequality must be satisfied for the approximation of quasi-neutrality to hold true, and it is therefore a necessary requirement for an ionized gas to be called a plasma.

Going back to the ratio $e\Phi/kT$, let's verify its magnitude in $r = \bar{d}$, still assuming $\bar{d} \ll \lambda$:

$$(e\Phi/kT)_{\bar{d}} = \frac{e^2}{kT} \frac{\exp(-\bar{d}/\lambda)}{\bar{d}} \simeq \frac{e^2}{kT} \frac{1}{\bar{d}} \simeq \frac{e^2}{kT} n^{1/3} \simeq \left(\frac{\bar{d}}{\lambda}\right)^2 \ll 1.$$

This justifies the consistency of using the approximation $e\Phi/kT \ll 1$ for a plasma.

## 1.3 Fundamental Plasma Parameters

The Debye length introduced in the previous paragraph is only one of a number of fundamental parameters describing the physical regime (to be better defined later) in which a plasma can be found. The rms or thermal speed of particles

$$v_T = \sqrt{3kT/m},$$

together with the Debye length, may be used to set a characteristic timescale $\tau_p$, by dividing the latter with the former: $\lambda_D/v_T$. The *plasma frequency* is proportional to $1/\tau_p$, and defined as:

$$\omega_p = \sqrt{\frac{4\pi e_0^2 n}{m}}, \tag{1.12}$$

where $e_0, m, n$ are the charge, mass and density of the species in question, so that even for a purely hydrogen plasma there will be a plasma frequency associated with electrons, $\omega_{pe}$, and one with the protons $\omega_{pi}$. For electrons, in cgs-Gaussian units as usual, we find

$$\omega_{pe} \simeq 5.64 \times 10^4 n_e^{1/2}.$$

The plasma frequency, obtained here on dimensional grounds, has an important physical meaning. Consider a plasma in which, in equilibrium, the densities of electrons and ions are the same $n_e = n_i = n_0 = const.$ If we perturb the electron density slightly, $n_e = n_0 + n_e'$ with $|n_e'| << n_0$, and assume the ions to be much heavier than the electrons, so that they do not move on the timescale over which the electrons react, the electron equation of motion may be written as (for simplicity, assume a 1D perturbation of the density only along the x-axis, so that all motion occurs along that axis):

$$m_e \frac{\partial v}{\partial t} = -eE \tag{1.13}$$

with the electric field being determined by Gauss' law:

$$\nabla \cdot \boldsymbol{E} = 4\pi q \rightarrow \frac{\partial E}{\partial x} = -4\pi e n_e'$$

Keeping in mind that all fluctuations are small so that only first order terms are retained in $n_e'$ and $v$, the continuity equation for electron density may be written as:

$$\frac{\partial n_e'}{\partial t} + n_0 \frac{\partial v}{\partial x} = 0.$$

Taking the derivative of the equation of motion with respect to both time and space, and using then Gauss' law and the continuity equation to eliminate the electric field and $v$ leads to :

$$\frac{\partial^2 n'_e}{\partial t^2} = -\frac{4\pi e^2 n_0}{m_e} n'_e = -\omega_{pe}^2 n'_e.$$

The same equation holds also for $v$, showing that both the electron speed and density execute harmonic oscillations with frequency $\omega_{pe}$. This *ordered* motion (as opposed to *random* thermal motions) is caused by the electric field that arises as a result of local violations to quasi-neutrality, and the plasma frequency therefore poses an upper limit to the frequencies of dynamic motions in the plasma satisfying quasi-neutrality.

Another important characteristic frequency in a plasma is the *collision frequency*. To estimate the collision frequency heuristically, consider the properties of different particle species interacting through the Coulomb (binary) force. We begin by defining binary collisions between particles as interactions occurring when the distance between particles decreases below a certain characteristic scale, $b$, given for example by the distance at which the kinetic energy of relative motion equals the electrostatic Coulomb potential energy. This condition is not in contrast with the previous one requiring that the electrostatic energy be much smaller than the thermal energy when the distance between particles is equal to the *mean interparticle* distance $\bar{d}$, provided $b << \bar{d}$. If the colliding particles have charges $Z_1 e$ and $Z_2 e$, $b$ is determined by:

$$\frac{Z_1 Z_2 e^2}{b} \simeq \frac{3}{2} kT,$$

The collision cross-section will therefore be given by:

$$\sigma_c = \pi b^2 = \frac{4\pi Z_1^2 Z_2^2 e^4}{(3kT)^2},$$

and the collision frequency:

$$\nu_c = n\sigma_c v_T = \frac{4\pi Z_1^2 Z_2^2 e^4 n}{m^{1/2}(3kT)^{3/2}}, \tag{1.14}$$

where now $m$ is the mass of the *impacting particles* while $n$ is the density of *target particles.*

Though the formulae above are only rough approximations, derived heuristically, they retain the correct dependence on the parameters of the colliding particles. A more rigorous treatment requires taking into account the fact that in a plasma, collisions can not be thought of as simply binary interactions; rather, inside the Debye sphere, each particle interacts with every other. This introduces a numerical constant multiplying the expressions above, as well as a factor known as the *Coulomb logarithm*, $\ln \Lambda$.

The final correct expression for electron-electron collisions is given by:

$$\begin{aligned}\nu_{ee} &= 1.43\,\frac{4\pi e^4 n_e}{m_e^{1/2}(3kT_e)^{3/2}}\ \ln\Lambda_{ee} \\ &\simeq 3.75 n_e T_e^{-3/2}\ \ln\Lambda_{ee}.\end{aligned} \tag{1.15}$$

The analagous expression for two identical ions of charge $Ze$ is:

$$\begin{aligned}\nu_{ii} &= 1.43\,\frac{4\pi Z^4 e^4 n_i}{m_i^{1/2}(kT_i)^{3/2}}\ \ln\Lambda_{ii} \\ &\simeq 8.76\times 10^{-3}\, n_i T_i^{-3/2}\ \ln\Lambda_{ii},\end{aligned} \tag{1.16}$$

where the final numerical expression refers to the proton-proton case.

The Coulomb logarithm is given by

$$\ln\Lambda = \ln(4\pi n\lambda_D^3) \simeq 8.33 - \frac{1}{2}\ln(n) + \frac{3}{2}\ln(T), \tag{1.17}$$

displaying only a weak logarithmic dependence on density, temperature (and particle charge). The numerical value given above corresponds to a plasma made up of protons and electrons.

For the general case of a plasma with ions of charge $Z$ quasi-neutrality implies $n_e = Zn_i$, leading to the following relation between electron-electron and ion-ion collision frequencies:

$$\frac{\nu_{ee}}{\nu_{ii}} \simeq \left(\frac{m_i}{m_e}\right)^{1/2}\left(\frac{T_i}{T_e}\right)^{3/2}\frac{1}{Z^3},$$

where however we have approximated $\Lambda_{ee} \simeq \Lambda_{ii}$.

For the case of ion-electron collisions the result depends on which of the particles is considered the target and which the impinging particle. In the case of electrons colliding on ion targets we have

$$\begin{aligned}\nu_{ei} &= 2\frac{4\pi Z^2 e^4 n_i}{m_e^{1/2}(3kT_e)^{3/2}}\ \ln\Lambda_{ei} \\ &\simeq 5.26\, n_i T_e^{-3/2}\ \ln\Lambda_{ei},\end{aligned} \tag{1.18}$$

while for ions colliding on target electrons

$$\begin{aligned}\nu_{ie} &= 2\left(\frac{24}{9\pi}\right)^{1/2}\left(\frac{m_e}{m_i}\right)^{1/2}\left(\frac{T_i}{T_e}\right)^{1/2}\frac{4\pi Z^2 e^4 n_e}{m_i^{1/2}(3kT_i)^{3/2}}\ \ln\Lambda_{ie} \\ &\simeq 2.63\times 10^{-3}\left(\frac{T_i}{T_e}\right)^{1/2} n_e T_i^{-3/2}.\end{aligned} \tag{1.19}$$

Again, the numerical values for both expressions are given for electron-proton collisions.[2]

Collisions are also the way through which a plasma thermalizes, i.e. through which a plasma containing particle populations with different temperatures reaches thermal equilibrium. In a collision, energy may be transferred from the particle of higher energy to the lower energy one. Consider then the case of populations of electrons and ions both out of thermodynamic equilibrium but with comparable energies. It may be shown that collisions lead to equilibrium among particles of the same species on a different timescale compared to that required for thermal equilibrium across species. For collision between identical particles, the characteristic timescale $\tau$ required to reach equilibrium is given by

$$\tau_{ee} \simeq (\nu_{ee})^{-1} \simeq \left(\frac{m_e}{m_i}\right)^{1/2} \tau_{ii}.$$

Thermal equilibrium across species implies collisions between electrons and ions: in one collision an electron can only lose a fraction of order $(m_e/m_i)$ of its energy. Reaching thermal equilibrium therefore requires a time

$$\tau_{ei} \simeq \left(\frac{m_i}{m_e}\right)^{1/2} \tau_{ii} \simeq \left(\frac{m_i}{m_e}\right) \tau_{ee}.$$

For an electron-proton plasma $\tau_{ei} \gg \tau_{ii} \gg \tau_{ee}$ and electron-proton thermalization requires substantially longer times than both electron-electron and proton-proton thermalization. This means that a plasma may survive for long times with electrons and protons at different temperatures. The timescale given by the inverse of the collision frequency defines a typical time between collisions, and is the characteristic timescale for the collisional phenomenon under consideration. In the previous paragraphs we have focused on the timescales of *energy exchange*. A different timescale is appropriate for *momentum exchange*. A detailed discussion may be found, for example, in Spitzer (1962) [4].

Collision frequencies are associated with *disordered* motions in a plasma, while as we have seen, motions at the plasma frequency due to departures from quasi-neutrality are *ordered* motions. For this ordered motion to occur, it is necessary that $\nu_{ei} \ll \omega_{pe}$. From Eq. (1.14) and the definition of $\omega_{pe}$ this condition is easily verified. Indeed

$$\frac{\nu_{ei}}{\omega_{pe}} \ll \frac{\nu_{ee}}{\omega_{pe}} \simeq (n\lambda_D^3)^{-1} \ll 1,$$

from Eq. (1.11).

[2] Numerical values given in different texts differ somewhat depending on the way in which the collisions are calculated, though they are all of the same order of magnitude. The values here are from Spitzer, "Physics of Fully Ionized Gases" [4].

Almost all natural plasmas are magnetized, in the sense that the charges move in a domain where a non-negligible magnetic field $B$ is present. Associated with the magnetic field $B$ are two further characteristic parameters, namely the *cyclotron* or *Larmor frequency*

$$\omega_c = \frac{|Ze|B}{mc}, \tag{1.20}$$

and the *Alfvén speed*:

$$c_a = \sqrt{\frac{B}{4\pi\rho}} = \sqrt{\frac{B}{4\pi m_i n_i}} \simeq 2.18 \times 10^{11} B\, n_i^{-1/2}. \tag{1.21}$$

The Alfvén speed depends only on the mass density of the plasma, while the cyclotron frequency depends explicitely on particle mass and charge and may be written for electrons and protons as

$$\omega_{ce} = \frac{|e|B}{m_e c} \simeq 1.76 \times 10^7 B,$$

and

$$\omega_{cp} = \frac{m_e}{m_p}\omega_{ce} \simeq 9.58 \times 10^3 B.$$

The plasmas studied in this textbook are classical plasmas, in the sense that the plasma dynamics is described by non-relativistic classical (non-quantum) mechanics. When does this approach lose its validity? Relativistic corrections become important when the thermal energy is such that $kT \gtrsim mc^2$. For electrons this implies the very extreme condition

$$T \gtrsim 6 \times 10^9 K,$$

implying that relativistic corrections are very rarely required for plasmas in *thermal equilibrium*. Often however there may be particle populations in a plasma with energies much greater than $kT$, responsible for so called *non thermal* processes, and for these species relativistic corrections may be important.

Quantum Effects are important when the characteristic length scales in the plasma approach the *De Broglie wavelength*, $\lambda_q = \hbar/p$, where $p$ is the particle momentum and $\hbar$ is the reduced Planck constant $\hbar = h/2\pi$. For thermal motions this implies

$$\lambda_q = \frac{\hbar}{(3mkT)^{1/2}},$$

and quantum-mechanical effects become important when:

$$\bar{d} \simeq n^{-1/3} \lesssim \lambda_q$$

or

$$Tn^{-2/3} \lesssim \frac{\hbar^2}{3mk} \simeq 2.95 \times 10^{-12}.$$

where the numerical value refers to electrons. Quantum effects therefore become important for low temperatures and/or extremely high densities.

## 1.4 The Classical Description of Plasmas

The classical, microscopic, description of a plasma considers the plasma as an ensemble of charged particles in a *vacuum*. Maxwell's equations are therefore written in the form

$$\begin{aligned} \nabla \times \boldsymbol{E} &= -\frac{1}{c}\frac{\partial \boldsymbol{B}}{\partial t} \\ \nabla \times \boldsymbol{B} &= \frac{1}{c}\frac{\partial \boldsymbol{E}}{\partial t} + \frac{4\pi}{c}\boldsymbol{J} \\ \nabla \cdot \boldsymbol{E} &= 4\pi q \\ \nabla \cdot \boldsymbol{B} &= 0 \end{aligned} \tag{1.22}$$

where $q$ and $\boldsymbol{J}$ are the *total* charge and current densities, including possible *external* charges and currents. Taking the divergence of the second equation in (1.22) and the time-derivative of Gauss' theorem, third equation in (1.22), we obtain the *equation of charge conservation or charge continuity*:

$$\nabla \cdot \boldsymbol{J} + \frac{\partial q}{\partial t} = 0.$$

Gauss' theorem for both the electric and magnetic fields (third and fourth equation in 1.22) in reality imposes only initial conditions on the dynamics:

$$0 = \nabla \cdot (\nabla \times \boldsymbol{E}) = -\frac{1}{c}\frac{\partial}{\partial t}\nabla \cdot \boldsymbol{B}$$

so that if $\nabla \cdot \boldsymbol{B} = 0$ is satisfied at any one time it will be satisfied for all subsequent times. Analogously, taking the time derivative of the equation for $\nabla \cdot \boldsymbol{E}$ and using conservation of charge we obtain:

$$\frac{\partial}{\partial t}(\nabla \cdot \boldsymbol{E} - 4\pi\, q) = c\,\nabla \cdot (\nabla \times \boldsymbol{B}) = 0.$$

This means that Gauss' theorems for the electric and magnetic fields should not be included when considering the number of independent equations describing plasma dynamics, though of course they are often used in applications to specific problems.

In addition, every particle $i = 1 \dots N$ (including ions and electrons) follows Newton's equations of motion:

$$m_i \ddot{\boldsymbol{r}}_i = e_i (\boldsymbol{E} + \frac{1}{c} \dot{\boldsymbol{r}}_i \times \boldsymbol{B}), \tag{1.23}$$

where $e_i$ is the charge of the $i$th particle, $\boldsymbol{r}_i(t)$ and $\dot{\boldsymbol{r}}_i(t)$ are the $i$th particle's position vector and velocity—also written $\boldsymbol{v}_i(t)$, respectively. At least in principle, complete knowledge of the microscopic plasma state for an ensemble of $N$ particles therefore requires solving the $3N$ equations of motion together with the 6 independent Maxwell equations (1.22), for a total of $3N+6$ equations in the $3N+10$ unknowns $\boldsymbol{E}$, $\boldsymbol{B}$, $\boldsymbol{j}$, $q$. The missing 4 equations are those determining charge and current densities from the motions of individual particles:

$$\begin{aligned} q(\boldsymbol{r}, t) &= \sum_{i=1}^{N} e_i \, \delta[\boldsymbol{r} - \boldsymbol{r}_i(t)] \\ \boldsymbol{J}(\boldsymbol{r}, t) &= \sum_{i=1}^{N} e_i \, \boldsymbol{v}_i(t) \, \delta[\boldsymbol{r} - \boldsymbol{r}_i(t)]. \end{aligned} \tag{1.24}$$

Apart from the complexity of solving such a huge nonlinear system of equations (clearly impossible in practice), these equations provide at least in principle a scheme for recovering *ab initio* the microscopic state of a plasma. However, even if the solution of such a system were attainable, the information content of the microscopic state would be far too great with respect to any possible experimental or observational verification. A step forward would be to move from this description to a statistical one, the best compromise between accuracy and manageability. This transition corresponds to a loss of detailed but often not essential information, such as, for example, the trajectories of individual particles. Even a statistical approach is in many cases too detailed, and one is content with a fluid-type closure, where in addition to trajectories all information on the detailed distribution of the velocities of particles in phase space is also lost in favor of a finite number of velocity moments, typically up to third order. The correct choice of the level of description depends on the nature of the problem and the type of information required to understand the dynamics. Each simplification has a cost in the sense that some "resolving power" is lost, and much of the "art" in solving plasma physics problems resides in the proper choice for understanding the question at hand.

There is at least one case where the complete description from Eqs. (1.22)–(1.24) is used, provided the electromagnetic fields in (1.22) are determined by *external* charges and currents. In this case the $3N$ particle motion equations become decoupled, and we are discussing single particle orbits in given electromagnetic fields. This is called "Orbit Theory" and belongs only marginally to plasma physics if, as we have said, by plasmas we mean a system of *collectively* interacting particles and fields. An understanding of particle orbit theory is however a pre-requisite for developing a kinetic plasma theory and gaining a deeper understanding of macroscopic plasma properties.

## Questions and Problems

**1.1.** Prove that in a head-on collision of an electron with an ion the electron loses an amount of energy of order $m_e/m_i$ in favor of the ion.
**1.2.** Prove that in a plasma permeated by a sufficiently strong electric field a regime of continuous or *runaway* acceleration of electrons develops and determine the critical minimum value for such a field, known as the Dreicer field.

## Solutions

**1.1.** Assume the ion to be originally at rest and call $v_e$ and $v_e'$ the electron speeds before and after the collision, $v_i'$ the ion speed after the collision. Write the equations of conservation for momentum and energy:

$$m_e v_e = m_e v_e' + m_i v_i',$$

$$\tfrac{1}{2} m_e v_e^2 = \tfrac{1}{2} m_e v_e'^2 + \tfrac{1}{2} m_i v_i'^2.$$

Solving for $v_e'$, $v_i'$ gives

$$v_e' = \frac{m_e - m_i}{m_e + m_i} v_e \quad ; \quad v_i' = \frac{2 m_e}{m_e + m_i} v_e,$$

and therefore

$$\frac{m_i v_i'^2}{m_e v_e^2} = 4 \frac{m_e/m_i}{1 + (m_e/m_i)^2} \simeq 4 \frac{m_e}{m_i}.$$

**1.2.** A *runaway* regime of continuous acceleration can develop when the electric force on an electron overcomes the "viscous" drag force due to collisions, in magnitude $e E > m_e \nu_{ei} v$. Using expression (1.17) for $\nu_{ei}$ and starting from a typical speed $v$ equal to an electron thermal speed, the result is:

$$E > E_D = \frac{8\pi Z^2 e^3 n_i}{3kT_e} \simeq 2 \times 10^{-7} \frac{n_i}{T_e} \; Volt/m. \; (Z = 1)$$

# Chapter 2
# Particle Orbit Theory

**Abstract** In which particle motion in assigned electric and magnetic fields is discussed and studied in detail in a few representative cases. Fundamental concepts such as guiding center motion, particle drifts, magnetic moment, and adiabatic invariants are introduced, and a brief discussion of the mirror force and magnetic bottles is presented.

This chapter is devoted to a description of the motion of individual charged particles in assigned magnetic and electric fields, which we assume to be given as a function of time and position. Such orbits can become quite complex even in this case, so gaining a general understanding of the main types of motions that may occur is an important pre-requisite to studying the dynamics of plasmas in their self-consistent electromagnetic fields. Let's begin with the simplest case, the motion of a charged particle in a constant, uniform, magnetic field.

## 2.1 Motion in a Uniform, Static Magnetic Field

Consider a magnetic field which is constant in space and time, so that with respect to a Cartesian system of coordinates the field reads $\boldsymbol{B} = (0, 0, B)$. The classical equation of motion for a particle of mass $m$ and charge $e_0$ in such a field may be written as

$$m\frac{\mathrm{d}\boldsymbol{v}}{\mathrm{d}t} = m\ddot{\boldsymbol{r}} = \frac{e_0}{c}\boldsymbol{v} \times \boldsymbol{B}. \tag{2.1}$$

Because the force on the particle is perpendicular to its velocity,

$$\ddot{z} = 0$$

and motion along the field occurs at constant speed, $\dot{z} = v_{\parallel}$ is constant. Another immediate consequence is found taking the scalar product of the equation of motion with the velocity $\dot{\boldsymbol{r}}$:

$$m\ddot{\boldsymbol{r}} \cdot \dot{\boldsymbol{r}} = 0,$$



C. Chiuderi and M. Velli, *Basics of Plasma Astrophysics*,
UNITEXT for Physics, DOI 10.1007/978-88-470-5280-2_2

implying that the total kinetic energy energy

$$\tfrac{1}{2}\, m\dot{r}^2 = W = W_{\parallel} + W_{\perp}$$

is constant. However, since $W_{\parallel} = \frac{1}{2}\, m v_{\parallel}^2$ is separately conserved, we find that the kinetic energy of motion perpendicular to the field $W_{\perp}$ is also constant.

The trajectory of the particle may be found by integrating the equation of motion, Eq. (2.1), which rewritten in its three cartesian components reads:

$$\frac{dv_x}{dt} = \Omega\, v_y \tag{2.2a}$$

$$\frac{dv_y}{dt} = -\Omega\, v_x \tag{2.2b}$$

$$\frac{dv_z}{dt} = 0 \tag{2.2c}$$

where

$$\Omega = \frac{e_0 B}{m\, c}.$$

The solutions to Eq. (2.2a)–(2.2c) are given by:

$$v_x = v_{\perp} \cos(\Omega t + \alpha) \tag{2.3a}$$

$$v_y = -v_{\perp} \sin(\Omega t + \alpha) \tag{2.3b}$$

$$v_z = v_{\parallel}. \tag{2.3c}$$

Integrating Eq. (2.3a)–(2.3c) once more we obtain the trajectory as a function of time:

$$x = x_0 + \left(\frac{v_{\perp}}{\Omega}\right) \sin(\Omega t + \alpha) \tag{2.4a}$$

$$y = y_0 + \left(\frac{v_{\perp}}{\Omega}\right) \cos(\Omega t + \alpha) \tag{2.4b}$$

$$z = z_0 + v_{\parallel} t. \tag{2.4c}$$

The trajectory is a helix which positively charged particles sweep in the clockwise sense and negative particles in a counter-clockwise sense as seen projected on the $(x, y)$ plane. The projected orbits are circular with radius $R_L = v_{\perp}/|\Omega|$, known as the *Larmor radius*. The gyration frequency $|\Omega|$ is known as the *cyclotron* frequency, also called the *gyrofrequency*.

The frequencies $\omega_{ce}$ and $\omega_{ci}$ defined in Chap. 1 are the values of $|\Omega|$ calculated for an electron and proton respectively. Let's proceed now to motion in combined electric and magnetic fields, starting from the case in which the fields are perpendicular (Figs. 2.1 and 2.2).

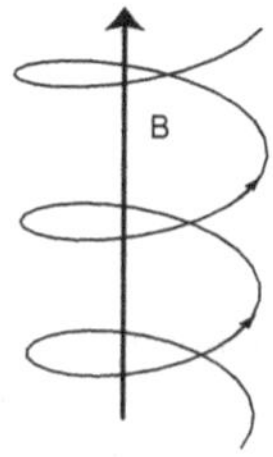


**Fig. 2.1** Trajectory of a negatively charged particle in a uniform magnetic field

## 2.2 Motion in Orthogonal Electric and Magnetic Fields

Starting from the same configuration as above, we add a constant, homogeneous electric field $\boldsymbol{E}$ orthogonal to $\boldsymbol{B}$:

$$\begin{aligned}\boldsymbol{E} &= (0, E, 0),\\ \boldsymbol{B} &= (0, 0, B),\end{aligned}$$

with $E$ and $B$ independent of position and time.

To solve Eq. (2.1) now we start from the $y$ component of the equation of motion, Eq. (2.3c) which becomes:

$$\frac{\mathrm{d}v_y}{\mathrm{d}t} = -\Omega\, v_x + \frac{e_0\, E}{m} = -\Omega(v_x - c\frac{E}{B}). \tag{2.5}$$

It is natural to consider therefore a frame of reference $S'$, moving at the speed $c\,E/B$ along the $x$ direction, which is possible as long as $E \ll B$: in this case the Lorentz transformation reduces to the Galilean transformation

$$\begin{aligned}x' &= x - c\,(E/B)t,\\ y' &= y,\\ z' &= z,\\ t' &= t,\end{aligned}$$

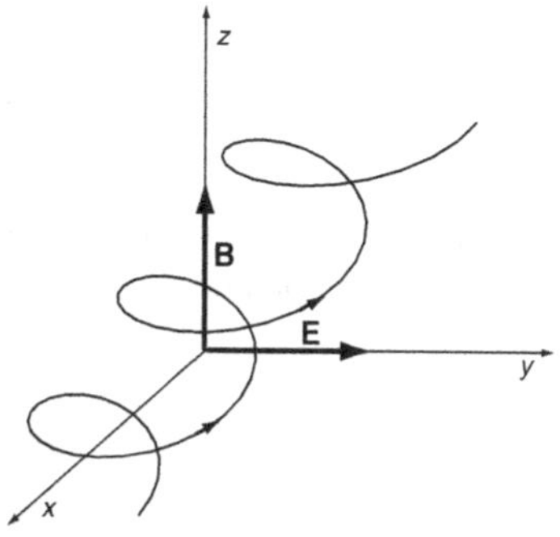


**Fig. 2.2** Motion in orthogonal and constant electric and magnetic fields

and in the frame $S'$ the equation of motion along $y'$ becomes

$$\frac{dv'_y}{dt} = -\Omega\, v'_x,$$

identical to Eq. (2.2c). The trajectory written in the new, primed coordinates therefore coincides with that given by Eq. (2.4a) where $y$ is replaced by $y'$. Returning to the original frame of reference we can see that the trajectory is just a superposition of the original helical trajectory and a uniform translation at speed $cE/B$ along the $x$-axis as shown in (Fig. 2.2). To interpret this result consider that the effect of the electric field in the $y$-direction is to accelerate (or decelerate, depending on the charge) a particle along that axis. In the presence of a magnetic field, the increase (decrease) in speed along that direction leads to an increasing (decreasing) centripetal force deflecting the motion orthogonally to both $\boldsymbol{E}$ and $\boldsymbol{B}$. This force gradually changes the direction of motion eventually leading to a change of sign of $v_y$, so that motion in the $y$-direction is confined to a finite displacement. On the other hand the different radius of curvature of the gyrating motion as the particle moves along and against the direction of the electric field leads to the average uniform displacement orthogonal to the fields. The velocity associated with this motion is known as the drift velocity $\boldsymbol{v}_E$ and may be written in a coordinate independent way as

$$\boldsymbol{v}_E = c\,\frac{\boldsymbol{E}\times\boldsymbol{B}}{B^2}. \tag{2.6}$$

When $E/B = O(1)$ relativistic corrections become important, and a different approach is required if $E > B$, since $v_x$ in this case would become:

$$v_x = c\frac{E}{B} + v_{0x}\cos(\Omega t),$$

which exceeds the speed of light. With $E > B$ one must therefore use the relativistic equations of motion,

$$\frac{d\boldsymbol{p}}{dt} = e_0\left(\boldsymbol{E} + \frac{\boldsymbol{v}}{c}\times\boldsymbol{B}\right)$$
$$\boldsymbol{p} = \gamma m\boldsymbol{v}$$
$$\gamma = \frac{1}{\sqrt{1 - v^2/c^2}}.$$

Consider the effects of a Lorentz transformation along the positive $x$-axis with speed $v$ on electric and magnetic fields:

$$E'_x = E_x \qquad B'_x = B_x \tag{2.7a}$$
$$E'_y = \gamma(E_y - \beta B_z) \qquad B'_y = \gamma(B_y + \beta E_z) \tag{2.7b}$$

$$E'_z = \gamma(E_z + \beta B_y) \qquad B'_z = \gamma(B_z - \beta E_y), \tag{2.7c}$$

where $\beta = v/c$.

For the configuration considered we have $E'_x = 0$, $E'_z = 0$, $E'_y = \gamma(E - \beta B)$. Therefore, for $E < B$, the choice $\beta = \frac{E}{B}$ makes the $y$ component of the electric field in the $S'$ frame vanish, and only a magnetic field along $z' = z$, $B'_z = \gamma(B - \beta E) = \gamma(B - E^2/B) = B/\gamma$ will remain, yielding the helical motion seen before. In the original frame $S$ the motion will therefore be a superposition of the helical motion and the translational one. More generally, which kind of Lorentz transformation may be used to simplify the equations of motion may be understood by recalling the two Lorentz invariants for the electromagnetic field, $\boldsymbol{E} \cdot \boldsymbol{B}$ and $E^2 - B^2$. If the first of these vanishes, then it will always be possible to find a frame in which either $E$ or $B$ vanishes, depending on the sign of the second invariant $E^2 - B^2$. So if the electric and magnetic fields are perpendicular, but $E > B$, a new frame can be found with a vanishing magnetic field, corresponding to a Lorentz boost along the $x$ direction with $\beta = B/E$. Once the equation of motion is solved in the new frame, one can transform the solution back to the original frame applying the inverse Lorentz transformation with $\beta = -B/E$ along the $x'$ axis (see Problem 2.4).

A very important property of the drift speed $\boldsymbol{v}_E$ is that it does not depend on the sign of the charge of the particle in motion. Both electrons and protons move with the same speed under this drift, and therefore no current is generated by the so called "E cross B" drift within the plasma. The above discussion may be generalized to the case of any constant force $\boldsymbol{F}$ orthogonal to the magnetic field $\boldsymbol{B}$: it will give rise to a drift which is orthogonal both to the force and the magnetic field, whose magnitude is obtained by substituting $\boldsymbol{E}$ with $\boldsymbol{F}/e_0$ in Eq. (2.6):

$$\boldsymbol{v}_F = c\,\frac{\boldsymbol{F} \times \boldsymbol{B}}{e_0 B^2} \tag{2.8}$$

As a consequence, the drift from a non-electric force will give rise to an average current in the plasma.

So far we have only considered cases in which $E \neq B$. However, the case $E = B$ has particular importance since it is related to the motion of a particle in the field of an electromagnetic wave. The method of a Lorentz boost, used to treat the cases with $E \neq B$, is no longer applicable and we must solve directly the equation of motion of special relativity in the laboratory frame. The more familiar non relativistic case can be easily obtained by taking the appropriate limit.

We shall consider a particle of mass $m$ and charge $e_0$, initially at rest at the origin of the coordinates, acted upon by a linearly polarized electromagnetic wave of frequency $\omega$. We may then choose a frame of reference in which the electric field lies on the $x$-axis and the magnetic field on the $y$-axis, so that the waves propagates along the $z$-direction. Thus,

$$\boldsymbol{E} = (E_0 \cos\phi)\, \boldsymbol{e}_x, \quad \boldsymbol{B} = (E_0 \cos\phi)\, \boldsymbol{e}_y,$$

with $\phi = \omega(t - z/c)$.

The equations of relativistic dynamics are

$$\frac{\mathrm{d}\,(m\gamma \boldsymbol{v})}{\mathrm{d}t} = e_0(\boldsymbol{E} + \boldsymbol{\beta} \times \boldsymbol{B}), \tag{2.9}$$

$$\frac{\mathrm{d}\,(m\gamma c)}{\mathrm{d}t} = e_0 \boldsymbol{E} \cdot \boldsymbol{\beta} \tag{2.10}$$

where $\boldsymbol{\beta} = (\boldsymbol{v}/c)$. Introducing the quantity $\omega_E = (e_0 E_0/mc)$, Eqs. (2.9) and (2.10) give

$$\frac{\mathrm{d}}{\mathrm{d}t}(\gamma\beta_x) = \omega_E(1 - \beta_z)\cos\phi, \tag{2.11}$$

$$\frac{\mathrm{d}}{\mathrm{d}t}(\gamma\beta_y 0) = 0, \tag{2.12}$$

$$\frac{\mathrm{d}}{\mathrm{d}t}(\gamma\beta_z) = \omega_E\, \beta_x \cos\phi, \tag{2.13}$$

$$\frac{\mathrm{d}\gamma}{\mathrm{d}t} = \omega_E\, \beta_x \cos\phi. \tag{2.14}$$

Equating Eqs. (2.13) and (2.14) we get

$$\frac{\mathrm{d}}{\mathrm{d}t}[\gamma(1 - \beta_z)] = 0$$

and therefore $\gamma(1 - \beta_z) = K$ is a constant of motion.

To determine the trajectory of the particle, it is convenient to use the variable $\phi$ in place of $t$. Then

$$\boldsymbol{\beta} = \frac{1}{c}\frac{\mathrm{d}\boldsymbol{r}}{\mathrm{d}t} = \frac{1}{c}\frac{\mathrm{d}\boldsymbol{r}}{\mathrm{d}\phi}\frac{\mathrm{d}\phi}{\mathrm{d}t} = \frac{\omega}{c}(1 - \beta_z)\frac{\mathrm{d}\boldsymbol{r}}{\mathrm{d}\phi} = \frac{\omega}{c}(1 - \beta_z)\boldsymbol{r}',$$

where the prime indicates the derivative with respect to $\phi$. Therefore

$$\frac{\mathrm{d}}{\mathrm{d}t}(\gamma\boldsymbol{\beta}) = \frac{\mathrm{d}}{\mathrm{d}t}[\gamma\frac{\omega}{c}(1 - \beta_z)\boldsymbol{r}'] = \frac{\omega^2 K}{c}(1 - \beta_z)\boldsymbol{r}''.$$

Comparing the last expression with Eq. (2.9) we finally get the following set of differential equations:

$$x'' = \frac{c}{\omega K}\frac{\omega_E}{\omega}\cos\phi$$

$$y'' = 0$$

$$z'' = \frac{1}{K}\frac{\omega_E}{\omega}x'\cos\phi.$$

Since $\boldsymbol{r}(0) = 0$ and $\dot{\boldsymbol{r}}(0) = 0$ (the dot indicates the time-derivative), we have

$$\phi(0) = 0; \quad \dot{\phi}(0) = \omega[1 - \beta_z(0)] = \omega; \quad K = K(0) = 1,$$

and the initial conditions for the system above are:

$$x(0) = 0, \quad x'(0) = \dot{x}(0)/\dot{\phi}(0) = 0 \ ; \quad y(0) = y'(0) = 0 \ ; \quad z(0) = z'(0) = 0.$$

The second equation shows that $y = 0$: no motion occurs in the $y$-direction. A straightforward integration of the other two equations gives

$$x = -\left(\frac{\omega_E}{\omega}\right)\left(\frac{c}{\omega}\right)(1 - \cos\phi), \tag{2.15}$$

$$z = \left(\frac{\omega_E}{2\omega}\right)^2\left(\frac{c}{\omega}\right)(\phi - \tfrac{1}{2}\sin 2\phi) = \xi(\phi - \tfrac{1}{2}\sin 2\phi),$$

with

$$\xi = \left(\frac{\omega_E}{2\omega}\right)^2\left(\frac{c}{\omega}\right).$$

Inserting the definition of $\phi$ into the the expression for $z$ we have

$$z = \omega\xi((t - z/c) - (\xi/2)\sin 2\phi,$$

or

$$z = \left(\frac{\omega\xi}{1 + \omega\xi/c}\right)t - \frac{\xi}{2(1 + \omega\xi/c)}\sin 2\phi. \tag{2.16}$$

The velocity along $z$ is therefore a superposition of a constant speed plus a periodic term. The constant term corresponds to a value of $\beta_z$

$$\beta_z = \frac{\omega\xi/c}{1 + \omega\xi/c} = \frac{(\omega_E/2\omega)^2}{1 + (\omega_E/2\omega)^2},$$

which tends to unity when $(\omega_E/\omega)$ is very large, namely when $E_0$ is very large or $\omega$ is very small. The first circumstance arises in the field of an intense laser beam or in linear acclerators. The second one has been considered in connection with the problem of the acceleration of cosmic rays in the vicinity of a pulsar. In fact, the pulsar emits a low frequency electromagnetic wave, which in principle could accelerate particles to very large energies. This model, however, presents a number of drawbacks and has been abandoned.

The periodic term in Eq. (2.16), coupled with the motion along $x$, Eq. (2.15), produces an eight-shaped trajectory in the $(x, z)$ plane of a frame moving along $z$ with a speed $c\beta_z$.

## 2.3 Motion in Slowly Variable Magnetic Fields

Because electric and magnetic fields in space are never uniform or constant in time, there are in general no exact integrals of motion to which one can resort to simplify the understanding of charged particle dynamics. However, when particle motion occurs in slowly variable magnetic fields, either in time or in space, the equation of particle motion, Eq. (2.1), may be solved approximately. In fact, the systematic expansion in terms of drift velocities of higher order is allowed if the the ratio of the Larmor radius $R_L$ to the gradient scale $L$ $R_L/L \ll 1$, or alternatively, if the characteristic frequency of time variation of the field $1/T$ is much smaller than the gyrofrequency $|\Omega|$, i.e. $|\Omega| T \gg 1$. When these conditions are satisfied, exact integrals of motion are replaced by approximate integrals or adiabatic invariants, quantities whose relative changes are bounded by inequalities more stringent than the ones above.

### 2.3.1 *Charged Particle Orbits in the Presence of a Magnetic Fields with a Weak Gradient*

Equation (2.8) may be used to understand particle motion by use of a local analysis in a magnetic field with gradients. Consider for example a magnetic field line having a certain curvature. The particle will locally gyrate around the magnetic field, which in view of the inequalities expressed above may be considered constant on the scale of the Larmor radius. However, the particle generally also has a motion parallel to the field $\boldsymbol{B}$, and, because of the curvature of the field line, the particle will feel in its own frame of reference an outward centrifugal force due to this motion $\boldsymbol{F} = (m\, v_{\parallel}^2\, \boldsymbol{R}_c)/(R_c^2)$, where $R_c$ is the local radius of curvature. Equation (2.8) then shows that as a consequence of curvature a drift motion will arise with velocity

$$\boldsymbol{v}_C = \frac{mc\, v_{\parallel}^2 (\boldsymbol{R}_c \times \boldsymbol{B})}{e_0\, R_c^2\, B^2} = \frac{2\, c\, W_{\parallel}}{e_0\, R_c\, B} (\hat{\boldsymbol{R}}_c \times \hat{\boldsymbol{B}}), \tag{2.17}$$

where $\hat{\boldsymbol{B}}$ is the unit vector tangent to the local magnetic field, while $\hat{\boldsymbol{R}}_c$ is unit vector along the radius of curvature. The factor in parentheses is therefore a purely geometrical term which describes the local geometry of magnetic field lines. This motion of charged particles is called the *curvature drift*.

Consider now instead a magnetic field directed along one direction, whose magnitude depends on a coordinate along an axis orthogonal to the magnetic field $\boldsymbol{B}$ itself. An example is given by a field $\boldsymbol{B} = [0, 0, B(y)]$, where $B(y)$ satisfies the weak gradient condition

$$\frac{\mathrm{d}B/\mathrm{d}y}{B} R_L \ll 1.$$

The Lorentz force around a given point $y_0$ is given by:

$$F(y) = F(y_0 + \delta y) = \frac{e_0}{c}\left[B\left(-v_x e_y + v_y e_x\right)\right]_{y_0+\delta y},$$

where we are considering motions only in a plane perpendicular to the magnetic field $\boldsymbol{B}$.

Denoting $y$-derivatives with a prime, the field in the neighborhood of this point may be written $B(y_0 + \delta y) = B(y_0) + \delta y\, B'(y_0) = B_0 + \delta y\, B_0'$ so that

$$\begin{aligned} F(y) &= \frac{e_0 B_0}{c}\left[-v_x e_y + v_y e_x\right] \\ &+ \frac{e_0 B_0}{c}\left[-v_x e_y + v_y e_x\right]\delta y\,(B_0'/B_0). \end{aligned} \tag{2.18}$$

For $\delta y$ of the same order as $R_L$ one has by assumption that $\delta y\,(B_0'/B_0) \ll 1$. Therefore the first term in the equation describes the unperturbed circular motion while the second term provides the first order correction. We can thus assume that $v_x, v_y, \delta y = y - y_0$ are the same as those obtained from the motion in a constant magnetic field, which are periodic functions of the phase $\Phi = \Omega t + \alpha$. Averaging the force $\boldsymbol{F}$ over a gyration period, $P = 2\pi/|\Omega|$, that is calculating

$$\langle F \rangle = \frac{1}{P}\int_0^P F\,\mathrm{d}t,$$

we see that the only non zero contribution comes from the term $v_x \delta y B_0'$ on the right hand side of Eq. (2.18). As a result

$$\langle F \rangle = -\frac{e_0}{c} v_\perp R_L \frac{1}{P}\int_0^P \cos^2(\Phi)\,\mathrm{d}t\, B_0'\, e_y = -\frac{e_0 v_\perp}{2c} R_L B_0'\, e_y = -\tfrac{1}{2}\, m v_\perp^2\, \frac{B_0'}{B_0}\, e_y.$$

In other words, on averaging over a gyroperiod, we are left with a constant force perpendicular to $\boldsymbol{B}_0$ which, according to Eq. (2.8), must give rise to a drift orthogonal to both $\boldsymbol{B}_0$ and $\langle \boldsymbol{F} \rangle$. Substituting the expression for $\langle \boldsymbol{F} \rangle$ in Eq. (2.8) we find:

$$v_G = \frac{c\, W_\perp\, B_0'}{e_0 B_0^2}\, e_x,$$

whose general vector expression may be written as:

$$v_G = \frac{c\, W_\perp (B_0 \times \nabla)|B_0|}{e_0 |B_0|^3}. \tag{2.19}$$

This is known as the *gradient drift*.

When, as in the preceding discussion, the gradients are small with respect to the cyclotron radius and the fastest time-scale is given by the gyro-motion one calls the averaged position of the particle over one gyroperiod the guiding center of the particle. In terms of our previous discussion of drifts, one could say the particle executes a gyromotion around its guiding center while the guiding center itself moves following the drifts in question. In order for the so-called guiding center approximation to be valid, an additional condition must be established concerning the parallel motion: the distance a particle moves along the field during one gyroperiod must also be small respect to the scale of gradients, or $v_{\parallel} P \ll L$.

In these circumstances we may always write for the equations of motion along and across the magnetic field the equations:

$$m\frac{\mathrm{d}\boldsymbol{v}_{\parallel}}{\mathrm{d}t} = \boldsymbol{F}_{\parallel} \qquad (2.20)$$
$$m\frac{\mathrm{d}\boldsymbol{v}_{\perp}}{\mathrm{d}t} = \boldsymbol{F}_{\perp} + \frac{e_0}{c}\left(\boldsymbol{v}_{\perp} \times \boldsymbol{B}\right).$$

We may then write $\boldsymbol{v}_{\perp} = \boldsymbol{v}_{\Omega} + \boldsymbol{v}_d$ where $\boldsymbol{v}_{\Omega}$ is the velocity of the cyclotron motion around the guiding center and orthogonal to the local magnetic field while $\boldsymbol{v}_d$ is the drift velocity. We now expand

$$\boldsymbol{v}_d = \boldsymbol{v}_d^0 + \boldsymbol{v}_d^1 + \boldsymbol{v}_d^2 + \cdots$$

and expand order by order in the small gradient parameter the equations for the drifts:

$$m\left(\frac{\mathrm{d}\boldsymbol{v}_d^0}{\mathrm{d}t} + \cdots\right) = \frac{e_0}{c}\left(\boldsymbol{v}_d^1 + \cdots\right) \times \boldsymbol{B}, \qquad (2.21)$$

so that, for example, considering that the time-dependence of the zero-order drift is a small quantity,

$$\boldsymbol{v}_d^1 = -\frac{mc}{e_0 B}\left(\frac{d\boldsymbol{v}_d^0}{dt} \times \frac{\boldsymbol{B}}{B}\right)$$

and so on and so forth. It is important to remark that quite generally the first order drift may differ in direction from the zero order drift, and must therefore be included to understand charged particle trajectories.

### *2.3.2 Magnetic Moment Conservation*

Consider a magnetic field configuration with cylindrical symmetry in which the magnetic field has a strong axial component and again satisfies the conditions for weak gradients defined above.

The field may be written in cylindrical coordinates $r, \theta, z$ as

$$\boldsymbol{B} = [B_r(r,z), 0, B_z(r,z)]; \quad B_z \gg B_r \quad \rightarrow \quad B = (B_r^2 + B_z^2)^{1/2} \simeq B_z.$$

On the other hand the divergence free condition for the magnetic field $\nabla \cdot \boldsymbol{B} = 0$ may be written as:

$$\frac{1}{r}\frac{\partial}{\partial r}(r\,B_r) + \frac{\partial B_z}{\partial z} = 0,$$

which may be integrated to yield

$$\int_0^r \frac{\partial}{\partial r}(r\,B_r)\mathrm{d}r = -\int_0^r r\frac{\partial B_z}{\partial z}\mathrm{d}r \simeq -\langle\frac{\partial B}{\partial z}\rangle\frac{r^2}{2}.$$

Here we have taken into account the fact that $r$ is of the same order as the Larmor radius $R_L$ and that the magnetic field and the magnetic field gradients vary on a much larger scale. We therefore obtain:

$$B_r \simeq -\frac{r}{2}\frac{\partial B}{\partial z},$$

where we have now omitted the brackets $\langle\rangle$. The equation of motion of a particle in the direction parallel to the magnetic field becomes:

$$m\frac{\mathrm{d}v_\parallel}{\mathrm{d}t} = -\frac{e_0}{c}(\boldsymbol{v}\times\boldsymbol{B})_z = -\frac{e_0}{c}v_\theta\,B_r.$$

Remembering that the sense of particle gyration is opposite in sign to the charge we may write

$$v_\theta = -\frac{e_0}{|e_0|}\,v_\perp,$$

and therefore

$$\frac{\mathrm{d}v_\parallel}{\mathrm{d}t} = -\frac{e_0}{c}\Big(-\frac{e_0}{|e_0|}\,v_\perp\Big)\Big(-\frac{r}{2}\frac{\partial B}{\partial z}\Big) \simeq -\Big(\frac{|e_0|}{c}R_L\,|\Omega|\,\frac{R_L}{2}\Big)\frac{\partial B}{\partial z} = -\Big(\frac{|e_0|}{c}\frac{\pi\,R_L^2}{P}\Big)\frac{\partial B}{\partial z},$$

where as before the gyroperiod $P = 2\pi/|\Omega|$.

The periodic gyromotion of a charged particle $e_0$ in the plane orthogonal to the magnetic field is electrically equivalent to a current carrying loop with $I = |e_0|/P$. According to Ampère's law, such a loop corresponds to a magnetic dipole with *magnetic moment* $\mu$, given by:

$$\mu = \frac{I\,S}{c} = \frac{|e_0|}{c}\frac{\pi\,R_L^2}{P} = \frac{1}{2B}mv_\perp^2 = \frac{W_\perp}{B}. \qquad (2.22)$$

We therefore get, along $\boldsymbol{B}$, the equation:

$$m\frac{\mathrm{d}v_{\parallel}}{\mathrm{d}t} = -\mu\frac{\partial B}{\partial z},$$

which, multiplying by $v_{\parallel}$ leads to

$$\frac{\mathrm{d}W_{\parallel}}{\mathrm{d}t} = mv_{\parallel}\frac{\mathrm{d}v_{\parallel}}{\mathrm{d}t} = -\mu v_{\parallel}\frac{\partial B}{\partial z} = -\mu\frac{\mathrm{d}B}{\mathrm{d}t},$$

where we have used

$$\frac{\mathrm{d}}{\mathrm{d}t} = \frac{\partial}{\partial t} + \boldsymbol{v}\cdot\boldsymbol{\nabla}.$$

On the other hand, since the magnetic field does no work on the particles and there is no explicit time-dependence of the field itself, the total energy of the particle is conserved:

$$\frac{\mathrm{d}W_{\perp}}{\mathrm{d}t} = -\frac{\mathrm{d}W_{\parallel}}{\mathrm{d}t} = \mu\frac{\mathrm{d}B}{\mathrm{d}t},$$

an expression which may also be found by writing the equation for perpendicular motion. We may now calculate the time- derivative of $\mu$ as follows:

$$\frac{\mathrm{d}\mu}{\mathrm{d}t} = \frac{\mathrm{d}}{\mathrm{d}t}(W_{\perp}/B) = \frac{1}{B}\frac{\mathrm{d}W_{\perp}}{\mathrm{d}t} - \frac{1}{B^2}W_{\perp}\frac{\mathrm{d}B}{\mathrm{d}t} = \frac{\mu}{B}\frac{\mathrm{d}B}{\mathrm{d}t} - \frac{1}{B^2}(\mu\, B)\frac{\mathrm{d}B}{\mathrm{d}t} = 0.$$

The magnetic moment is therefore a conserved quantity, provided gradient length-scales of the average field are small compared to the particle gyro-radii: this is what is meant by an *adiabatic invariant*. In the particle's frame of reference, the magnetic field changes, albeight slowly, with time, because of the motion of the particle. We can therefore expect that conservation of magnetic moment also holds for slow time-variations of the magnetic field.

Consider then such a slowly varying field, directed along the axis in a cylindrical coordinate system: $\boldsymbol{B} = [0, 0, B(t)]$, whose typical time variation $T$ is much longer than the gyroperiod $2\pi/|\Omega|$. The time-variation induces an azimuthal electric field $\boldsymbol{E}$ which appears in the perpendicular components of the equation of motion. Taking the scalar product of the equation of motion with $\boldsymbol{v}_{\perp}$ we find:

$$m\,\frac{\mathrm{d}\boldsymbol{v}_{\perp}}{\mathrm{d}t}\cdot\boldsymbol{v}_{\perp} = e_0\boldsymbol{E}\cdot\boldsymbol{v}_{\perp},$$

or

$$\frac{\mathrm{d}}{\mathrm{d}t}(\tfrac{1}{2}m\,v_{\perp}^2) = e_0\boldsymbol{E}\cdot\boldsymbol{v}_{\perp}.$$

Integrating this over the gyroperiod, from time 0 to $P = 2\pi/|\Omega|$, the left hand side gives the variation of the kinetic energy of perpendicular motion over one gyroperiod $\Delta W_\perp$:

$$\Delta W_\perp = e_0 \int_0^P \boldsymbol{E} \cdot \boldsymbol{v}_\perp \, \mathrm{d}t = e_0 \oint \boldsymbol{E} \cdot \mathrm{d}\boldsymbol{r}_\perp = e_0 \int_S (\nabla \times \boldsymbol{E}) \cdot \mathrm{d}\boldsymbol{S},$$

where $\mathrm{d}\boldsymbol{r}_\perp = \boldsymbol{v}_\perp \mathrm{d}t$ while $\mathrm{d}\boldsymbol{S}$ is an oriented element of the surface upon which the cyclotron motion orbit rests.

From Maxwell's equations, we may rewrite:

$$\Delta W_\perp = -e_0 \int_S \frac{\partial \boldsymbol{B}}{\partial t} \cdot \mathrm{d}\boldsymbol{S} \simeq \pi R_L^2 |e_0| \dot{B},$$

where we have been able to bring $< \partial \boldsymbol{B}/\partial t >= \dot{B}$ out of the integral because of the hypothesis of slow variations of the magnetic field and we have taken into account the sign coming from the orientation of particle orbits of different sign. Substituting $R_L = |\boldsymbol{v}_\perp|/|\Omega|$ we obtain:

$$\Delta W_\perp = W_\perp \frac{2\pi}{|\Omega|} \frac{\dot{B}}{B} .$$

On the other hand, $(2\pi/|\Omega|)\dot{B}$ is the change of $B$ over a gyroperiod, i.e. $\Delta B$, so we can write:

$$\Delta W_\perp = W_\perp \frac{\Delta B}{B} ,$$

or, recalling Eq. (2.22),

$$\Delta(W_\perp/B) = \Delta(\mu) = 0, \tag{2.23}$$

Therefore, we have shown that the magnetic moment is an adiabatic invariant even in the case of slowly time-varying magnetic fields.

### *2.3.3 Magnetic Mirrors and Magnetic Bottles*

The concept of the conservation of magnetic moment finds application in the study of magnetic mirrors and magnetic bottles as plasma confinement devices. Let's consider again a magnetic field, for example with axial symmetry, which increases in intensity along the positive direction of the axial direction $z$ as shown in Fig. 2.3.

Since the magnetic field magnitude $B$ increases with $z$, the constancy of $\mu$ also implies that the kinetic energy in perpendicular motion $W_\perp$ must increase. Conservation of energy, $W = W_\parallel + W_\perp = constant$, then means that any increase

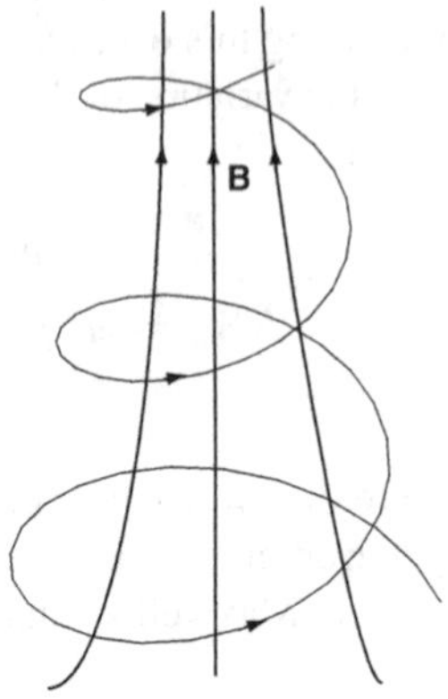


**Fig. 2.3** Trajectory in a spatially varying magnetic field

in $W_\perp$ must be accompanied by a decrease in $W_\parallel$. Thus, it may happen that for some sufficiently large value of $B = B_R$, $W_\parallel$ vanishes. In such case, the charged particle, which cannot continue its trajectory along positive $z$, is reflected. This kind of configuration is called a *magnetic mirror*.

Introducing the pitch angle $\vartheta$, defined as the angle formed by the velocity vector $\boldsymbol{v}$ with the axial direction $z$, that is with the direction of the dominant magnetic field component $B$, we find $v_\perp = v \sin\vartheta$ so that the magnetic moment invariance may be written in the form:

$$\tfrac{1}{2}mv^2 \frac{\sin^2\vartheta}{B} = \text{constant} \quad \text{i.e.} \quad \frac{\sin^2\vartheta}{B} = \text{constant}$$

because of the conservation of total energy $\frac{1}{2}mv^2$. By evaluating the constants at the reflection point we obtain:

$$\sin^2\vartheta = \frac{B}{B_R}. \tag{2.24}$$

Now if the field has a maximum somewhere, $B_{\max}$, obviously $B_R < B_{\max}$ must hold: indeed, if the particle is able to reach the position where $B = B_{\max}$ with $v_\parallel \neq 0$ it will continue in the same direction, as $B$ decreases beyond the point $B = B_{\max}$ and therefore $W_\parallel$ increases at the expense of $W_\perp$. From Eq. (2.24) it follows that we can write the condition for reflection as

$$\sin^2\vartheta = \frac{B}{B_R} \geqslant \frac{B}{B_{\max}}.$$

Particles that start off with a pitch angle $\sin^2\vartheta < B/B_{\max}$ will not be reflected by the magnetic mirror.

If we now consider the illustrated magnetic field structure, comprised of two separate magnetic mirrors encompassing the same axial field, we see that it is possible

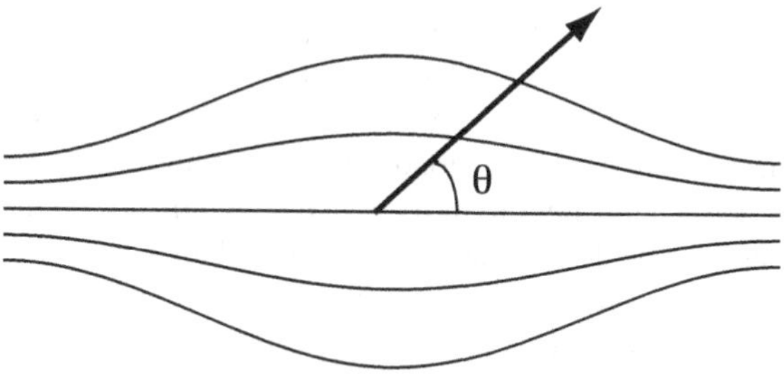


**Fig. 2.4** A magnetic bottle configuration

to confine particles with sufficiently large pitch angles in the region between the two mirrors (Fig. 2.4).

Denoting with $B = B_0$ the lowest value of the magnetic field, the particles that will be confined will be those with

$$\sin^2 \vartheta_0 \geqslant \frac{B_0}{B_{\max}} = 1/R,$$

the quantity $R = B_{\max}/B_0$ being called the *mirror ratio*. The other particles, those with angles falling witihn a cone defined by $\vartheta_0$, known as the *loss cone*, will not be confined. The probability $\mathcal{P}$ of losing a particle depends on the ratio between the solid angles swept by the loss cone and $2\pi$, i.e.

$$\mathcal{P} = \tfrac{1}{2\pi} \int_{\Omega_0} \sin\vartheta \, d\vartheta \, \mathrm{d}\varphi = \int_0^{\vartheta_0} \sin\vartheta \, \mathrm{d}\vartheta = 1 - \sqrt{1 - 1/R} \simeq 1/2R \quad \text{for} \quad R \gg 1.$$

In a real situation, the loss process will never stop, because collisions between charged particles will continually feed particles into the loss cone. For this reason, the *magnetic bottle* is not an efficient confinement device with regards to possible future nuclear fusion machines.

A magnetic bottle configuration in which the intensity of the magnetic field varies slowly with time possesses another adiabatic invariant, the so-called *longitudinal adiabatic invariant* $J$, defined as

$$J = \int_{s_1}^{s_2} v_\parallel \, \mathrm{d}s,$$

where $s_1$ and $s_2$ are the reflection points for the parallel motion whose position is assumed to change on time-scales much slower than the transit time inside the bottle. One may show (see, e.g. Boyd and Sanderson: *The Physics of Plasmas*, p. 28) that $J$ is also an adiabatic invariant.

An interesting and important application of the existence of this invariant is connected to understanding the acceleration of cosmic rays. A simple model may be developed as follows: an ensemble of charged particles finds itself in an environment where magnetic clouds are present and whose behaviour may be schematized as being equivalent to moving magnetic mirrors. In the regions between clouds, the magnetic field has values much smaller than in the clouds, so that particles which start out with a sufficiently large pitch angle are confined in the region between clouds. If $s_0$ is a measure of the distance between clouds at a certain time $t$ and $s_0'$ that at a successive moment in time $t'$, the invariance of $J$ implies that

$$v_{\|}\, s_0 \simeq v_{\|}'\, s_0' \quad \text{or} \quad v_{\|}' \simeq v_{\|} \frac{s_0}{s_0'},$$

where $v_{\|}$ and $v_{\|}'$ are the average velocities respectively at times $t$ an $t'$. It follows that the energy $W_{\|}'$ is equal to $W_{\|}' = W_{\|}\,(s_0/s_0')^2$ and increases if clouds approach each other while it decreases if they increase their separation. Since $W_{\perp} = \mu\, B = constant$
($\mu$ is a constant because it is also an adiabatic invariant and B does not change much as the clouds and particles move around), we see that the total energy $W = W_{\|} + W_{\perp}$ also increases as clouds approach and vice-versa. Since the relative motion of clouds must be considered a random variable, one might think that on average energy gains and losses might balance out. This is wrong however, since over a given interval of time the number of head on collisions between clouds is greater than the number of overtaking ones. Their relative frequencies are, in fact, proportional to $(v_{\|} + v_C)/(v_{\|} - v_C)$, where $v_C$ is the speed of the cloud. The net effect is therefore a gain of energy. This acceleration mechanism for cosmic rays was proposed by Enrico Fermi in 1949 and is known as the second order Fermi mechanism. It is not very efficient because of the particles that are scattered into the loss cone. Today it is thought that a much more efficient mechanism is given by the so-called first order Fermi mechanism where particles are accelerated as they repeatedly cross magnetic shock waves.

## Problems

**2.1.** Consider a magnetic field configuration with a slight curvature: $\boldsymbol{B} \equiv (0, B_y(z), B_{0z})$, and $B_{0z}$ constant while $B_y$ and $\mathrm{d}B_y/\mathrm{d}z$ are small quantities. Show that there is a drift along the $x$ direction with speed

$$v_c = -\frac{m v_{\|}^2 c}{e_0 B_{0z}^2} (\mathrm{d}B_y/\mathrm{d}z).$$

and show that this coincides with what one finds from Eq. (2.17).

**2.2.** Consider the motion of a proton and an electron at the surface of the Earth in the presence of a horizontal uniform magnetic field with an intensity of $B_0 \simeq 0.3$ G. Calculate the drift speed and directions knowing that the magnetic field is directed from South to North. Assuming that the electron density $n_e$ is equal to that of protons $n_p$ calculate thecurrent $\boldsymbol{J}$ resuting from this drift. Finally, show that

$$\frac{\boldsymbol{J}}{c} \times \boldsymbol{B} + (n_e m_e + n_p m_p)\boldsymbol{g} = 0,$$

where $\boldsymbol{g}$ is the gravitational acceleration vector at the surface of the Earth.

This example shows how, in the presence of gravity, a horizontal magnetic field induces a current in the plasma whose effect is to suspend the plasma in the magnetic field, cancelling out gravity. In reality, for a plasma volume of finite extent, there is a separation of charge induced at the lateral boundaries by the current $\boldsymbol{J}$. Show that the generated polarization electric field provides a downward E cross B drift motion, whose time derivative leads the plasma to fall through the magnetic field with acceleration $\boldsymbol{g}$. This points to the fundamental role of *boundary conditions* in establishing the correct behavior of a plasma.

**2.3.** The Earth's magnetic field may be considered with good approximation to be a dipole field, whose intensity in the magnetic equatorial plane is given by $B = B_0(R_E/R)^3$ where $B_0 = 0.3$ G, $R_E$ is the radius of the Earth and R the Earth's radius. Show that because of the gradient drift in the dipole field, Eq. (2.19), a particle with pitch angle $\theta = 90°$carries out a circular orbit around the Earth and find an expression for this orbit in terms of the Energy of the particle $E$ and its charge $q$. Calculate the period of the orbit for a proton and an electron with an energy of 1 keV at a distance of $R = 2R_E$. Compare this drift speed with the one due to the gravitational field directly and with the gravitational orbital speed at the same height.

**2.4.** Consider the motion of a charged particle in the presence of orthogonal uniform electric (along the $y$-direction) and magnetic (along the $z$-direction) fields where the electric field is greater than the magnetic field $E > B$. Solve the problem by moving to a frame moving along the $x$ direction with $\beta = B/E$. Show that in this frame the magnetic field vanishes and solve the equation of motion explicitely. Move back to the original frame of reference and describe the particle orbit in this frame.

## Solutions

**2.1.** Calling $\Omega = e_0 B_{0z}/mc$ e $\Omega_y = e_0 B_y/mc$, the equations of motion become:

$$\ddot{x} = \Omega\dot{y} - \Omega_y\dot{z},$$
$$\ddot{y} = -\Omega\dot{x},$$
$$\ddot{z} = \Omega_y\dot{x}.$$

Taking the time derivative and neglecting terms that are quadratic in small quantities we find:

$$\ddot{x} + \Omega^2 \dot{x} = -\Omega_y' v_\parallel^2,$$

$$\dddot{y} + \Omega^2 \dot{y} = \Omega \Omega_y v_\parallel.$$

A particular (non oscillating solution) of this equation is given by

$$\dot{x} = -\frac{\Omega_y'}{\Omega^2} = -\frac{m v_\parallel^2 c}{e_0 B_{0z}^2}(\mathrm{d}B_y/\mathrm{d}z),$$

providing the requested drift speed.

From the definition of the radius of curvature

$$\frac{\partial \boldsymbol{B}}{\partial s} = -\frac{\boldsymbol{R}_C}{R_C^2},$$

and taking into acocunt that $\partial \boldsymbol{B}/\partial s \simeq (\mathrm{d}B_y/\mathrm{d}z)\boldsymbol{e}_y$, while $\boldsymbol{B} \simeq B_z \boldsymbol{e}_z$, one finds that the above result is equivalent to that found using Eq. (2.17).

**2.2.** The drift speed is given by Eq. (2.8) where the gravitational acceleration is orthogonal to the magnetic field, so that

$$|v_{p,e}| = \frac{c m_{p,e} g}{eB} = 1.86 \times 10^{-4}\ \mathrm{cm/s},\ \text{(electron)} \quad = \quad 3.42 \times 10^{-1} \mathrm{cm/s},\ \text{(proton)}.$$

The velocity is towards the East (West) for a proton (electron). The current from this drift is also directed Eastwards and has a magnitude $|\boldsymbol{J}| = c n_p (m_p + m_e) g/B$.

**2.3.** The gradient drift for this case reduces to

$$|v_G| = \frac{cE}{q}\frac{1}{B^2}|dB/dr| = \frac{cE}{qB_0}\frac{3R^2}{R_E^3}$$

directed towards the West IEast) for protons (electrons). For a particle with the elctron or proton charge and an energy of 1 keV at a height of $R = 2R_E$ we find $|v_G| = 6.27 \times 10^6$ cm/s.

**2.4.** In the reference frame moving with $v_x = cB/E$ one finds from Eq. (2.7) that $B_z' = \Gamma(B_z - \beta E_y) = 0$ while $E_y' = \Gamma(E_y - \beta B_z) = E/\Gamma$ where here $\beta = B/E$ and $\Gamma = 1/\sqrt{1 - B^2/E^2}$. In the new, primed frame, the relativistic equation of motions are given by [see Eqs. 2.9) and (2.10)]

$$\frac{\mathrm{d}(\gamma' \beta_x')}{\mathrm{d}t'} = \frac{\mathrm{d}(\gamma' \beta_z')}{\mathrm{d}t'} = 0; \quad \frac{\mathrm{d}(\gamma' \beta_y')}{\mathrm{d}t'} = \omega_E'; \quad \frac{\mathrm{d}\gamma'}{\mathrm{d}t'} = 0,$$

with $\omega'_E = (e_0 E'/mc) = \Omega_E/\Gamma$. If the particle starts from rest at the origine $\beta'_x = \beta'_z = 0$, $\beta' = \beta'_y$ and $\gamma' = (1-\beta'^2)^{-1/2}$. From the remaining two equations we get:

$$\beta' \frac{d\gamma'}{dt'} + \gamma' \frac{d\beta'}{dt'} = \omega'_E = \omega'_E \beta'^2 + \gamma' \frac{d\beta'}{dt'},$$

or

$$\omega'_E (1-\beta'^2) = \gamma' \frac{d\beta'}{dt'} \quad \Longrightarrow \quad \frac{d\beta'}{dt'} = \frac{\omega'_E}{\gamma'^3}.$$

Integrating the last equation we obtain

$$\beta' = \frac{\tau'}{\sqrt{1+\tau'^2}} \quad \Longrightarrow \gamma'^2 = 1 + \tau'^2,$$

where $\tau' = \omega'_E t'$. Writing

$$v'_y = \frac{dy'}{dt'} = \omega'_E \frac{dy'}{d\tau'} = c\beta' = \frac{c\tau'}{\sqrt{1+\tau'^2}},$$

we see that the particle's speed increases in the $y'$-direction, i.e. along $\boldsymbol{E}'$, approaching $c$ when $t' \to \infty$.

Integrating the above equation we finally find

$$y' = \frac{c}{\omega'_E}\left(\sqrt{1+\tau'^2} - 1\right),$$

which, coupled with $x' = z' = 0$, gives the trajectory of the particle in the primed frame. Transforming back to the original frame by means of a Lorentz transformation vith $V = -B/E$, we easily find

$$x = Vt; \quad y = \left(\frac{c\Gamma}{\omega_E}\right)\left[\left(1 + \frac{{\omega_E}^2 t^2}{\Gamma^4}\right)^{1/2} - 1\right].$$

# Chapter 3
# Kinetic Theory of Plasmas: An Outline

**Abstract** A short summary of the kinetic theory of plasmas is presented as a basis for the development of fluid models. The general kinetic equation is derived and followed by a discussion of the hierarchy of moment equations and the fluid closure problem. The Vlasov equation for collisionless plasmas is presented and Jeans' theorem concerning its solution is proved.

As already remarked in the introductory chapter, when the number of particles becomes large, a *statistical approach* to the description of plasma dynamics plasma provides a good compromise between accuracy and tractability. A complete discussion of plasma kinetic theory is definitely beyond the scope of this book: we limit ourselves here to a brief outline of the basic elements of the theory, necessary prerequisites to the derivation of the fluid models subject of the next chapter.

## 3.1 The Distribution Function

The complete description of the state of a single particle in classical mechanics requires knowledge of its position and velocity at all times. Thus, the natural framework to deal with particle dynamics is a six-dimensional space defined by three spatial coordinates $(x, y, z)$ and three velocity coordinates $(v_x, v_y, v_z)$ known as *phase space*. At any given time $t$, the dynamical state of a particle is represented by a point in phase space. The ensemble of all representative points of the particles thus describes the state of the entire system. The first step towards a statistical approach is the definition of a density function in phase space, telling us which is the average number of particles present, at any given time, in an infinitesimal cell of such a space. This density, generally a function of the six phase space coordinates and of time, is called the *distribution function*, $f(\boldsymbol{r}, \boldsymbol{v}, t)$. The average number of particles with positions lying between $\boldsymbol{r}$ and $\boldsymbol{r} + d\boldsymbol{r}$ and velocities between $\boldsymbol{v}$ and $\boldsymbol{v} + d\boldsymbol{v}$, will then be given by:

$$\mathrm{d}N = f(\boldsymbol{r}, \boldsymbol{v}, t)\, \mathrm{d}\boldsymbol{r}\, \mathrm{d}\boldsymbol{v}. \tag{3.1}$$



C. Chiuderi and M. Velli, *Basics of Plasma Astrophysics*,
UNITEXT for Physics, DOI 10.1007/978-88-470-5280-2_3

When the number of particles is sufficiently large, the distribution function can be considered a *continuous* function of its variables and the total number of particles $N$ can be computed as

$$N = \int_{V_6} f(\boldsymbol{r}, \boldsymbol{v}, t)\, \mathrm{d}\boldsymbol{r}\, \mathrm{d}\boldsymbol{v},$$

where $V_6$ is the phase space volume. To derive the equation describing the dynamical evolution of the distribution function, consider the situation in which the total number $N$ of particles does not change during the evolution of the system. This implies that we are neglecting, for instance, all ionization or recombination processes, that, however, could be easily reintroduced at a later stage. Whenever in physics we have a conserved quantity ($N$ in our case) it is possible to write down a continuity equation. For example, in non-relativistic mechanics, mass is a conserved quantity. The mass contained in a volume $V$ is $M = \int_V \rho\, dV$ and therefore

$$\frac{\mathrm{d}M}{\mathrm{d}t} = \int_V \frac{\partial \rho}{\partial t}\, \mathrm{d}V + \int_S \rho\, (\boldsymbol{v} \cdot \boldsymbol{n}) \mathrm{d}S = 0,$$

where the first integral gives the variation of the mass inside $V$ and the second represents the mass flux through the surface delimiting $V$, $\boldsymbol{n}$ being the unit vector normal to the surface element $\mathrm{d}S$, pointing towards the *exterior*.[1] By transforming the surface integral into a volume integral, the preceding equation becomes:

$$\int_V \left[ \frac{\partial \rho}{\partial t} + \nabla \cdot (\rho \boldsymbol{v}) \right] \mathrm{d}V = 0.$$

Since $V$ is an arbitrary volume, we necessarily have

$$\frac{\partial \rho}{\partial t} + \nabla \cdot (\rho \boldsymbol{v}) = 0.$$

The same procedure can be adopted for any conserved quantity, in particular for $N$. In this case, however, we must modify the definition of the operator $\nabla$ to take into account the dependence of the distribution function on the velocities. This can be achieved by generalizing

$$\nabla \to \sum_{i=1}^{3} \frac{\partial}{\partial x_i} + \sum_{i=1}^{3} \frac{\partial}{\partial v_i} = \nabla + \nabla_{\boldsymbol{v}}. \tag{3.2}$$

[1] from now on $\boldsymbol{n}$ will be referred to as *the external normal*.

The conservation law for the number of particles can now be expressed as

$$\frac{\partial f}{\partial t} + \nabla \cdot (f\boldsymbol{v}) + \nabla_{\boldsymbol{v}} \cdot (f\boldsymbol{a}) = 0, \tag{3.3}$$

where $\boldsymbol{a}$ is the acceleration calculated in the corresponding elementary cell in phase space. The preceding equation can be simplified by noting that

- In phase space $r$ and $v$ are *independent* variables. Thus, $\nabla \cdot \boldsymbol{v} = 0$ and the second term of (4.30) simplifies into $\boldsymbol{v} \cdot \nabla f$,
- The term $f\, \nabla_{\boldsymbol{v}} \cdot \boldsymbol{a} = f\, \nabla_{\boldsymbol{v}} \cdot (\boldsymbol{F}/m) = 0$. In fact, the forces normally coming into play are velocity independent, with the notable exception of the Lorentz force, $\boldsymbol{F} = (e_0/c)(\boldsymbol{v} \times \boldsymbol{B})$. However, in this case too we have

$$\nabla_{\boldsymbol{v}} \cdot \boldsymbol{F} = \frac{e_0}{c} \sum_i \frac{\partial (\boldsymbol{v} \times \boldsymbol{B})_i}{\partial v_i} = 0$$

since the $i$-th component of $\boldsymbol{v} \times \boldsymbol{B}$ does not depend on $v_i$.

Keeping all this in mind Eq. (4.30) becomes:

$$\frac{\partial f}{\partial t} + \boldsymbol{v} \cdot \nabla f + \frac{\boldsymbol{F}}{m} \cdot \nabla_{\boldsymbol{v}} f = 0. \tag{3.4}$$

So far we have implicitly assumed that all the particles occupying the same cell of phase space are subject to the same acceleration. This is certainly true as far as *collective* effects are concerned, i.e. those due to the simultaneous action of all the particles of the plasma, but may not be true for *collisions*, namely for interactions involving only two particles at a time. This suggests separating the collective from collisional forces. Note that the phase space trajectories generated by these two types of forces are also very different from one another: collective effects give rise to forces that are weakly position-dependent and generate smooth trajectories, while collisions produce abrupt variations of the speed of the interacting particles at the collision location. We shall therefore write:

$$\boldsymbol{F} = \boldsymbol{F}_{sv} + \boldsymbol{F}_{coll},$$

where the index *sv* stands for "slowly varying" and we shall distinguish the collisional contribution by moving it to the right hand side of the Eq. (3.4). Introducing the notation:

$$-\frac{\boldsymbol{F}_{coll}}{m} \cdot \nabla_{\boldsymbol{v}} f = \left(\frac{\partial f}{\partial t}\right)_{coll},$$

we obtain the final form of the equation describing the dynamical evolution of the distribution function, the so-called *kinetic equation*,

$$\frac{\partial f}{\partial t} + \boldsymbol{v} \cdot \nabla f + \frac{\boldsymbol{F}}{m} \cdot \nabla_{\boldsymbol{v}} f = \left(\frac{\partial f}{\partial t}\right)_{coll}, \tag{3.5}$$

where it is understood that the forces $\boldsymbol{F}$ appearing on the lhs are only those due to the collective effects.

Everything said so far is equally valid for a neutral gas and for a plasma. In the latter case the force term can be written as

$$\boldsymbol{F} = e_0(\boldsymbol{E} + \frac{1}{c}\boldsymbol{v} \times \boldsymbol{B}) + \boldsymbol{f},$$

where $\boldsymbol{f}$ are the forces of non-electromagnetic origin acting on the plasma, the most common additional force in astrophysics being gravitational. Equation (9.7) has to be coupled with the set of Maxwell's equations (1.22), that determine the electric and magnetic fields, and with the other equations entering the definition of $\boldsymbol{f}$. Neglecting for a moment non-electromagnetic forces write:

$$\frac{\partial f}{\partial t} + \boldsymbol{v} \cdot \nabla f + \frac{e_0}{m}(\boldsymbol{E} + \frac{1}{c}\boldsymbol{v} \times \boldsymbol{B}) \cdot \nabla_{\boldsymbol{v}} f = \left(\frac{\partial f}{\partial t}\right)_{coll}. \tag{3.6}$$

At this point we must specify the nature of the collisions that determine the r.h.s of the Eqs. (9.7) and (3.6). For each type of collision, or better, for each collision *model*, we shall obtain a different kinetic equation. If, for instance, we consider a rarefied plasma dominated by collective effects, we may neglect collisions and equate the rhs of Eq. (9.7) to zero, leading to the *Vlasov equation*:

$$\frac{\partial f}{\partial t} + \boldsymbol{v} \cdot \nabla f + \frac{e_0}{m}(\boldsymbol{E} + \frac{1}{c}\boldsymbol{v} \times \boldsymbol{B}) \cdot \nabla_{\boldsymbol{v}} f = 0. \tag{3.7}$$

If we adopt the Boltzmann collision model (binary elastic collisions) we will get the *Boltzmann equation*, where the collisional term is given in terms of an integral that involves the product of two distribution functions. The Boltzmann equation is particularly important for neutral gases, where binary collisions are dominant, but it is not the most appropriate to describe plasmas. In fact, as we have already remarked, inside the Debye sphere every particle interacts with many other particles at the same time and the deflection suffered by the particle is the result of many small deviations rather than of a single interaction. The collisional term must therefore be suitably modified and the resulting equation is called the *Fokker-Planck equation*.

## 3.2 The Moments of the Distribution Function

The analytical solution of the kinetic equations is generally impossible, even in the simplest case of the Vlasov equation. Unfortunately, their numerical solution also presents enormous difficulties because $f$ depends on a high number of variables (seven, in the general case). It is not surprising therefore that the few analytical solutions and the majority of the numerical solutions concern configurations showing some symmetry property, with the consequent decrease of the number of independent variables. Another cause of complexity is due to the fact that in a plasma are present at least two types of charged particles of opposite signs, typically electrons and positive ions. As a consequence, we have to deal with a number of kinetic equations equal to the number of species present. Finally, observe that the linear character of the lhs of the kinetic equations with respect to $f$ is guaranteed *only if* the fields $\boldsymbol{E}$ and $\boldsymbol{B}$ are *external*, namely if their origin is not tied to the dynamical behaviour of the plasma particles. In all other cases, the kinetic equations have to be complemented with Maxwell's equations, that couple the electromagnetic fields to the distributions of charges and currents which, in turn, are expressed in terms of the motions of the charged particles forming the plasma, described by means of the respective distribution functions. The fields will therefore depend on the distribution functions and the kinetic equations will have a *nonlinear* character.

For the sake of discussion, let's assume that we have managed to solve one of the kinetic equations, obtaining an explicit analytical form for the distribution function. Is there any way we can verify that our solution does indeed correspond to some physically realistic situation? In general, the answer is negative: the distribution function does not appear to be a directly observable quantity. To determine $f$ experimentally we should *measure* the number of particles in a given volume having a well defined value of the velocity and repeat this measure for all values of the speed and for all directions. This type of measurement is possible, although not very precisely, for a limited number of directions and for very rarefied plasmas, conditions that are hard to realize in a laboratory. In interplanetary space such conditions are indeed fulfilled and many spacecraft have succeeded in determining the shape of distribution functions for the plasma particles that flow from the Sun (mostly for electrons, protons, and some for alpha particles) and fill all of interplanetary space, the so-called *solar wind*. These data convincingly demonstrate that the space external to the Earth should be considered as an extraordinary cosmical laboratory, where experiments unfeasible in a terrestrial lab are actually performed. If the distribution function cannot be measured, why is it useful? May we use it to determine other quantities that are more susceptible to a direct measurement? To answer these questions, we must first ask ourselves if we really need all the information contained in $f$. Once again, we are facing a situation where we must decide whether we can give up as redundant certain types of information. For instance, if we judge that the essential information concerns the behavior of our system in ordinary space and not in the whole phase space, we may decide to give up the information concerning the distribution of the velocities, by introducing some sort of averaging process over the velocities themselves. As an example, let's consider the integral

$$\int f(\boldsymbol{r}, \boldsymbol{v}, t) \mathrm{d}\boldsymbol{v} \equiv \iiint f(\boldsymbol{r}, v_v, v_y, v_z; t) \mathrm{d}v_x \mathrm{d}v_y \mathrm{d}v_z. \quad (3.8)$$

Taking into account the definition of $f$, Eq. (3.1), this integral defines the average number of particles whose position vector lies between $\boldsymbol{r}$ and $\boldsymbol{r} + d\boldsymbol{r}$, *regardless of their velocity*, or, in other words the numerical density of particles in ordinary space, that we shall call $n(\boldsymbol{r}, t)$

$$n(\boldsymbol{r}, t) = \int f(\boldsymbol{r}, \boldsymbol{v}, t) \mathrm{d}\boldsymbol{v}. \quad (3.9)$$

In the same way, the average value of any other velocity-dependent quantity $\Phi(\boldsymbol{v})$ can be defined as

$$\langle \Phi \rangle = \frac{\int \Phi(\boldsymbol{v}) f(\boldsymbol{r}, \boldsymbol{v}, t) \mathrm{d}\boldsymbol{v}}{\int f(\boldsymbol{r}, \boldsymbol{v}, t) \mathrm{d}\boldsymbol{v}} = \frac{1}{n(\boldsymbol{r}, t)} \int \Phi(\boldsymbol{v}) f(\boldsymbol{r}, \boldsymbol{v}, t) \mathrm{d}\boldsymbol{v}. \quad (3.10)$$

The factor $1/n$ appearing in the definition of average values is due to the fact that $f$ does nor represent the *probability* of finding a particle in the given phase space cell, but the *average number* of particles within the cell. In other words, $f$ is not normalized to unity, but to the total number of particles, $N$. Clearly, all average values are functions of $\boldsymbol{r}$ and $t$. A special class of functions $\Phi$ is particularly interesting: those that are multilinear products of components of the velocity $\boldsymbol{v}$, namely $\Phi = v_i v_j \ldots v_k$. They are connected to the definition of the $n$-th *order moment* of $f$ as the integral of the velocity distribution function times the product of $n$ velocity components:

$$n\text{-th } \textit{order moment} = \int (v_i v_j \ldots v_k) f(\boldsymbol{r}, \boldsymbol{v}, t) \mathrm{d}\boldsymbol{v}, \quad (3.11)$$

where the product of components entering the integrand has $n$ factors. Recalling the definition of an average value, Eq. (3.10), it is obvious that:

$$\int (v_i v_j \ldots v_k) f(\boldsymbol{r}, \boldsymbol{v}, t) \mathrm{d}\boldsymbol{v} = n(\boldsymbol{r}, t) \langle (v_i v_j \ldots v_k) \rangle \quad (3.12)$$

Therefore, the numerical density is a moment of order zero (in this case $\Phi = 1$). The average value of the velocity, $\boldsymbol{u}(\boldsymbol{r}, t)$, will be given by

$$\boldsymbol{u}(\boldsymbol{r}, t) = \frac{1}{n(\boldsymbol{r}, t)} \int \boldsymbol{v} f(\boldsymbol{r}, \boldsymbol{v}, t) \mathrm{d}\boldsymbol{v}. \quad (3.13)$$

The components of the average velocity are therefore connected with the first-order moments of the distribution function. Notice that $\boldsymbol{u}(\boldsymbol{r}, t)$ represents the average speed of the particles located in the "point $\boldsymbol{r}$" and not the velocity of a single (microscopic) particle, a quantity that has been indicated by $\boldsymbol{v}$.

In a similar way, it is possible to define the charge density, $q(\boldsymbol{r}, t)$, and the current density $\boldsymbol{j}(\boldsymbol{r}, t)$. In fact, if $f_e$ and $f_i$ are, respectively, the electrons' and ions' distribution functions, that for simplicity we shall assume to be singly ionized, we have:

$$q(\boldsymbol{r}, t) = \int e(f_i - f_e)\mathrm{d}\boldsymbol{v}, \quad \text{and} \quad \boldsymbol{j}(\boldsymbol{r}, t) = \int e\,\boldsymbol{v}(f_i - f_e)\mathrm{d}\boldsymbol{v}. \tag{3.14}$$

These examples suggest that the moments of the distribution function have a well defined physical meaning and that they represent quantities that can be measured. However, as we have already stressed, the averaging process causes a loss of information, that related to the behavior of the particles in velocity space. This loss is not essential in all the cases where the velocities of the single particles are not substantially different from the average speed. If, however, a subset of particles exists which behaves in a peculiar way with respect to the others, either because of a clumping in velocity space, such as a particle beam, or because there is a region in velocity space where particles can become resonant with the oscillations of the electromagnetic field, the potential effect of these particles might disappear in the averaging process, with the risk of overlooking important physical effects. We shall return to this point later on, when discussing the phenomenon known as *Landau damping*.

As the distribution function evolves in time, the moments of the distribution functions must also evolve, and their evolution equation, called the *general moment equation* is obtained by taking moments of the kinetic equation. In other words, multiply (3.5) by $\psi(\boldsymbol{v}) = v_i, v_j \dots v_k$ and integrate over all velocities. The first term, recalling Eq. (3.12), becomes:

$$\int \psi(\boldsymbol{v})\frac{\partial f}{\partial t}\mathrm{d}\boldsymbol{v} = \frac{\partial}{\partial t}\int \psi f \mathrm{d}\boldsymbol{v} = \frac{\partial}{\partial t}(n\langle\psi\rangle),$$

since $\psi$ is time independent.

Analogously, since $\psi$ is also independent of $r$, the second term becomes:

$$\int \psi\,\boldsymbol{v}\cdot\nabla f \mathrm{d}\boldsymbol{v} = \nabla\cdot\int f\boldsymbol{v}\,\psi\mathrm{d}\boldsymbol{v} = \nabla\cdot(n\langle\boldsymbol{v}\,\psi\rangle).$$

In the third term, involving $\boldsymbol{F}/m$, separate the velocity-independent forces from the Lorentz force and write (see comments after Eq. (3.3)) $\boldsymbol{F} = \boldsymbol{g} + (e_0/c)\boldsymbol{v}\times\boldsymbol{B}$. For the first group of forces we have:

$$\begin{aligned}\frac{1}{m}\int \psi(\boldsymbol{v})(\boldsymbol{g}\cdot\nabla_{\boldsymbol{v}} f)\mathrm{d}\boldsymbol{v} &= \frac{\boldsymbol{g}}{m}\cdot\int \psi\,(\nabla_{\boldsymbol{v}} f)\mathrm{d}\boldsymbol{v}\\ &= \frac{\boldsymbol{g}}{m}\cdot\int\Big[\nabla_{\boldsymbol{v}}(\psi\, f) - f\,(\nabla_{\boldsymbol{v}}\psi)\Big]\mathrm{d}\boldsymbol{v} = -\frac{n}{m}\boldsymbol{g}\cdot\langle\nabla_{\boldsymbol{v}}\psi\rangle,\end{aligned}$$

where we have used the fact that the fist term in square brackets vanishes if

$$\lim_{|v|\to\infty} (\psi\, f) = 0.$$

Acting in the same way on the term $(e_0/c)\boldsymbol{v} \times \boldsymbol{B}$ and recalling that the $i$-th component of $\boldsymbol{v} \times \boldsymbol{B}$ does not contain $v_i$ we get:

$$\frac{e_0}{m\,c}\int \psi(\boldsymbol{v})\,(\boldsymbol{v}\times\boldsymbol{B})\cdot(\nabla_{\boldsymbol{v}} f)\mathrm{d}\boldsymbol{v} = -\frac{n\,e_0}{m\,c}\langle(\boldsymbol{v}\times\boldsymbol{B})\cdot\nabla_{\boldsymbol{v}}\psi\rangle.$$

Collecting the preceding results, we obtain the general moment equation:

$$\frac{\partial}{\partial t}(n\langle\psi\rangle) + \nabla\cdot(n\langle\boldsymbol{v}\,\psi\rangle) - \frac{n}{m}\boldsymbol{g}\cdot\langle\nabla_{\boldsymbol{v}}\psi\rangle - \frac{n\,e_0}{m\,c}\langle(\boldsymbol{v}\times\boldsymbol{B})\cdot\nabla_{\boldsymbol{v}}\psi\rangle = \\ = \int\psi(\boldsymbol{v})\left(\frac{\partial f}{\partial t}\right)_{coll}\mathrm{d}\boldsymbol{v}. \qquad (3.15)$$

The form of the collisional term on the rhs depends on the properties of particle collisions, i.e. the collisional model adopted. However, remembering that $(\partial f/\partial t)_{coll}$ represents the temporal variation of $f$ due to collisions, we will assume it may be written in a similar way to the term $\partial f/\partial t$ representing the temporal variation of $f$ due to collective forces:

$$\int\psi(\boldsymbol{v})\left(\frac{\partial f}{\partial t}\right)_{coll}\mathrm{d}\boldsymbol{v} = \left(\frac{\partial}{\partial t}(n\langle\psi\rangle)\right)_{coll}.$$

Equation 3.15 with this last form for the collisional term becomes the basis for the development of fluid plasma models, which are the subject of the next chapter.

## 3.3 Vlasov Equation and Jeans' Theorem

While in the case of neutral gases it seldom happens that collisions are negligible, the Vlasov equation (3.7) has many applications in plasma physics, and in some sense provides the simplest model for the kinetic behavior of a plasma. The reason is that while in a system of neutral particles collective effects due to average long range scale self-consistent forces are generally unimportant (with some important exceptions discussed further below), they are fundamental to plasma dynamics.

In spite of its simplicity, the solution of the Vlasov equation generally constitutes a difficult problem, because of the non-linearity arising from the coupling of the electromagnetic fields to the particles, the latter being sources for the fields which determine their motion. An extremely important simplification occurs when the fields $\boldsymbol{E}$ and $\boldsymbol{B}$ entering Eq. (3.7) are *external* fields. In this case the Vlasov equation becomes linear in the distribution function $f$, and a general solution can be found when the system admits one or more constants of motion, a result known as Jeans' theorem.

To prove Jean's theorem, let's introduce the concept of *derivative along a curve* and, for the sake of simplicity, let's consider the case of ordinary geometrical space. Let $g = g(x, y)$ be the equation of a surface. The differential of $g$, namely the variation of $g$ for *arbitrary* variations of $x$ and $y$, is given by:

$$\mathrm{d}g = \frac{\partial g}{\partial x}\,\mathrm{d}x + \frac{\partial g}{\partial y}\,\mathrm{d}y.$$

Let's now assume that the variations of $x$ and $y$ are no longer arbitrary, but that the representative point in the $(x, y)$ plane is constrained to move along the curve $y = y(x)$, so that $\mathrm{d}y = (\mathrm{d}y/\mathrm{d}x)\,\mathrm{d}x$. Then

$$dg = \frac{\partial g}{\partial x}\,\mathrm{d}x + \frac{\partial g}{\partial y}\frac{\mathrm{d}y}{\mathrm{d}x}\,\mathrm{d}x.$$

The quantity

$$\frac{\mathrm{d}g}{\mathrm{d}x} = \frac{\partial g}{\partial x} + \frac{\partial g}{\partial y}\frac{\mathrm{d}y}{\mathrm{d}x}, \tag{3.16}$$

is called the derivative of $g$ along the curve $y = y(x)$.

Let's now apply this definition to the time derivative of $f$, a function of $(\boldsymbol{r}, \boldsymbol{v}, t)$, along the trajectory of the representative point in phase space. The trajectory is given in parametric form by the functions

$$\boldsymbol{r} = \boldsymbol{r}(c_j, t), \quad \boldsymbol{v} = \boldsymbol{v}(c_j, t) \quad j = 1 \ldots 6, \tag{3.17}$$

where the quantities $c_j$ are the six integration constants necessary to completely define the trajectory. By inverting the system of Eq. (3.17), we obtain

$$c_j = c_j(\boldsymbol{r}, \boldsymbol{v}, t), \quad j = 1 \ldots 6. \tag{3.18}$$

Any constant of motion can be obviously expressed in terms of the $c_j$ and therefore in terms of the primary variables $\boldsymbol{r}, \boldsymbol{v}$ using the (3.18). Recalling the definition of derivative along a curve, we write:

$$\frac{\mathrm{d}f}{\mathrm{d}t} = \frac{\partial f}{\partial t} + \sum_{i=1}^{3} \frac{\partial f}{\partial x_i}\frac{\mathrm{d}x_i}{\mathrm{d}t} + \sum_{i=1}^{3} \frac{\partial f}{\partial v_i}\frac{\mathrm{d}v_i}{\mathrm{d}t}.$$

Since $\mathrm{d}x_i/\mathrm{d}t = v_i$ and $\mathrm{d}v_i/\mathrm{d}t = F_i/m$, the preceding equation becomes:

$$\frac{\mathrm{d}f}{\mathrm{d}t} = \frac{\partial f}{\partial t} + \boldsymbol{v} \cdot \nabla f + \frac{\boldsymbol{F}}{m} \cdot \nabla_{\boldsymbol{v}} f.$$

Therefore, if $f$ is a solution of Vlasov equation we may conclude that $(\mathrm{d}f/\mathrm{d}t) = 0$, i.e. that the distribution function remains constant along a trajectory in phase space.

Let's now check that the general solution of Vlasov equation can be expressed as

$$f(\boldsymbol{r}, \boldsymbol{v}, t) = \mathcal{F}(c_1, c_2 \ldots c_6),$$

where $\mathcal{F}$ is an arbitrary function of its variables. We thus have:

$$\begin{aligned}\frac{\partial \mathcal{F}}{\partial t} + \boldsymbol{v} \cdot \nabla \mathcal{F} + \frac{\boldsymbol{F}}{m} \cdot \nabla_v \mathcal{F} &= \sum_j \frac{\partial \mathcal{F}}{\partial c_j}\left[\frac{\partial c_j}{\partial t} + \boldsymbol{v} \cdot \nabla c_j + \frac{\boldsymbol{F}}{m} \cdot \nabla_v c_j\right] \\ &= \sum_j \frac{\partial \mathcal{F}}{\partial c_j}\left[\frac{\mathrm{d}c_j}{\mathrm{d}t}\right]_{trajectory} = 0,\end{aligned}$$

since, by definition, the quantities $c_j$ are constants on the trajectory. This proves **Jeans' theorem**: *Any function of the constants of motion is a solution of the Vlasov equation*

Of course, the explicit form of the solution is obtained by expressing the $c_j$ in terms of the primary physical variables $\boldsymbol{r}$ and $\boldsymbol{v}$ using Eq. (3.18), once it has been verified that the solution obtained has the necessary convergence properties for $v \to \infty$.

A simple application of Jeans' theorem is given by the motion of a particle of mass $m$ and charge $e_0$, subject to the action of an electrostatic electric field, whose potential is $\Phi(\boldsymbol{r})$. The total energy of the particle,

$$E = \frac{1}{2} m v^2 + e_0 \Phi(\boldsymbol{r}),$$

is a constant of motion and therefore any function of the energy is a solution of Vlasov equation. In particular, if for $r \to \infty$, where the potential $\Phi$ is assumed to vanish, the system is in thermodynamical equilibrium, with a density $n_0$ and a temperature $T_0$, the distribution function will asymptotically tend to a maxwellian:

$$f_0 = n_0 \left(\frac{m}{2\pi k T_0}\right)^{3/2} \exp\left[-(\frac{1}{2} m v^2 / k T_0)\right] = n_0 \left(\frac{m}{2\pi k T_0}\right)^{3/2} \exp(-E/kT_0).$$

Jeans' theorem now allows us to write down the solution at a generic point as:

$$f(\boldsymbol{r}, \boldsymbol{v}) = n_0 \left(\frac{m}{2\pi k T_0}\right)^{3/2} \exp\left[-(\frac{1}{2} m v^2 + e_0 \Phi(\boldsymbol{r}))/kT_0\right].$$

This form of the distribution function has been already used (without justification) in the discussion of the Debye length, in Sect. 1.3 of Chap. 1.

The Vlasov equation can sometimes also be used for systems composed of neutral particles, the most interesting case being that of stellar systems. A system composed by a very large number of stars, like a galaxy, could be consider as a sort of "gas", whose "molecules" are represented by the stars. The collective effects are those produced by the gravitational interactions of the ensemble of stars, the direct collisions between being in general negligible. The equation describing the dynamical evolution of the system is again Eq. (3.4), where $\boldsymbol{F}/m$ now represent the gravity acceleration produced at the given point by the collective action of all the stars of the system. Actually, the description of a stellar system by means of Eq. (3.4) preceded the formulation of the Vlasov equation for plasmas (1938) [31], thanks to the works of Jeans (1915) [14] and Chandrasekhar (1942) [15].

We thus see that a very deep analogy exists between the dynamics of a system of mass points and that of a system of charged particles. The law expressing the gravitational interaction between two mass points (Newton) is identical to the one for the electrostatic interaction between two charges (Coulomb), while the Coriolis force in a rotating reference system is perfectly analogous to the Lorentz force exerted on a particle by a magnetic field. There are also deep differences, mostly stemming from the fact that the gravitational interaction is always attractive, while the electrostatic one can be either attractive or repulsive. As a consequence, in "gravitational plasmas" the screening effect that gives rise to the concept of Debye length does not exist.

## Questions and Problems

**3.1.** In interplanetary space distribution functions are often observed, both for protons and electrons, known as *bi-maxwellians*, given by:

$$f_{B0} = n_c\left(\frac{m}{2\pi kT_c}\right)^{3/2}\exp\left(-\frac{mv^2}{2kT_c}\right) + n_h\left(\frac{m}{2\pi kT_h}\right)^{3/2}\exp\left(-\frac{mv^2}{2kT_h}\right). \tag{3.19}$$

$n_c + n_h = n_0$ is the total density of particles: $n_c$, $T_c$ are the parameters defining the "cold" dominant part of the distribution, while $n_h$, $T_h$ refer to a population of "hot" particles of low density but with a much higher value of the temperature. Show that the temperature of the combined distribution, defined by the second-order moment

$$kT = \frac{m}{3n_0}\int v^2 f_{B0}\mathrm{d}\boldsymbol{v},$$

is given by:

$$T_0 = \frac{n_c}{n_0}T_c + \frac{n_h}{n_0}T_h.$$

Using Jeans' theorem verify, assuming spherical symmetry, with $r$ distance from sun center, $r_0$ the distance at which density is measured, and a static atmosphere, that the total density $n(r)$ varies with distance from the Sun as:

$$n(r) = n_c\exp\left[\left(-\frac{GmM_\odot}{r_0kT_c}\right)\left(1-\frac{r_0}{r}\right)\right] + n_h\exp\left[\left(-\frac{GmM_\odot}{r_0kT_h}\right)\left(1-\frac{r_0}{r}\right)\right].$$

Then show that the temperature varies with heliocentric distance as:

$$T(r) = \frac{n_c T_c}{n(r)} \exp\Big[\Big(-\frac{GmM_\odot}{r_0 k T_c}\Big)\Big(1 - \frac{r_0}{r}\Big)\Big] + \frac{n_h T_h}{n(r)} \exp\Big[\Big(-\frac{GmM_\odot}{r_0 k T_h}\Big)\Big(1 - \frac{r_0}{r}\Big)\Big],$$

where $M_\odot$ is the solar mass.

**3.2.** The visible surface of the Sun, the photosphere, has a much lower temperature than the solar corona and solar wind. Because the sources of solar energy, nuclear reactions at the center of the Sun, are below the photosphere, the temperature inversion seems to violate the second law of thermodynamics. A possible solution to this "coronal heating" problem consists in assuming that bi-maxwellian distribution functions are present already within the solar atmosphere at heights corresponding to the transition zone between photosphere and corona, with a dominant component at a temperature $T_c \simeq 10^4$ K and a minor component say with a density $n_h/n_0 \simeq 10^{-3}$ and $T_h \simeq 10^6$ K. According to the formula of the preceding problem, the temperature at a distance of $r = r_0 \simeq 7\ 10^5$ km, corresponding to the transition zone, turns out to be $T_0 \simeq 1.1 \times 10^4$ K. Taking first into account only the protons, $m = m_p$, show that the temperature at a distance of only $r = 1.01\ r_0$ has already risen to a value of $T \simeq 10^6$ K. The electrons, because of their lower mass, reach such a temperature only at a much greater distance: how much greater? In the absence of collisions, which are the effects that may prevent the formation of free charges inside the coronal plasma and the solar wind?

# Chapter 4
# Fluid Models

**Abstract** The formal derivation of fluid models is presented, starting from the hierarchy of moment equations stemming from kinetic theory. Starting from the example of the ideal, neutral gas to establish the procedure, a derivation of the two-fluid plasma equation is carried out in detail. Finally, the one-fluid plasma equations are derived and their meaning discussed, with particular emphasis on energy equation and Ohm's law.

In Chap. 3 we have shown that the moments of the distribution functions are related to measurable macroscopic physical quantities. The moments themselves, however, are defined in terms of the distribution function, so that a knowledge of the latter would seem to be required to recover such quantities. A considerable step forward would be achieved if it were possible to find a system of differential equations containing *only* the moments. In this case, the moments would be defined as the solutions of the of such a system, without further reference to the distribution function, whose knowledge would become unnecessary. This type of approach is the basis for the development of fluid models, that we are now going to discuss.

## 4.1 The Case of Neutral Gases

To establish the procedure, let's start by considering the simple case of a neutral gas, composed by identical particles of mass $m$. We shall therefore need just one distribution function, that we shall still denote by $f$. The starting point of our program will obviously be the general moment equation, Eq. (3.15):

$$\frac{\partial}{\partial t}(n\langle\psi\rangle) + \nabla\cdot(n\langle\boldsymbol{v}\,\psi\rangle) - \frac{n}{m}\boldsymbol{F}\cdot\langle\nabla_{\boldsymbol{v}}\psi\rangle = \left(\frac{\partial}{\partial t}(n\langle\psi\rangle)\right)_{coll}. \tag{4.1}$$

There will be one equation for each moment, which means that we are moving from a *single kinetic equation* to a *system of equations*. It is easy to realize that actually we shall be faced with an *infinite* system. In fact, let the function $\psi$ entering Eq. (4.1) be an $n$-th order moment, namely the product of $n$ velocity components. Equation (4.1)



C. Chiuderi and M. Velli, *Basics of Plasma Astrophysics*,
UNITEXT for Physics, DOI 10.1007/978-88-470-5280-2_4

would then be the equation for the $n$-th moment. But, the second term of Eq. (4.1) contains the quantity $\langle \boldsymbol{v}\, \psi \rangle$ that involves moments of order $n+1$. We conclude therefore that the equation for the moment of order $n$ inevitably contains moments of order $n+1$ and therefore the system of moment equations is actually an infinite chain. It is clear that, unless some way is found to truncate the chain of equations, our program will turn out to be totally unfeasible. This is the so-called *closure problem* that we shall discuss later on.

For the time being, we shall write down the appropriate equation for each moment. At **zeroth-order** we put $\psi = m$ in Eq. (4.1). Thus

$$\frac{\partial \rho}{\partial t} + \nabla \cdot (\rho\, \boldsymbol{u}) = m\left(\frac{\partial n}{\partial t}\right)_{coll}, \tag{4.2}$$

where we have used the definition of the bulk speed $\boldsymbol{u}$, Eq. (3.13) .

At **first order** we use $\psi = m\, v_i$ and write, for the $i$-th velocity component:

$$\begin{aligned} &\frac{\partial}{\partial t}(nmu_i) + \frac{\partial}{\partial x_k}(nm\langle v_i v_k\rangle) - n\, F_k\, \langle \frac{\partial v_i}{\partial v_k}\rangle \\ =&\frac{\partial}{\partial t}(nmu_i) + \frac{\partial}{\partial x_k}(nm\langle v_i v_k\rangle) - nF_i = \left(\frac{\partial (nm\langle v_i\rangle)}{\partial t}\right)_{coll}, \end{aligned} \tag{4.3}$$

where a sum over repeated indexes must be understood. As expected, in Eq. (4.3) we notice the presence of second order moments, $nm\langle v_i v_k\rangle$. To make the physical meaning of those terms clear, let's write the velocity of a single particle as:

$$\boldsymbol{v} = \boldsymbol{u} + \boldsymbol{w}, \tag{4.4}$$

splitting it into an *average (or flow) speed*, $\boldsymbol{u}$, and a *peculiar speed*, $\boldsymbol{w}$. Since $\boldsymbol{u} = \langle \boldsymbol{v} \rangle$, the average of Eq. (4.4) gives:

$$\langle \boldsymbol{w} \rangle = 0. \tag{4.5}$$

Thus, the average of $\boldsymbol{w}$ vanishes, a typical feature of random motions. This suggests that $\boldsymbol{w}$ could be identified with the velocity of the chaotic thermal motion of the particles. Inserting the expression Eq. (4.4) in $\langle v_i v_k\rangle$ and noticing that $\langle u_i\rangle = u_i$, since $\boldsymbol{u}$ is a function only of time and spatial coordinates, and the velocity-averaging process represented by $\langle\,\rangle$ leaves $\boldsymbol{u}$ unaltered, we obtain:

$$\langle v_i v_k\rangle = \langle (u_i + w_i)(u_k + w_k)\rangle = u_i u_k + \langle w_i w_k\rangle. \tag{4.6}$$

Thus:

$$nm\langle v_i v_k\rangle = nmu_i u_k + nm\langle w_i w_k\rangle = nmu_i u_k + \mathsf{P}_{ik}.$$

The tensor $\mathsf{P}_{ik} = nm\langle w_i w_k\rangle$ is called the *pressure tensor*.

Since $nmu_k$ is simply the momentum density in the $k$-direction, $nmu_i u_k = u_i(nmu_k)$ gives the flux of the $k$-th component of the momentum across the unit surface whose normal is parallel to the $i$-axis. Therefore, the first term of Eq. (4.6) represents the momentum flux connected with *ordered* motion, whereas the second term represents the momentum flux connected with *disordered or random* motion, that we have already identified with the thermal motion. The diagonal terms of the $\mathsf{P}_{ik}$ tensor, which have the same value in the case of an isotropic medium, correspond to the normal definition of pressure of a fluid, while the off-diagonal ones are different from zero only in the presence of viscous forces, which, in fact, act in a direction perpendicular to the fluid velocity. These terms can be identified with *shear-stresses* in the flow.

The pressure tensor is clearly a symmetric one. It may be useful to separate its diagonal part from the off-diagonal one, by writing, always in an isotropic medium,

$$\mathsf{P}_{ik} = P\,\delta_{ik} + \Pi_{ik},$$

where the tensor $\Pi_{ik}$ is a symmetric tensor with vanishing diagonal terms. Reverting to Eq. (4.3) we shall therefore write:

$$\frac{\partial}{\partial t}(nmu_i) + \frac{\partial}{\partial x_k}(nmu_iu_k + \mathsf{P}_{ik}) - n\,F_i = \left(\frac{\partial(nm\langle v_i\rangle)}{\partial t}\right)_{coll}. \tag{4.7}$$

Moving now to the **second-order** moments, we can limit ourselves to $\psi = \frac{1}{2}mv_iv_i = \frac{1}{2}mv^2$ and write:

$$\frac{\partial}{\partial t}(n\langle\tfrac{1}{2}mv^2\rangle) + \nabla\cdot(n\langle \boldsymbol{v}\tfrac{1}{2}mv^2\rangle) - n\boldsymbol{F}\cdot\boldsymbol{u} = \left(\frac{\partial}{\partial t}(n\langle\tfrac{1}{2}mv^2\rangle)\right)_{coll}. \tag{4.8}$$

Making use once more of the decomposition given by Eq. (4.4), we may express the term $n\langle v_k\frac{1}{2}mv^2\rangle$ as

$$n\langle v_k\tfrac{1}{2}mv^2\rangle = n(\tfrac{1}{2}m(u^2 + \langle w^2\rangle)u_k + u_i\,\mathsf{P}_{ik} + n\langle w_k(\tfrac{1}{2}mw^2)\rangle.$$

Moreover, since

$$n(\tfrac{1}{2}m\langle v^2\rangle) = \tfrac{1}{2}\rho u^2 + \tfrac{1}{2}\sum_i \mathsf{P}_{ii} = \tfrac{1}{2}\rho u^2 + \tfrac{1}{2}(3P),$$

we get:

$$n\langle v_k\tfrac{1}{2}mv^2\rangle = (\tfrac{1}{2}\rho u^2 + \tfrac{3}{2}P)u_k + u_i\,\mathsf{P}_{ik} + n\langle w_k(\tfrac{1}{2}mw^2)\rangle.$$

If the gas obeys the perfect gas equation, $P = nkT$, the term $\frac{3}{2}P$ gives the internal energy per unit volume, $\epsilon$.[1]

[1] We did implicitly assume a monoatomic gas with 3 degrees of freedom; in the general case of $f$ degrees of freedom we have $\epsilon = P/(\gamma - 1)$, with $\gamma = (f+2)/f$.

Summarizing, the first term on the rhs represents the total energy flux (kinetic energy of the ordered motion + internal energy) transported by the ordered motions, the third gives the flux of the internal energy transported by the thermal motions, namely the heat flux , while the second gives the work done by the pressure forces. The preceding expression allows us to write Eq. (4.8) as:

$$\frac{\partial}{\partial t}\left(\tfrac{1}{2}\rho u^2 + \tfrac{3}{2}P\right) + \frac{\partial}{\partial x_k}\left[\left(\tfrac{1}{2}\rho u^2 + \tfrac{3}{2}P\right)u_k + u_i\,\mathsf{P}_{ik} + q_k\right] - n\,F_i u_i = \left(\frac{\partial}{\partial t}(n\langle\tfrac{1}{2}mv^2\rangle)\right)_{coll}, \tag{4.9}$$

where we have explicitly introduced the heat flux vector, $\boldsymbol{q} = n\langle \boldsymbol{w}(\frac{1}{2}mw^2)\rangle$.

So far, we have avoided discussing the collisional terms appearing on the rhs of Eqs. (4.2), (4.7) and (4.9). A rigorous theory should start from a specific collision model, for instance that of binary elastic collisions by Boltzmann. Adopting such a collisional term it is possible to show the the laws of conservation, in a single collision, of the number of particles, of the total momentum and of the total energy, imply the fulfillment of the following conditions:

$$\begin{aligned} \int \left(\frac{\partial f}{\partial t}\right)_{coll} &= 0 \\ \int m\boldsymbol{v}\left(\frac{\partial f}{\partial t}\right)_{coll} &= 0 \\ \int \tfrac{1}{2}mv^2\left(\frac{\partial f}{\partial t}\right)_{coll} &= 0. \end{aligned} \tag{4.10}$$

The importance of these conditions, that reflect the fundamental conservation laws, is such that their validity is considered to be a necessary condition for adopting a collisional term different from that of Boltzmann. In other words, the relationships expressed by Eqs. (4.10), must be verified *by any model for the collisional term*. The fulfillment of conditions (4.10) implies that all the terms at the rhs of Eqs. (4.2), (4.7) and (4.9) are actually zero.

This result can be made plausible by considering, for instance, equation (4.2). The collisional term represents, apart from the factor $m$, the temporal variation of the number of particles in a given phase-space volume caused by collisions. But the particle number is a conserved quantity in a collision if, as assumed, we neglect the processes that modify the number of interacting particles, such as ionization or recombination. Hence, the collisional term of Eq. (4.2) vanishes and we are entitled to write:

$$\frac{\partial \rho}{\partial t} + \nabla \cdot (\rho\,\boldsymbol{u}) = 0, \tag{4.11}$$

where we recognize the mass continuity equation of fluid mechanics.

In a similar way, the collisional terms of Eqs. (4.7) and (4.9) give, respectively, the temporal variations of momentum and energy caused by collisions, that both vanish in elastic encounters.

In Eq. (4.7) we are thus entitled to put the rhs equal to zero and to write, taking into account Eq. (4.11), the first two terms as:

$$\frac{\partial}{\partial t}(\rho u_i) + \frac{\partial}{\partial x_k}(\rho u_i u_k) = \rho\left(\frac{\partial u_i}{\partial t} + u_k \frac{\partial u_i}{\partial x_k}\right) + u_i\left(\frac{\partial \rho}{\partial t} + \frac{\partial(\rho u_k)}{\partial x_k}\right)$$
$$= \rho\left(\frac{\partial u_i}{\partial t} + u_k \frac{\partial u_i}{\partial x_k}\right).$$

But the expression on the rhs is the derivative of $u_i$ along the trajectory $\boldsymbol{u} = (\mathrm{d}\boldsymbol{r}/\mathrm{d}t)$, a quantity that in fluid mechanics is known as the *lagrangian derivative* of $u_i$

$$\frac{\mathrm{d}}{\mathrm{d}t} = \frac{\partial}{\partial t} + (\boldsymbol{u} \cdot \boldsymbol{\nabla}). \tag{4.12}$$

The final form of Eq. (4.7) is therefore

$$\rho \frac{\mathrm{d}u_i}{\mathrm{d}t} = -\frac{\partial \mathsf{P}_{ik}}{\partial x_k} + n\, F_i . \tag{4.13}$$

In the preceding expression we recognize the *equation of motion* of a fluid. In the case of a *perfect* fluid, $\mathsf{P}_{ik} = P\,\delta_{ik}$, and Eq. (4.13) becomes *Euler's equation*:

$$\rho \frac{\mathrm{d}\boldsymbol{u}}{\mathrm{d}t} = -\boldsymbol{\nabla} P + n\, \boldsymbol{F}, \tag{4.14}$$

while for a *viscous* fluid ($\Pi_{ik} \neq 0$) we get the *Navier-Stokes equation*:

$$\rho \frac{\mathrm{d}u_i}{\mathrm{d}t} = -\frac{\partial P}{\partial x_i} - \frac{\partial \Pi_{ik}}{\partial x_k} + n\, F_i . \tag{4.15}$$

In both the above equations a distinction has been made between pressure forces and other types of forces.

Turning now to Eq. (4.9) and putting again the collisional term equal to zero, we get:

$$\frac{\partial}{\partial t}\left(\tfrac{1}{2}\rho u^2 + \tfrac{3}{2}P\right) = -\frac{\partial}{\partial x_k}\left([\tfrac{1}{2}\rho u^2 + \tfrac{3}{2}P]u_k + P u_k + u_i\,\Pi_{ik} + q_k\right) + n\, F_i u_i . \tag{4.16}$$

This is the *energy equation*, which states that the temporal variation of the total energy contained in a given volume equals the flux of the same quantity across the surface enclosing the volume, plus the work done by *all* forces, pressure forces and forces of a different nature, plus the effect of the heat flux. It is important to notice that all the terms, with the exception of $q_k$, are proportional to one component of the fluid speed. Hence, these terms refer to energy fluxes connected with macroscopic motions of matter, i.e. they are *convective* terms. On the contrary, the term containing $\boldsymbol{q}$ survives even in static conditions, $\boldsymbol{u} = 0$, and represents *heat conduction*.

A useful alternative expression for the energy equation is the following. Let's rewrite Eq. (4.16) in the form:

$$\left(\frac{\partial}{\partial t}\left(\tfrac{1}{2}\rho u^2\right) + \frac{\partial}{\partial x_k}\left(\tfrac{1}{2}\rho u^2 u_k\right)\right) + \tfrac{3}{2}\left(\frac{\partial P}{\partial t} + \tfrac{5}{3}\frac{\partial}{\partial x_k}\left(P u_k\right)\right)$$
$$= -\frac{\partial}{\partial x_k}\left(u_i \Pi_{ik} + q_k\right) + n\, F_i u_i .$$

Expanding the derivatives and using the equation of continuity, Eq. (4.11), and the equation of motion, Eq. (4.13), we get:

$$\tfrac{3}{2}\left(\frac{\partial P}{\partial t} + u_k \frac{\partial P}{\partial x_k} + \tfrac{5}{3} P \frac{\partial u_k}{\partial x_k}\right) = -\Pi_{ik}\frac{\partial u_i}{\partial x_k} - \frac{\partial q_k}{\partial x_k}.$$

Using once again Eq. (4.11) to eliminate $\partial u_k/\partial x_k$ and recalling the definition of lagrangian derivative, Eq. (4.12), we obtain:

$$\tfrac{3}{2}\left(\frac{\mathrm{d}P}{\mathrm{d}t} - \frac{5}{3}\frac{P}{\rho}\frac{\mathrm{d}\rho}{\mathrm{d}t}\right) = -\Pi_{ik}\frac{\partial u_i}{\partial x_k} - \frac{\partial q_k}{\partial x_k}.$$

Multiplying and dividing the lhs by $\rho^{-5/3}$ we finally arrive at:

$$\tfrac{3}{2}\rho^{5/3}\frac{\mathrm{d}}{\mathrm{d}t}\left(P\,\rho^{-5/3}\right) = -\Pi_{ik}\frac{\partial u_i}{\partial x_k} - \frac{\partial q_k}{\partial x_k}. \tag{4.17}$$

Equation (4.17) is valid for a monoatomic gas. In the general case we have:

$$\frac{1}{\gamma - 1}\rho^{\gamma}\frac{\mathrm{d}}{\mathrm{d}t}\left(P\,\rho^{-\gamma}\right) = -\Pi_{ik}\frac{\partial u_i}{\partial x_k} - \frac{\partial q_k}{\partial x_k}. \tag{4.18}$$

Notice that for a perfect gas, with vanishing viscosity and thermal conduction ($\Pi_{ik} = 0, q_k = 0$), the preceding equation reduces to

$$\frac{\mathrm{d}}{\mathrm{d}t}\left(P\,\rho^{-\gamma}\right) = 0,$$

which is the *adiabatic* equation of a prefect gas.

It is easy to generalize Eq. (4.11) to include processes implying a variation of the total number of particles. The rhs will no longer vanish and will consist of a sum of positive terms, representing the processes that increase the number of particles, forcing the density to increase as well, and of negative terms representing a decrease in particle number.

It is also possible to include in Eq. (4.18), beyond the effects of viscosity and heat conduction, other dissipative effects, such as radiative losses, of fundamental importance in astrophysics. In fact, introducing the expression of the entropy *per unit mass* of a perfect gas, $s = c_V \ln(P\rho^{-\gamma})$, and recalling that

$$c_V = \frac{\mathcal{R}}{\mu}\frac{1}{\gamma - 1},$$

where $c_V$ is the specific heat at constant volume, $\mathcal{R}$ the gas constant and $\mu$ the mean molecular weight, we may write the lhs of Eq. (4.18) as $\rho T \mathrm{d}s/\mathrm{d}t$, that we recognize as the expression for the total heat exchanged by the system. This suggests writing Eq. (4.18) in the symbolic form

$$\rho T \frac{\mathrm{d}s}{\mathrm{d}t} = \mathcal{G} - \mathcal{P}, \tag{4.19}$$

where $\mathcal{G}$ and $\mathcal{P}$ represent, respectively, the gains and losses of energy.

We have managed to deduce formally the equations of fluid mechanics, but still we have not completed our program of obtaining a series of relationships among the moments that don't require the previous knowledge of the distribution function. In fact, it is easily seen that, even in the absence of non- pressure forces ($\boldsymbol{F} = 0$), Eqs. (4.11), (4.15) and (4.18) form a system of 5 scalar equations in the 11 unknowns $P,\ \rho,\ \boldsymbol{u},\ \boldsymbol{q},\ \Pi_{ik}$. Hence, 6 quantities are still defined in terms of the distribution function. It is clear that nothing can be gained by writing the equations for the moments of order higher than second, since it is easily seen that the number of unknowns increases more rapidly than that of equations. Therefore, a scheme must be established allowing to express higher-order moments in terms of lower-order ones, to finally end up with an equal number of equations and unknowns.

The simplest case is that of a collision-dominated, neutral gas. Such a system evolves toward a state of thermodynamical equilibrium described by a maxwellian distribution function:

$$\begin{aligned} f_0 &= n\left(\frac{m}{2\pi kT}\right)^{3/2} \exp\left[-\frac{\frac{1}{2}m(\boldsymbol{v}-\boldsymbol{u})^2}{kT}\right] \\ &= n\left(\frac{m}{2\pi kT}\right)^{3/2} \exp\left[-\frac{\frac{1}{2}mw^2}{kT}\right] \end{aligned} \tag{4.20}$$

A distinction must be made between the case in which the macroscopic quantities $n$, $T$ and $\boldsymbol{u}$ are constants and that in which they depend on space and time. In the first case we have a state of thermodynamical equilibrium, in the second of a *local* thermodynamical equilibrium, where $n(\boldsymbol{r}, t)$, $T(\boldsymbol{r}, t)$ and $\boldsymbol{u}(\boldsymbol{r}, t)$ are slowly varying functions over distances of the order of the mean free path and times of the order of the collision time. In both cases it's easy to check that $\Pi_{ik} = 0$ and that $q_k = 0$. In fact, the definition of moments involves integrations over the velocity space only and the circumstance that $T$, $n$ and $\boldsymbol{u}$ depend on spatial as well as temporal coordinates becomes irrelevant. Thus $\mathsf{P}_{ik} = P\,\delta_{ik}$ and Eqs. (4.11), (4.15) and (4.18) are sufficient to determine the 5 unknowns $P,\ \rho,\ \boldsymbol{u}$.

If the gas finds itself in a non-equilibrium condition, it is still possible to make some progress by adopting a perturbative technique. This amounts to the assumption that the distribution function can be represented as $f = f_0 + f_1$ with $|f_1| \ll |f_0|$. The system of fluid equation can be linearized and first-order expressions for the off-diagonal terms of the pressure tensor and of the heat flux can be found. A discussion of this method, called the *Chapman-Enskog* method, is beyond the scope of this book (see, however, Exercise 4.1).

## 4.2 The Plasma Case: Two-Fluid Models

To apply the procedure developed for neutral gases to a plasma we must take into account that the number of distribution functions increases. Restricting ourselves for simplicity to the case of a completely ionized hydrogen plasma, composed by electrons and protons, we will have to deal with two kinetic equations, one for each species: we shall indicate them by $f_p$ and $f_e$. If the only forces coming into play are those of electromagnetic origin, the two kinetic equations can be written as:

$$\frac{\partial f_s}{\partial t} + \boldsymbol{v}\cdot\nabla f_s + \frac{e_s}{m_s}(\boldsymbol{E} + \frac{1}{c}\boldsymbol{v}\times\boldsymbol{B})\cdot\nabla_{\boldsymbol{v}} f_s = \left(\frac{\partial f_s}{\partial t}\right)_{coll}, \quad s = e, p. \tag{4.21}$$

Following the procedure adopted for neutral gases, we obtain the equation equivalent to Eq. (4.2):

$$\frac{\partial n_s}{\partial t} + \nabla\cdot(n_s\,\boldsymbol{u}^{(s)}) = \left(\frac{\partial n_s}{\partial t}\right)_{coll}, \tag{4.22}$$

where, obviously,

$$n_s = \int f_s \mathrm{d}\boldsymbol{v} \quad \mathrm{e} \quad \boldsymbol{u}^{(s)} = \frac{1}{n_s}\int \boldsymbol{v}\, f_s\, \mathrm{d}\boldsymbol{v}.$$

The equation of motion analogous to Eq. (4.3) becomes:

$$\frac{\partial}{\partial t}(n_s m_s u_i^{(s)}) + \frac{\partial}{\partial x_k}(n_s m_s \langle v_i v_k\rangle_s) - e_s n_s\, E_i - \frac{e_s n_s}{c}(\boldsymbol{u}^{(s)}\times\boldsymbol{B})_i = R_i^{(s)},$$

where the notation $\langle\rangle_s$ indicates that the average must be taken with respect to the distribution function $f_s$ and we have introduced the compact notation $R_i^{(s)}$ for the collisional term,

$$\boldsymbol{R}^{(s)} = \int m_s \boldsymbol{v}\left(\frac{\partial f_s}{\partial t}\right)_{coll} \mathrm{d}\boldsymbol{v}.$$

If we now define a pressure tensor for each species:

$$\mathsf{P}_{ik}^{(s)} = n_s m_s \langle w_i w_k \rangle = P^{(s)}\, \delta_{ik} + \Pi_{ik}^{(s)},$$

we arrive at the equation of motions for electrons and protons:

$$\frac{\partial}{\partial t}\left(n_s m_s u_i^{(s)}\right) + \frac{\partial}{\partial x_k}\left(n_s m_s u_i^{(s)} u_k^{(s)} + \mathsf{P}_{ik}^{(s)}\right) - e_s n_s\, E_i - \frac{e_s n_s}{c}\left(\boldsymbol{u}^{(s)} \times \boldsymbol{B}\right)_i = R_i^{(s)}. \tag{4.23}$$

In a similar manner, the two energy equations can be derived:

$$\frac{\partial}{\partial t}\left(\tfrac{1}{2} m_s n_s \left(u^{(s)}\right)^2 + \tfrac{3}{2} P^{(s)}\right) + \frac{\partial}{\partial x_k}\left(\left[\tfrac{1}{2} m_s n_s \left(u^{(s)}\right)^2 + \tfrac{5}{2} P^{(s)}\right] u_k + u_i^{(s)}\, \Pi_{ik}^{(s)} + q_k^{(s)}\right) - e_s n_s E_i u_i^{(s)} = Q^{(s)}, \tag{4.24}$$

with

$$Q^{(s)} = \int \tfrac{1}{2} m_s v^2 \left(\frac{\partial f_s}{\partial t}\right)_{coll} \mathrm{d}\boldsymbol{v}.$$

What we did so far is a straightforward extension of the procedure used for neutral gases, but the collisional terms require closer scrutiny. In fact, unlike the neutral case, where it has been assumed that all the particles have the same mass, here we must face cases in which the colliding particles have vastly different masses. For each species, therefore, the collisional term will be the sum of the contributions coming from both same species particle and different species particle collisions. To analyze these cases and to simplify the notation let's put:

$$C(s, s') = \left(\frac{\partial f_s}{\partial t}\right)_{coll} \qquad (s, s' = e, p).$$

Making use of this notation, the conservation law for total particle number can be written in the form

$$\int C(s, s')\mathrm{d}\boldsymbol{v} = 0, \tag{4.25}$$

valid both for $s = s'$ and $s \neq s'$.

The global momentum and energy conservation give:

$$\int m_s \boldsymbol{v}\, C(s, s)\mathrm{d}\boldsymbol{v} = 0$$

$$\int \tfrac{1}{2} m_s v^2\, C(s, s)\mathrm{d}\boldsymbol{v} = 0,$$

for collisions between particles of the same species, while for collisions between particles of different species we shall have:

$$\int m_s \boldsymbol{v}\, C(s,s')\mathrm{d}\boldsymbol{v} + \int m'_s \boldsymbol{v}\, C(s',s)\mathrm{d}\boldsymbol{v} = 0$$

$$\int \tfrac{1}{2} m_s v^2\, C(s,s')\mathrm{d}\boldsymbol{v} + \int \tfrac{1}{2} m'_s v^2\, C(s',s)\mathrm{d}\boldsymbol{v} = 0. \tag{4.26}$$

Thus, the collisional term given by Eq. (4.22) always vanishes, while in those of Eqs. (4.23) and (4.24), only contributions from particles of different species enter,

$$\left(\frac{\partial n_s}{\partial t}\right)_{coll} = 0,$$

$$\boldsymbol{R}^{(s)} = \int m_s \boldsymbol{v}\, C(s,s')\mathrm{d}\boldsymbol{v} \quad (s' \neq s),$$

$$Q^{(s)} = \int \tfrac{1}{2} m_s v^2\, C(s,s')\mathrm{d}\boldsymbol{v} \quad (s' \neq s).$$

On the other hand, since the momentum and energy lost by one species are gained by the other, we have

$$\boldsymbol{R}^{(e)} = -\boldsymbol{R}^{(p)} \quad e \quad Q^{(e)} = -Q^{(p)}. \tag{4.27}$$

The system given by Eq. (4.22) (with no collisional term), Eqs. (4.23) and (4.24) is the basic system of equations of the so-called *two-fluid model*. Notice that in general the temperatures and the heat fluxes of the two fluids will be different:

$$T^{(s)} = \frac{p^{(s)}}{k\, n_s}; \; q_i^{(s)} = \left\langle \tfrac{1}{2} m_s\, w_i \sum_k w_k w_k \right\rangle_s.$$

The closure problem for the two-fluid model is similar to that of neutral gases. Here as well, the closure can be obtained by assuming that each fluid is in a state of local thermodynamical equilibrium *at its own temperature*, an assumption that allows the neglect of the viscous and thermal conduction terms. The coupling between the two species is maintained by the terms that represent the mutual exchange of momentum and energy and, to achieve closure, they must be expressed in terms of the macroscopic variables. The two-fluid model is a good description of the electro-proton system when the two species have not yet attained a common thermal equilibrium. As observed in the Introduction, such a situation may occur in a rarefied plasma, because of the relative inefficiency of the collisions between different species, $\tau_{ep} \gg \tau_{pp} \gg \tau_{ee}$. A typical example of a two-fluid system is given by the solar wind, where, at Earth orbit, the electron temperature is higher than the proton one.

## 4.3 The One-Fluid Model

Even in the case where a closure can be obtained, the two-fluid model is still a complex one. This prompts us to look for a simpler description, at least when the temperatures of the two fluids are not substantially different. This further simplification can be obtained by introducing a *fictitious* fluid, somehow representative of the entire plasma. We expect that the equations of such a model, if it can be found, look similar to those of a neutral gas, with some extra terms to take into account the electromagnetic effects.

The *total* numerical density of the plasma, $n(\boldsymbol{r}, t)$, i.e. the total number of particles per unit volume, irrespective of the sign of the charge, will be given by:

$$n = n_e + n_p.$$

We can further define the following quantities for our one-fluid model;

$$\textit{mass density} \quad \rho(\boldsymbol{r}, t) = n_p m_p + n_e m_e \tag{4.28a}$$

$$\textit{charge density} \quad q(\boldsymbol{r}, t) = e(n_p - n_e) \tag{4.28b}$$

$$\textit{current density} \quad \boldsymbol{J}(\boldsymbol{r}, t) = e(n_p \boldsymbol{u}^{(p)} - n_e \boldsymbol{u}^{(e)}) \tag{4.28c}$$

An equation for $\rho$ can be easily obtained simply by multiplying the continuity equations, Eq. (4.22), of each species by the respective mass and adding them. Thus,

$$\frac{\partial \rho}{\partial t} = \nabla \cdot \left(m_e n_e \boldsymbol{u}^{(e)} + m_p n_p \boldsymbol{u}^{(p)}\right).$$

This expression can be cast in the familiar form of the mass continuity equation; by introducing the vector

$$\boldsymbol{U} = \frac{m_e n_e \boldsymbol{u}^{(e)} + m_p n_p \boldsymbol{u}^{(p)}}{m_e n_e + m_p n_p} = \frac{m_e n_e \boldsymbol{u}^{(e)} + m_p n_p \boldsymbol{u}^{(p)}}{\rho}. \tag{4.29}$$

Keeping in mind that $n_e \simeq n_p$ (charge quasi-neutrality!), we see that $\boldsymbol{U}$, called the *fluid velocity*, is a sort of local center-of-mass speed. Using this definition, we finally obtain

$$\frac{\partial \rho}{\partial t} + \nabla \cdot (\rho \boldsymbol{U}) = 0. \tag{4.30}$$

If we now multiply the equations Eq. (4.22) of the two species by the respective charges, and recall the definitions Eq. (4.28a)–(4.28c), we obtain the charge-conservation equation:

$$\frac{\partial q}{\partial t} + \nabla \cdot \boldsymbol{J} = 0. \tag{4.31}$$

Having introduced the velocity $\boldsymbol{U}$, it is natural to attempt to write the equation of motion in terms of such a velocity, by adding the two equations of motions Eq. (4.23). However, this procedure gives rise to a problem connected to the fact that the pressure tensors $\mathsf{P}_{ik}^{(s)}$ are defined in terms of the velocities $\boldsymbol{w} = \boldsymbol{v} - \boldsymbol{u}^{(s)}$, whereas, by analogy to the case of neutral gases that we use as a paradigm, it would be preferable to define the random speed as $\boldsymbol{w}' = \boldsymbol{v} - \boldsymbol{U}$. Moreover, since the pressures of the two species have to be added, it would be more logical to refer their peculiar velocities to the same fluid speed, $\boldsymbol{U}$, rather than to two different average speeds, $\boldsymbol{u}^{(e)}$ e $\boldsymbol{u}^{(p)}$. However, the new definition of the peculiar velocities implies

$$\langle \boldsymbol{w}' \rangle_s = \boldsymbol{u}^{(s)} - \boldsymbol{U} \neq 0.$$

The term $m_s n_s \langle v_i v_k \rangle_s$ then needs to be modified as follows:

$$m_s n_s \langle v_i v_k \rangle = m_s n_s \langle (w_i' + U_i)(w_k' + U_k) \rangle_s = \mathsf{P}_{ik}^{(s)} + m_s n_s (u_i^{(s)} U_k + u_k^{(s)} U_i - U_i U_k),$$

where it has to be understood that the tensors $P_{ik}^{(s)}$ are now defined in terms of the respective $\boldsymbol{w}'$. As a result, new terms appear in Eq. (4.23), that now reads

$$\frac{\partial}{\partial t}(n_s m_s u_i^{(s)}) + \frac{\partial}{\partial x_k}\left[n_s m_s (u_i^{(s)} U_k + u_k^{(s)} U_i - U_i U_k) + \mathsf{P}_{ik}^{(s)}\right] \\ - e_s n_s E_i - \frac{e_s n_s}{c}(\boldsymbol{u}^{(s)} \times \boldsymbol{B})_i = R_i^{(s)}, \tag{4.32}$$

where it has been assumed that the only forces acting on the system are those of electromagnetic origin.

At this stage we define a total pressure tensor as

$$\mathsf{P}_{ik} = \mathsf{P}_{ik}^{(e)} + \mathsf{P}_{ik}^{(p)},$$

add the two equations (4.32) for the two species and, taking into account Eqs. (4.27), (4.28b), (4.28c) e (4.29), we get:

$$\rho \frac{\partial U_i}{\partial t} + \rho U_k \frac{\partial U_i}{\partial x_k} = -\frac{\partial \mathsf{P}_{ik}}{\partial x_k} + q\, E_i + \frac{1}{c}(\boldsymbol{J} \times \boldsymbol{B})_i, \tag{4.33}$$

or, in vector notation,

$$\rho \frac{\mathrm{d}\boldsymbol{U}}{\mathrm{d}t} = -\nabla \cdot \mathsf{P} + q\boldsymbol{E} + \frac{1}{c}(\boldsymbol{J} \times \boldsymbol{B}). \tag{4.34}$$

The kinetic temperature of the one-fluid model can be defined by generalizing the definition used for neutral gases:

$$\tfrac{3}{2}nkT = \sum_s \int \tfrac{1}{2}m_s w'^2 \, f_s \mathrm{d}\boldsymbol{v} = \sum_s \tfrac{1}{2}\mathsf{P}_{ii} = \tfrac{3}{2}P,$$

obtaining

$$T = \frac{\mathsf{P}_{ii}}{3nk}. \tag{4.35}$$

A similar procedure can be adopted with the two energy equations (4.24), provided that the heat flux vectors, $\boldsymbol{q}^{(s)}$, are defined in terms of the peculiar velocities $\boldsymbol{w}'$. The term $\frac{1}{2}m_s n_s \langle v_i\, v^2\rangle_s$ now becomes

$$\begin{aligned}\tfrac{1}{2}m_s n_s \langle v_i\, v^2\rangle &= \tfrac{3}{2}P^{(s)}U_i + \mathsf{P}^{(s)}_{ik}U_k \\ &\quad + \tfrac{1}{2}m_s n_s\left(u_i^{(s)}U_kU_k + 2u_k^{(s)}U_kU_i + u_i^{(s)}U^2 - 2U_iU^2\right) + q_i^{(s)}\end{aligned} \tag{4.36}$$

with these new definitions Eqs. (4.24) reads

$$\begin{aligned}\frac{\partial}{\partial t}\left[\tfrac{1}{2}m_s n_s(2\,u_k^{(s)}U_k - U^2) + \tfrac{3}{2}P^{(s)}\right] &+ \frac{\partial}{\partial x_i}\Big[\tfrac{3}{2}P^{(s)}U_i + \mathsf{P}^{(s)}_{ik}U_k \\ &+ \tfrac{1}{2}m_s n_s\left(u_i^{(s)}U_kU_k + 2u_k^{(s)}U_kU_i + u_i^{(s)}U^2 - 2U_iU^2\right) \\ &+ q_i^{(s)}\Big] - e_s n_s E_k u_k^{(s)} = Q^{(s)}.\end{aligned} \tag{4.37}$$

Finally, defining the total heat flux vector as $\boldsymbol{q} = \boldsymbol{q}^{(e)} + \boldsymbol{q}^{(p)}$ and adding the two energy equations (4.37) we obtain the energy equation for the one-fluid model:

$$\frac{\partial}{\partial t}\left(\tfrac{1}{2}\rho U^2 + \tfrac{3}{2}P\right) + \frac{\partial}{\partial x_i}\left[U_i\left(\tfrac{1}{2}\rho U^2 + \tfrac{5}{2}P\right) + \Pi_{ik}U_k + q_i\right] - J_k E_k = 0. \tag{4.38}$$

The meaning of the term $\boldsymbol{J}\cdot\boldsymbol{E}$ in the preceding equation becomes clearer if we write it in the form:

$$\boldsymbol{J}\cdot\boldsymbol{E} = \left(\frac{c}{4\pi}(\nabla\times\boldsymbol{B}) - \frac{c}{4\pi}\frac{\partial\boldsymbol{E}}{\partial t}\right)\cdot\boldsymbol{E} =$$

$$= -\frac{c}{4\pi}\Big[\nabla\cdot(\boldsymbol{E}\times\boldsymbol{B}) - \boldsymbol{B}\cdot(\nabla\times\boldsymbol{E})\Big] - \frac{c}{4\pi}\frac{\partial\boldsymbol{E}}{\partial t}\cdot\boldsymbol{E},$$

where use has been made of the vector identity

$$\nabla \cdot (\boldsymbol{F} \times \boldsymbol{G}) = \boldsymbol{G} \cdot (\nabla \times \boldsymbol{F}) - \boldsymbol{F} \cdot (\nabla \times \boldsymbol{G}).$$

If we now use the Maxwell equation for $\nabla \times \boldsymbol{E}$ we get the equation describing electromagnetic energy density

$$\boldsymbol{J} \cdot \boldsymbol{E} = -\frac{c}{4\pi}\left[\nabla \cdot \left(\boldsymbol{E} \times \boldsymbol{B}\right) + \frac{1}{c}\left(\boldsymbol{B} \cdot \frac{\partial \boldsymbol{B}}{\partial t} + \boldsymbol{E} \cdot \frac{\partial \boldsymbol{E}}{\partial t}\right)\right]$$
$$= -\nabla \cdot \boldsymbol{S} - \frac{\partial}{\partial t}\left(\frac{B^2}{8\pi} + \frac{E^2}{8\pi}\right),$$

where the Poynting vector $\boldsymbol{S} = (c/4\pi)(\boldsymbol{E} \times \boldsymbol{B})$, representing electromagnetic energy flux, has been introduced. This particular form of the term $\boldsymbol{J} \cdot \boldsymbol{E}$ is completely general and is valid both for ideal and dissipative plasmas. Indeed, the coupling between the electromagnetic energy equation and the plasma energy equation occurs precisely via $\boldsymbol{J} \cdot \boldsymbol{E}$, which when positive provides a loss for the electromagnetic field and a corresponding source for plasma energy, so that total energy is conserved. Finite conductivity effects are implicitly contained in $\boldsymbol{E}$, as we shall see when discussing the generalized form of Ohm's law.

Equation (4.38) now becomes

$$\frac{\partial}{\partial t}\left(\tfrac{1}{2}\rho U^2 + \tfrac{3}{2}P + \frac{B^2}{8\pi} + \frac{E^2}{8\pi}\right) + \frac{\partial}{\partial x_i}\left[U_i\left(\tfrac{1}{2}\rho U^2 + \tfrac{5}{2}P\right) + \Pi_{ik}U_k + q_i + S_i\right] = 0. \tag{4.39}$$

The physical meaning of this equation is clear: the time variation of the total energy (including the contributions coming from electric and magnetic fields) is balanced by the flux terms that also contain the dissipative effects, some of which appear explicitly while others are implicitly contained in the Poynting vector, as already remarked. Therefore, even if a magnetic diffusivity term $\eta = c^2/4\pi\sigma$ does not appear explicitly, the transformation of magnetic energy into different energy forms, such as thermal energy or kinetic energy of accelerated particles, can still take place. We may now follow the procedure that allowed us to transform Eq. (4.16) into Eq. (4.18). Taking into account the equation of motion (4.33) it can be easily verified that the energy equation can be written as:

$$\frac{1}{\gamma - 1}\rho^{\gamma}\frac{\mathrm{d}}{\mathrm{d}t}\left(P\rho^{-\gamma}\right) = -\Pi_{ik}\frac{\partial U_i}{\partial x_k} - \frac{\partial q_k}{\partial x_k} + \left(qU_k - J_k\right)\left(E_k + \frac{1}{c}(\boldsymbol{U} \times \boldsymbol{B})_k\right). \tag{4.40}$$

Notice that the term $q\boldsymbol{U} \cdot (\boldsymbol{U} \times \boldsymbol{B})$ identically vanishes. It has been introduced only to write the above equation in a form suitable for further developments.

The system of equations formed by Eqs. (4.30), (4.31), (4.33) and (4.38) [or (4.40)], coupled to Maxwell's equations for $\nabla \times \boldsymbol{E}$ and $\nabla \times \boldsymbol{B}$, is the basis of the one-fluid model. It consists of 12 scalar equations in the 21 unknowns $\rho, P, \boldsymbol{U}, q, \boldsymbol{J}, \boldsymbol{E}, \boldsymbol{B}, \Pi_{ik}, \boldsymbol{q}$. It is quite evident that here too we will have to face the closure problem, namely to find ways to express the last 6 unknowns in terms of the previous ones. But, even leaving this problem aside, we still need three extra scalar equations (or one vector equation) to equate the number of equations and unknowns.

To find the missing vector equation, we notice that the equation of motion for the one-fluid model was obtained by adding the two vector equations of motions for the two species. A second independent equation can be obviously obtained by subtracting them. To proceed, we first multiply each of the equations by $e_s/m_s$ and then add the resulting equations, which is practically equivalent to subtracting them because of the opposite sign of the electrical charges. But now the collisional terms $\boldsymbol{R}^{(s)}$ no longer cancel each other and to close the system we must express their difference in terms of the fundamental variables of the fluid model. Explicitly,

$$\begin{aligned}
&\frac{\partial}{\partial t}\Big(en_p u_i^{(p)} - en_e u_i^{(p)}\Big) \\
&\quad + \frac{\partial}{\partial x_k}\Big[en_p\Big(u_i^{(p)}U_k + u_k^{(p)}U_i\Big) - en_e\Big(u_i^{(e)}U_k + u_k^{(e)}U_i\Big) + e\Big(\frac{\mathsf{P}_{ik}^{(p)}}{m_p} - \frac{\mathsf{P}_{ik}^{(e)}}{m_e}\Big)\Big] \\
&\quad - e^2\Big(\frac{n_p}{m_p} + \frac{n_e}{m_e}\Big)E_i - \frac{e^2}{c}\Big(\frac{n_p}{m_p}(\boldsymbol{u}^{(p)} \times \boldsymbol{B})_i + \frac{n_e}{m_e}\boldsymbol{u}^{(e)} \times \boldsymbol{B})_i\Big) \\
&= e\Big(\frac{R_i^{(p)}}{m_p} - \frac{R_i^{(e)}}{m_e}\Big) = eR_i^{(p)}\Big(\frac{1}{m_p} + \frac{1}{m_e}\Big),
\end{aligned}$$

where use has been made of Eq. (4.27). As already stated, $e_p = e$, $e_e = -e$ .

Recalling the definition Eq. (4.28b) and considering that $m_e \ll m_p$ e $n_e \simeq n_p$, the preceding expression becomes

$$\begin{aligned}
&\frac{\partial J_i}{\partial t} + \frac{\partial}{\partial x_k}\Big(J_i U_k + J_k U_i\Big) - \frac{e}{m_e}\frac{\partial \mathsf{P}_{ik}^{(e)}}{\partial x_k} \\
&\quad - \frac{e^2 n_e}{m_e}E_i - \frac{e^2 n_e}{m_e m_p c}\Big[\Big(m_e\boldsymbol{u}^{(p)} + m_p\boldsymbol{u}^{(e)}\Big) \times \boldsymbol{B}\Big]_i = \frac{eR_i^{(p)}}{m_e}
\end{aligned} \tag{4.41}$$

The factor $(m_e\boldsymbol{u}^{(p)} + m_p\boldsymbol{u}^{(e)})$ entering the term that contains the magnetic field $\boldsymbol{B}$ can be rewritten, taking into account Eqs. (4.28b) and (4.29), as:

$$m_e\boldsymbol{u}^{(p)} + m_p\boldsymbol{u}^{(e)} = (m_e + m_p)\boldsymbol{U} + (m_e - m_p)\frac{\boldsymbol{J}}{en_e} \simeq m_p(\boldsymbol{U} - \frac{\boldsymbol{J}}{en_e}).$$

We still need to define the relationship between the collisional terms $\boldsymbol{R}^{(s)}$ and the fluid variables. A reasonable assumption is to express $\boldsymbol{R}^{(s)}$ in terms of *the difference between the fluid speeds* of the two species:

$$\boldsymbol{R}^{(s)} = -n_s m_s \nu_{s,s'}\Big(\boldsymbol{u}^{(s)} - \boldsymbol{u}^{(s')}\Big),$$

where the parameter $\nu_{s,s'}$ represents an average frequency for collisions of the particles of type $s$ with the particles of type $s'$. This approximation basically corresponds to envisaging that the force acting between the two species during a collision is of a viscous nature, so that it vanishes when the fluid speeds of the the species coincide. Since $n_e \simeq n_p$, Eq. (4.27) implies that

$$m_e \nu_{e,p} = m_p \nu_{p,e}.$$

Inserting the various expressions we have found into Eq. (4.41) and multiplying it by $(m_e/n_e e^2)$, we finally obtain:

$$E_i + \frac{1}{c}(\boldsymbol{U} \times \boldsymbol{B})_i - \frac{J_i}{\sigma} = \frac{m_e}{e^2 n_e}\left[\frac{\partial J_i}{\partial t} + \frac{\partial}{\partial x_k}\left(J_i U_k + J_k U_i\right)\right] + \frac{1}{e n_e c}\left(\boldsymbol{J} \times \boldsymbol{B}\right)_i - \frac{1}{e n_e}\frac{\partial \mathsf{P}^{(e)}_{ik}}{\partial x_k}, \quad (4.42)$$

where we have introduced the quantity

$$\sigma = \frac{e^2 n_e}{m_e \nu_{ep}},$$

the plasma electric conductivity [See Eq. (1.1) in the introductory chapter].[2] Equation (4.42) is known as the *generalized Ohm equation*. In fact, whenever all the terms of the rhs can be neglected, the classic form of Ohm's law in a moving, conducting medium is recovered. Equations (4.30), (4.31), (4.33) and (4.38) [or (4.40)] together with Maxwell's equations for $\nabla \times \boldsymbol{E}$ and $\nabla \times \boldsymbol{B}$ and Eq. (4.42) now provide a consistent system of equations with an equal number of equations and unknowns provided that we are able to solve the closure problem, i.e. to express the fluid quantities $\Pi_{ik}$, $\mathsf{P}^{(e)}_{ik}$, $\boldsymbol{q}$ in terms of the others. There is generally no unique way to achieve closure, however there are two situations where the system may be closed without considering any moment higher than third order.

The most radical procedure is to assume that all components of the pressure tensor $\mathsf{P}_{ik}$, as well as those of the heat flux vector $\boldsymbol{q}$, vanish. This approximation is known as the *cold plasma model*. Clearly, it is a highly idealized situation where all thermal effect are neglected. In a cold plasma the number of unknowns reduces to 14, while the number of equations is still 15. However, by inspecting the energy equation in the form (4.40), we see that the rhs identically vanishes, so that the energy equation reduces to

$$(q\boldsymbol{U} - \boldsymbol{J}) \cdot \left(\boldsymbol{E} + \frac{1}{c}(\boldsymbol{U} \times \boldsymbol{B})\right) = 0.$$

[2] We must however observe that the quantity $\nu_{ep}$ that appears in the definition of $\boldsymbol{R}^{(s)}$ relates to the transfer of **momentum** between species, while the collisional frequencies discussed in the Introduction relate to the transfer of **energy**. For a more detailed illustration of these aspects, see Lifshitz and Pitaevskii [8], Chap. 4.

This equation is automatically satisfied provided that Ohm's law takes the form:

$$\boldsymbol{E} + \frac{1}{c}(\boldsymbol{U} \times \boldsymbol{B}) = 0.$$

Thus a cold plasma turns out also to be a necessarily ideal plasma. This is not too surprising since in the cold plasma approximation we have neglected all effects associated with the existence of collisions between the microscopic particles of the plasma.

The other situation in which the system can be easily closed is that of a *collisional plasma*, namely a plasma which is always in local thermodynamical equilibrium and whose distribution function is a maxwellian. In this case we may apply the same reasonings and approximations developed for neutral gases and conclude that, as a first approximation, it is possible to write

$$\Pi = 0 \;\; ; \quad \boldsymbol{q} = 0 \;\; ; \quad P \neq 0 \;\; ; \quad P^{(e)} \simeq P^{(p)} = \tfrac{1}{2}P.$$

In the presence of a strong magnetic field and a low degree of collisionality, plasmas may retain anisotropic thermodynamic properties even within the fluid approximation, in the sense that temperatures and pressures in the directions parallel and orthogonal to the magnetic field will be different. In this limit a fluid closure was developed by Chew et al. [14], in which the pressure tensor of the plasma, considered in a one-fluid approximation, is written as

$$P_{ij} = P_{\perp}\delta_{ij} + (P_{\parallel} - P_{\perp})b_i b_j$$

where $b_i = B_i/B$ is the $i$-th component of a unit vector along the mean magnetic field $\boldsymbol{B}$ and subscripts $_{\parallel,\perp}$ indicate directions parallel and perpendicular to the magnetic field respectively. The equations describing the CGL model may be derived by repeating the procedure described in Sect. 4.3. As far as the equation of continuity is concerned, there is no change. For the single fluid equation of motion, one can proceed without separating parallel and perpendicular components for the velocity, but retaining the general tensor nature of the pressure in Eq. (4.34). Taking the the scalar product of that equation with $\boldsymbol{b}$ leads to

$$\rho \frac{\mathrm{d}\boldsymbol{U}}{\mathrm{d}t}_{\parallel} = -\nabla_{\parallel} P_{\parallel} - (P_{\perp} - P_{\parallel})\left(\frac{\nabla B}{B}\right)_{\parallel}.$$

Upon subtraction of the parallel component from Eq. (4.34) we find for the momentum equation orthogonal to the mean field

$$\rho \frac{\mathrm{d}\boldsymbol{U}}{\mathrm{d}t}_{\perp} = -\nabla_{\perp}\left(P_{\perp} + \frac{B^2}{8\pi}\right) + \left(\frac{\boldsymbol{B} \cdot \nabla \boldsymbol{B}}{4\pi}\right)_{\perp}\left(1 + \frac{P_{\perp} - P_{\parallel}}{B^2/4\pi}\right) = 0.$$

Particular care is required in deriving the equation describing internal energy (4.40), which becomes

$$\frac{1}{2}\frac{DP_\parallel}{Dt} + \frac{DP_\perp}{Dt} + \left(\frac{1}{2}P_\parallel + P_\perp\right)\frac{\partial U_j}{\partial x_j} + P_{ij}\frac{\partial U_j}{\partial x_i} = 0. \tag{4.43}$$

A separate independent equation is required for $P_\parallel$, and it is obtained from the general moment equations by considering the function $\psi = \frac{1}{2}mv_\parallel^2$, with $v_\parallel = \boldsymbol{v} \cdot \boldsymbol{b}$. The difficulty comes from the fact that $\psi$ now depends not only on velocity, but also on position and time, through the time and spatial dependence of $\boldsymbol{b}$. As a consequence, we find that

$$\frac{\partial \psi}{\partial t} = mv_\parallel \boldsymbol{v} \cdot \frac{\partial \boldsymbol{b}}{\partial t},$$

$\nabla\psi = mv_\parallel(\boldsymbol{v} \cdot \nabla\boldsymbol{b})$ and $\nabla_{\boldsymbol{v}}\psi = mv_\parallel\boldsymbol{b}$. For simplicity we will consider here only one particle species. Recall that $v_\parallel = w_\parallel + U_\parallel$ where $U$ is the mean velocity so that by definition $P_\parallel = \rho < w_\parallel^2 >$. Taking the scalar product between $U_\parallel\boldsymbol{b}$ and the momentum equation (4.33) we obtain:

$$\frac{\rho}{2}\frac{DU_\parallel^2}{Dt} - \rho U_\parallel U_i\frac{Db_i}{Dt} = -U_\parallel b_i\frac{\partial P_{ik}}{\partial x_k} + qU_\parallel E_\parallel. \tag{4.44}$$

Subtracting this equation from Eq. (3.15) for the second order moment, we obtain the equation for the parallel pressure

$$\frac{DP_\parallel}{Dt} + P_\parallel\frac{\partial U_j}{\partial x_j} + 2P_\parallel b_i b_j\frac{\partial U_i}{\partial x_j} = 0 \tag{4.45}$$

and subtracting this from Eq. (4.43), we obtain the equation describing the perpendicular pressure

$$\frac{DP_\perp}{Dt} + 2P_\perp\frac{\partial U_j}{\partial x_j} - P_\perp b_i b_j\frac{\partial U_i}{\partial x_j} = 0. \tag{4.46}$$

These equations replace the adiabatic equation of state valid for an isotropic rarified plasma and lead to interesting changes in the dynamics in the presence of a strong magnetic field.

## Questions and Problems

**4.1.** Starting from the kinetic equation (3.6), put $f = f_0 + f_1$ with $|f_1| \ll |f_0|$, and write the collisional term in the approximate form

$$\left(\frac{\partial f}{\partial t}\right)_{coll} = -\nu_c(f - f_0) = -\nu_c f_1 = -\frac{f_1}{\tau_c},$$

where $\nu_c = 1/\tau_c$ (assumed to be constant) represents the average collision freequency (see Eq. (1.14)). $f_0$ is a maxwellian function with constant $n$ and $T$ and $\boldsymbol{u} = 0$, while $f_1$ is a function of the velocity only.

(a) Discuss the meaning of the form adopted for the collisional term.

(b) Consider the stationary case ($\partial f/\partial t = 0$), with the only force acting on the system being that due to an electric field, $\boldsymbol{E}$. If $\boldsymbol{E}$ is a first order quantity, namely of the same order of $f_1$, show that the electrical conductivity $\sigma$, defined by $\boldsymbol{j} = \sigma \boldsymbol{E}$ can be written in the form given by Eq. (1.1):

$$\sigma = \frac{ne^2}{m\nu_c}.$$

Finally, taking Eq. (1.14) into account, deduce the dependence of $\sigma$ on the density and the temperature.

**4.2.** Using the same assumptions of the preceding problem, let $n$ and $T$ be weakly dependent on the position, so that their spatial gradients could be considered as first order quantities. The coefficient of thermal conductivity $\kappa$ is defined by the relation $\boldsymbol{q} = -\kappa \nabla T$, where $\boldsymbol{q}$ is the heat flux vector. Assuming that the pressure $P = nkT$ is constant, show that

$$\kappa = \frac{5Pk}{2m\nu_c}$$

and evaluate the dependence of $\kappa$ on the density and temperature.

**4.3.** In a viscous fluid the off-diagonal terms of the pressure tensor do not vanish, $\Pi_{ik} \neq 0$. If the mean flow is $\boldsymbol{u} = u(z)\boldsymbol{e}_x$, the viscosity $\nu$ is defined by

$$\rho\,\nu \frac{\mathrm{d}U}{\mathrm{d}z} = -\Pi_{xz}.$$

Using the approximate collisional term of the preceding problems show that

$$\nu = \frac{P}{\rho}\tau_c$$

and evaluate the dependence of $\nu$ on the density and temperature

## Solutions

**4.1.** (a) Since the collisional term vanishes when $f = f_0$ and collisions cause the distribution function to evolve towards the equilibrium distribution $f_0$ on a timescale characterized by $\tau_c = 1/\nu_c$, the form adopted for the collisional term is an acceptable first order representation.

(b) Within our assumptions, the first order kinetic equation reduces to

$$\frac{e\boldsymbol{E}}{m}\cdot\frac{\partial f_0}{\partial \boldsymbol{v}} = -\nu_c f_1.$$

If the electric field is written as $\boldsymbol{E} = E\boldsymbol{e}_z$, from the definition of $\sigma$ we have:

$$\sigma E = j = e\int v_z f \mathrm{d}\boldsymbol{v} = e\int v_z f_1 \,\mathrm{d}\boldsymbol{v}.$$

Multiplying the kinetic equation by $ev_z$ and integrating over the velocities we obtain

$$\sigma E = E\frac{e^2}{\nu_c kT}\int v_z^2 f_0 \,\mathrm{d}\boldsymbol{v}.$$

Performing the integral we find the required result. The electrical conductivity is independent of $n$ and scales with the temperature as $T^{3/2}$.

**4.2.** Though the pressure is constant, both $n$ and $T$ depend on the position. The equilibrium distribution function $f_0$ then reads

$$f_0 = n\Big(\frac{m}{2\pi kT}\Big)^{3/2}\exp\Big(-\frac{mv^2}{2kT}\Big) = \frac{P}{k}\Big(\frac{m}{2\pi k}\Big)^{3/2}T^{-5/2}\exp\Big(-\frac{mv^2}{2kT}\Big),$$

while the kinetic equation is simply given by

$$\boldsymbol{v}\cdot\boldsymbol{\nabla} f_0 = -\nu_c f_1.$$

If $T = T(z)$ we have $\boldsymbol{q} = q\boldsymbol{e}_z, q = -\kappa(\mathrm{d}T/\mathrm{d}z)$, with $q = \frac{1}{2}m\int v^2 v_z f_1 \,\mathrm{d}\boldsymbol{v}$, and the preceding equation becomes

$$f_0 v_z\Big(\frac{5}{2T} - \frac{mv^2}{2kT^2}\Big) = \nu_c f_1.$$

Multiplying the above equation by $v^2 v_z$ and integrating in $\mathrm{d}\boldsymbol{v}$ we obtain the required result. The thermal conductivity is independent of $n$ and scales with the temperature as $T^{5/2}$.

**4.3.** For $\boldsymbol{u} = u(z)\boldsymbol{e}_x$, $f_0$ takes up the form

$$f_0 = n\Big(\frac{m}{2\pi kT}\Big)^{3/2}\exp\{-\frac{m}{2kT}[(v_x - u(z))^2 + v_y^2 + v_z^2]\},$$

and the kinetic equation becomes

$$\frac{mv_z}{kT}\frac{\mathrm{d}u}{\mathrm{d}z}(v_x - u) f_0 = -\frac{f_1}{\tau_c}.$$

On the other hand,

$$\rho \nu \frac{\mathrm{d}u}{\mathrm{d}z} = -m \int v_x v_z f \mathrm{d}\boldsymbol{v} = -m \int v_x v_z (f_0 + f_1) \mathrm{d}\boldsymbol{v} = -m \int v_x v_z f_1 \mathrm{d}\boldsymbol{v},$$

since $\Pi_{ik} = 0$ for a maxwellian distribution. Multiplying the kinetic equation above by $v_x v_z$ and integrating in $\mathrm{d}\boldsymbol{v}$ we obtain

$$\nu = \frac{kT}{m} \tau_c = \frac{P}{\rho} \tau_c.$$

The kinematic viscosity scales as $T^{5/2}/n$.

# Chapter 5
# Magnetohydrodynamics

**Abstract** From the one-fluid plasma model the simplifications leading to magnetohydrodynamics (MHD) are discussed in detail. The general form of the MHD equations, both ideal and resistive, is presented and their consequences analyzed, in particular in connection with the time evolution of magnetic fields. MHD equilibria are then described, both when magnetic forces are absent (force-free fields) and when they are present. Finally, the concept of perturbed equilibrium states is introduced.

Although the collisional one-fluid model is the simplest of fluid models for a plasma, it is still complex enough to make general solutions difficult to recover in all but the simplest problems. In the fluid description of a plasma one typically attempts to make further approximations by limiting the range of possible parameters, namely densities, temperatures, magnetic fields and velocities to those of the specific plasma under examination, introducing the concept of a plasma *regime*. The choice of a regime generally allows one to neglect terms in the equations which are only important when either the time or spatial scales are far from those of the plasma in consideration. A very important regime derived from the collisional one-fluid model is the (classical) *magnetohydrodynamic, or MHD, regime*.

## 5.1 MHD Equations

To develop the scalings for magnetohydrodynamics, begin by defining characteristic length-scales $\mathcal{L}$ and timescales $\tau$ for electric and magnetic field dynamics, so that $\mathcal{L}$ is the spatial scale over which the fields show a significant variation and $\tau$ is the typical timescale over which the fields change. Let $\mathcal{U}$ be a representative value of the fluid velocity of the plasma. The MHD regime is then defined by the following conditions:

$$\mathcal{U} \simeq \frac{\mathcal{L}}{\tau} \quad \text{and} \quad \mathcal{U} \ll c. \tag{5.1}$$

The first requirement means that "the typical speed" of *electromagnetic* phenomena, that we identify with $\mathcal{L}/\tau$, is of the same order of magnitude as the typical speed of



C. Chiuderi and M. Velli, *Basics of Plasma Astrophysics*,
UNITEXT for Physics, DOI 10.1007/978-88-470-5280-2_5

*hydrodynamical* phenomena as defined by $\mathcal{U}$. Therefore, since the two classes of phenomena "proceed with the same speed", the potential for mutual interaction is maximized. This regime is intermediate between a situation dominated by electromagnetic effects, where hydrodynamical aspects appear as perturbations, and the symmetrical one, where hydrodynamics dominates and electromagnetic interactions bring in only small corrections. The second requirement is that the macroscopic plasma motions be non-relativistic. If this holds systematically, since individual speeds of the bulk of particles must also be close to the macroscopic fluid speed, the thermal speed of the plasma must also be non-relativistic.

To see how the one-fluid equations may be simplified in the MHD regime, we carry out a dimensional analysis, starting from Maxwell's equations. Call $\mathcal{E}$, $\mathcal{B}$, $\mathcal{Q}$, $\mathcal{J}$ the characteristic values for the electric and magnetic fields and for the charge and current densities, respectively. Begin with

$$\nabla \times \boldsymbol{E} = -\frac{1}{c}\frac{\partial \boldsymbol{B}}{\partial t},$$

which we write as:

$$\frac{\mathcal{E}}{\mathcal{L}} \simeq \frac{1}{c}\frac{\mathcal{B}}{\tau}.$$

This implies

$$\frac{\mathcal{E}}{\mathcal{B}} \simeq \frac{1}{c}\frac{\mathcal{L}}{\tau} \simeq \frac{\mathcal{U}}{c} \ll 1. \tag{5.2}$$

Similarly, the equation for $\nabla \times \boldsymbol{B}$ gives:

$$\nabla \times \boldsymbol{B} = \frac{4\pi}{c}\boldsymbol{J} + \frac{1}{c}\frac{\partial \boldsymbol{E}}{\partial t} \quad \Rightarrow \quad \frac{\mathcal{B}}{\mathcal{L}} \simeq \frac{4\pi}{c}\mathcal{J} + \frac{\mathcal{E}}{\tau}.$$

Dividing both members by $\mathcal{B}/\mathcal{L}$ we get:

$$1 \simeq \frac{4\pi}{c}\mathcal{J}\frac{\mathcal{L}}{\mathcal{B}} + \frac{\mathcal{E}}{\mathcal{B}}\frac{\mathcal{L}}{\tau} \simeq \frac{4\pi}{c}\mathcal{J}\frac{\mathcal{L}}{\mathcal{B}} + \left(\frac{\mathcal{U}}{c}\right)^2.$$

The last term is negligible with respect to unity and the equation for $\nabla \times \boldsymbol{B}$ reduces to

$$\nabla \times \boldsymbol{B} = \frac{4\pi}{c}\boldsymbol{J}. \tag{5.3}$$

In the MHD regime it is thus possible to neglect the displacement current, which makes it clear that we are in a *low frequency* regime. The displacement current, in fact, becomes important only when $\boldsymbol{E}$ varies rapidly with time, namely in a *high frequency* regime. The elimination of the displacement current implies that the continuity equation for the electrical charge must also change. A dimensional analysis of the latter gives

$$\frac{\mathcal{Q}}{\tau} + \frac{\mathcal{J}}{\mathcal{L}} = 0,$$

and, replacing $\mathcal{J}$ with the value obtained by a dimensional analysis of (5.3) and $\mathcal{Q}$ with a similar analysis of the equation for $\nabla \cdot \boldsymbol{E} \Rightarrow \mathcal{E}/\mathcal{L} \simeq 4\pi\,\mathcal{Q}$, we obtain

$$\frac{\mathcal{E}}{4\pi\mathcal{L}}\frac{1}{\tau} + \frac{c\mathcal{B}}{4\pi\mathcal{L}}\frac{1}{\mathcal{L}} = 0.$$

Therefore, the ratio of of the first term to the second one is

$$\frac{\mathcal{E}}{\mathcal{B}}\frac{\mathcal{L}}{\tau} \simeq \left(\frac{\mathcal{U}}{c}\right)^2 \ll 1,$$

implying that in the MHD regime the term with the temporal derivative of the charge density is negligible: the equation for the charge conservation can be written simply as $\nabla \cdot \boldsymbol{J} = 0$, in perfect agreement with Eq. (5.3).

Coming now to the momentum equation (4.43) and performing the usual dimensional analysis, one obtains:

$$\rho\frac{\mathcal{U}}{\tau} \simeq -\frac{P}{\ell} + \mathcal{Q}\mathcal{E} + \frac{1}{c}\mathcal{J}\mathcal{B}.$$

Using the estimates found previously, the ratio of the electric to the magnetic parts of the Lorentz force becomes

$$\frac{\mathcal{Q}\mathcal{E}}{\mathcal{J}\mathcal{B}/c} \simeq \left(\frac{\mathcal{E}}{\mathcal{B}}\right)^2 \simeq \left(\frac{\mathcal{U}}{c}\right)^2 \ll 1,$$

so that the electric part of the force may be neglected, leading to the form of momentum equation in MHD:

$$\rho\frac{\mathrm{d}\boldsymbol{U}}{\mathrm{d}t} = -\nabla P + \frac{1}{c}\boldsymbol{J} \times \boldsymbol{B}. \tag{5.4}$$

To estimate the relative importance of the various terms entering Ohm's equation (5.5), that for convenience we repeat here,

$$E_i + \frac{1}{c}\Big(\boldsymbol{U} \times \boldsymbol{B}\Big)_i - \frac{J_i}{\sigma} = \frac{m_e}{e^2 n_e}\left[\frac{\partial J_i}{\partial t} + \frac{\partial}{\partial x_k}\Big(J_i U_k + J_k U_i\Big)\right] + \frac{1}{e n_e c}\Big(\boldsymbol{J} \times \boldsymbol{B}\Big)_i - \frac{1}{e n_e}\frac{\partial \mathsf{P}_{ik}^{(e)}}{\partial x_k}, \tag{5.5}$$

begin by dividing all terms by the representative value of the electric field $E$. Notice that the two terms in square bracket have the same order of magnitude. Defining

$$\omega \simeq \tau^{-1}, \quad c_s \simeq \left(P/\rho\right)^{\frac{1}{2}},$$

and comparing the various terms in the same order in which they enter the above equation we obtain:

$$1:1:\left(\omega/\omega_{pe}\right)\left(\nu_{ep}/\omega_{pe}\right)\left(c/\mathcal{U}\right)^2:\left(\omega/\omega_{pe}\right)^2\left(c/\mathcal{U}\right)^2:$$
$$\left(\omega/\omega_{pe}\right)\left(\omega_{ce}/\omega_{pe}\right)\left(c/\mathcal{U}\right)^2:\left(\omega/\omega_{cp}\right)^2\left(c_s/\mathcal{U}\right)^2,$$

where $\omega_{pe}$ is the electron plasma frequency, $\omega_{ce}$ and $\omega_{cp}$ are, respectively, the electron and proton cyclotron frequencies and $\nu_{ep}$ is the electron-proton collision frequency. We thus see that, in order to neglect the terms in square bracket, we must have

$$\left(\omega/\omega_{pe}\right) \ll \mathcal{U}/c\,, \tag{5.6}$$

to neglect the term proportional to $\boldsymbol{J} \times \boldsymbol{B}$, connected with the so-called *Hall effect* we must have

$$\left(\omega\omega_{ce}/\omega_{pe}^2\right) \ll \left(\mathcal{U}/c\right)^2, \tag{5.7}$$

which may also be rewritten in terms of the proton cyclotron frequency $\omega_{cp}$ and the Alfvén speed $c_a = B/\sqrt{4\pi m_p n_p}$ as:

$$\omega/\omega_{cp} \ll \left(\mathcal{U}/c_a\right)^2. \tag{5.8}$$

Finally, the electron pressure term can be neglected when

$$\left(\omega/\omega_{cp}\right) \ll \left(\mathcal{U}/c_s\right)^2. \tag{5.9}$$

When all these conditions are satisfied, which is most easily achieved in the *low frequency* regime typical of MHD, Ohm's equation reduces simply to

$$\boldsymbol{E} + \frac{1}{c}\boldsymbol{U} \times \boldsymbol{B} = \frac{\boldsymbol{J}}{\sigma}, \tag{5.10}$$

which is what is usually known as Ohm's equation for a *resistive plasma*. If the further condition

$$\left(\omega \nu_{ep}/\omega_{pe}^2\right) \ll \left(\mathcal{U}/c\right)^2, \tag{5.11}$$

is satisfied, the term on the rhs of equation (5.10) can also be neglected. This happens in the presence of a very high electrical conductivity, in which case we obtain the form of Ohm's equation appropriate to an *ideal plasma*:

$$\boldsymbol{E} + \frac{1}{c}\boldsymbol{U} \times \boldsymbol{B} = 0. \tag{5.12}$$

Finally, noticing that the term $\mathcal{Q}\mathcal{U}$ is of the order of $(\mathcal{U}/c)^2$ with respect to $\mathcal{J}$, the energy equation (4.40) can be written as

$$\frac{1}{\gamma - 1}\rho^\gamma \frac{\mathrm{d}}{\mathrm{d}t}\left(P\,\rho^{-\gamma}\right) = \frac{J^2}{\sigma}. \tag{5.13}$$

The fundamental equations of the one-fluid model in the MHD regime are therefore Eqs. (4.30), (5.4), (5.13) and Eq. (5.3), with the addition of the equations that still contain the electric field, namely:

$$\nabla \times \boldsymbol{E} = -\frac{1}{c}\frac{\partial \boldsymbol{B}}{\partial t}$$

$$\boldsymbol{E} + \frac{1}{c}\boldsymbol{U} \times \boldsymbol{B} = \frac{\boldsymbol{J}}{\sigma}.$$

However by taking the curl of Ohm's law and eliminating the the electric field via the equation for $\nabla \times \boldsymbol{E}$ one obtains an equation containing only the fluid velocity and the magnetic field:

$$\begin{aligned}
\nabla \times \boldsymbol{E} + \frac{1}{c}\nabla \times (\boldsymbol{U} \times B) &= -\frac{1}{c}\frac{\partial \boldsymbol{B}}{\partial t} + \frac{1}{c}\nabla \times (\boldsymbol{U} \times \boldsymbol{B}) \\
&= \nabla \times \left(\frac{\boldsymbol{J}}{\sigma}\right) \\
&= \nabla \times \left(\frac{c}{4\pi\sigma}\nabla \times \boldsymbol{B}\right).
\end{aligned}$$

Defining the *magnetic diffusivity*,

$$\eta = \frac{c^2}{4\pi\sigma}, \tag{5.14}$$

expanding the lhs and remembering that $\nabla \cdot \boldsymbol{B} = 0$, we obtain

$$\frac{\partial \boldsymbol{B}}{\partial t} = \nabla \times (\boldsymbol{U} \times \boldsymbol{B}) + \eta \nabla^2 \boldsymbol{B} - \nabla \eta \times (\nabla \times \boldsymbol{B}). \tag{5.15}$$

This equation is known as the *magnetic induction equation*, sometimes also called *Faraday's equation*, and allows a further reduction of the number of equations of the MHD model.

Summarizing, the *resistive MHD equations*, are:

$$\frac{\partial \rho}{\partial t} + \nabla \cdot (\rho \boldsymbol{U}) = 0 \tag{5.16a}$$

$$\rho \frac{\mathrm{d}\boldsymbol{U}}{\mathrm{d}t} = -\nabla P + \tfrac{1}{c} \boldsymbol{J} \times \boldsymbol{B} = -\nabla P + \frac{1}{4\pi} (\nabla \times \boldsymbol{B}) \times \boldsymbol{B} \tag{5.16b}$$

$$\frac{1}{\gamma - 1} \rho^\gamma \frac{\mathrm{d}}{\mathrm{d}t} \left( P \rho^{-\gamma} \right) = \frac{4\pi}{c^2} \eta J^2 \tag{5.16c}$$

$$\frac{\partial \boldsymbol{B}}{\partial t} = \nabla \times (\boldsymbol{U} \times \boldsymbol{B}) + \eta \nabla^2 \boldsymbol{B} - \nabla \eta \times (\nabla \times \boldsymbol{B}). \tag{5.16d}$$

In the momentum equation we have expressed the current density in terms of $\boldsymbol{B}$ using Eq. (5.3). When the electrical conductivity is constant the last term of Eqs. (5.15) and (5.16d) vanishes. The equations for an ideal plasma are easily obtained by taking $\eta = 0$, or, equivalently, by letting $\sigma$ tend to infinity.

The system (5.16a–5.16d) is a closed system that completely determines the 8 *primary* unknowns $\rho$, $P$, $\boldsymbol{U}$, $\boldsymbol{B}$. The other quantities can then be deduced in terms of the primary ones. The current density is obviously given by

$$\boldsymbol{J} = \frac{c}{4\pi} (\nabla \times \boldsymbol{B}),$$

the electric field is obtained from Ohm's law

$$\boldsymbol{E} = -\frac{1}{c} \boldsymbol{U} \times \boldsymbol{B} + \frac{\boldsymbol{J}}{\sigma},$$

and the charge density from

$$q = \frac{1}{4\pi} (\nabla \cdot \boldsymbol{E}).$$

The electrical conductivity $\sigma$ is assumed to be a known quantity and depends in general on density and temperature via the collision frequency, though it will often be taken as a constant in the text.

By comparing the MHD equations with the corresponding ones for a neutral gas, we realize that they represent the minimum possible number of equations describing a plasma as a conducting medium. In spite of the fact that inclusion of electromagnetic effects introduces three vector quantities ($\boldsymbol{E}$, $\boldsymbol{B}$, $\boldsymbol{J}$) and one scalar quantity ($q$) when compared to neutral flow, the primary equations in MHD have only been increased by three, thanks to the choice of an appropriate regime.

### 5.1.1 Magnetic Pressure

By means of the MHD equations it is possible to investigate some aspects of the interactions between the plasma and the magnetic field that are not immediately apparent.

Let's start from the momentum equation (5.16b) whose last term can be modified making use of the vector identity:

$$\nabla(\boldsymbol{F}\cdot\boldsymbol{G}) = (\boldsymbol{F}\cdot\nabla)\boldsymbol{G} + (\boldsymbol{G}\cdot\nabla)\boldsymbol{F} + \boldsymbol{F}\times(\nabla\times\boldsymbol{G}) + \boldsymbol{G}\times(\nabla\times\boldsymbol{F}). \tag{5.17}$$

Using $\boldsymbol{F} = \boldsymbol{G} = \boldsymbol{B}$ we obtain

$$\frac{1}{4\pi}(\nabla\times\boldsymbol{B})\times\boldsymbol{B} = \frac{1}{4\pi}(\boldsymbol{B}\cdot\nabla)\boldsymbol{B} - \frac{1}{8\pi}\nabla\left(B^2\right),$$

and Eq. (5.16b) can be written as

$$\rho\frac{\mathrm{d}\boldsymbol{U}}{\mathrm{d}t} = -\nabla\left(P + \frac{B^2}{8\pi}\right) + \frac{1}{4\pi}(\boldsymbol{B}\cdot\nabla)\boldsymbol{B}. \tag{5.18}$$

The $i$-th component of the preceding equation is

$$\rho\frac{\mathrm{d}U_i}{\mathrm{d}t} = -\frac{\partial}{\partial x_i}\left(P + \frac{B^2}{8\pi}\right) + \frac{1}{4\pi}B_k\frac{\partial B_i}{\partial x_k} = -\frac{\partial}{\partial x_i}\left(P + \frac{B^2}{8\pi}\right) + \frac{1}{4\pi}\frac{\partial(B_i B_k)}{\partial x_k},$$

namely

$$\rho\frac{\mathrm{d}U_i}{\mathrm{d}t} = -\frac{\partial}{\partial x_k}\mathsf{T}_{ik}, \tag{5.19}$$

where use has been made of $\nabla\cdot\boldsymbol{B} = 0$ and a new tensor has been introduced:

$$\mathsf{T}_{ik} = \left(P + \frac{B^2}{8\pi}\right)\delta_{ik} - \frac{1}{4\pi}B_i B_k. \tag{5.20}$$

Assuming that the z-axis of our reference system is aligned with the magnetic field, which is always achievable *locally* by an appropriate rotation of the axes, the tensor $\mathsf{T}_{ik}$ reduces to

$$\begin{pmatrix} P + \dfrac{B^2}{8\pi} & 0 & 0 \\ 0 & P + \dfrac{B^2}{8\pi} & 0 \\ 0 & 0 & P - \dfrac{B^2}{8\pi} \end{pmatrix}.$$

We thus see that the presence of a magnetic field introduces an extra isotropic pressure, $B^2/8\pi$, and an *anisotropic negative pressure* (i.e. a tension) along the field,

$-B^2/4\pi$. To better appreciate the effect of the tension term, let's examine the case of a magnetic field given by $\boldsymbol{B} = [0, B_y(x), B_z(x)]$, whose field lines are straight lines (for $B_y = 0$, the field lines are straight lines parallel to the z-axis, while if $B_y \neq 0$ the field lines are straight in any given plane $x = const.$, but their inclination changes when $x$ is varied). For this field the tension term vanishes, which suggests that the tension is active only when the field lines are curved. In a certain sense, the field lines of $\boldsymbol{B}$ behave as they were composed of an elastic material: any deformation induces a tension that tends to restore a configuration with unbent, straight field lines.

The relative importance of plasma kinetic to magnetic pressures is measured by the parameter $\beta$, defined by

$$\beta = \frac{P}{B^2/8\pi}. \tag{5.21}$$

Whenever $\beta \gg 1$ the system's dynamics is dominated by hydrodynamical effects, while when $\beta \ll 1$ magnetic effects are dominant. This is confirmed by the fact that $\beta$ can also be written in the form

$$\beta = \frac{2}{3}\frac{E_{th}}{E_{mag}},$$

where $E_{th}/V = \frac{3}{2}P$ is the thermal energy density and $E_{mag}/V = B^2/8\pi$ is the magnetic energy density.

### *5.1.2 The Conservative Form of MHD Equations*

The continuity equation (5.16a)

$$\frac{\partial \rho}{\partial t} + \nabla \cdot (\rho\, \boldsymbol{U}) = 0,$$

describes mass conservation in differential form. It states that the amount of mass contained in a given volume varies with time due to mass flux crossing the surface that encloses the volume. More generally, an equation is said to be conservative if it is possible to cast it in a form similar to the continuity equation, namely:

$$\frac{\partial \Sigma}{\partial t} + \nabla \cdot \boldsymbol{\Phi} = 0,$$

where $\Sigma$ is the density of some quantity, and $\boldsymbol{\Phi}$ can be thought as the flux of the same quantity. As written, the equation refers to the density of a *scalar* quantity. Its generalization to a *vector* quantity $\boldsymbol{\Sigma}$ is:

$$\frac{\partial \Sigma_i}{\partial t} + \frac{\partial \Phi_{ik}}{\partial x_k} = 0,$$

where $\Phi$, a tensor, is now the flux of the vector density. The equations of magnetohydrodynamics may all be cast in such conservative form. The conservative form of MHD is useful both for the physical intuition it provides, and, more technically, for its importance when attempting to solve the equations numerically. The lhs of the momentum equation can be written as

$$\begin{aligned}\rho\frac{\partial U_i}{\partial t}+\rho U_k\frac{\partial U_i}{\partial x_k} &= \frac{\partial(\rho U_i)}{\partial t}-U_i\frac{\partial\rho}{\partial t}+\rho U_k\frac{\partial U_i}{\partial x_k}\\ &= \frac{\partial(\rho U_i)}{\partial t}+U_i\frac{\partial(\rho U_k)}{\partial x_k}+\rho U_k\frac{\partial U_i}{\partial x_k}\\ &= \frac{\partial(\rho U_i)}{\partial t}+\frac{\partial(\rho U_i U_k)}{\partial x_k},\end{aligned}$$

where use has been made of the continuity equation. It is now easy to see that it is possible to write the momentum equation in conservative form:

$$\frac{\partial}{\partial t}(\rho\, U_i)+\frac{\partial}{\partial x_k}\left[\rho\, U_i U_k+(P+\frac{B^2}{8\pi})\delta_{ik}-\frac{1}{4\pi}B_i B_k\right]=0. \tag{5.22}$$

In this case:

$$\boldsymbol{\Sigma}=\rho\,\boldsymbol{U};\quad \Phi_{ik}=\mathsf{T}_{ik}+\rho\, U_i U_k,$$

with the tensor flux $\mathsf{T}_{ik}$ given by Eq. (5.20).

The energy equation in the form given by (5.16c) cannot be cast in conservative form, because it describes how the *internal* energy varies with time, and it is not internal energy density that is conserved, but the total energy density. The internal energy per unit mass of a perfect gas is given by

$$W=c_v T=c_v\frac{R}{\mu}\frac{P}{\rho}=\frac{1}{\gamma-1}\frac{P}{\rho},$$

and the internal energy per unit volume is therefore $\epsilon=\rho\, W=P/(\gamma-1)$. It is easily verified that Eq. (5.16c) can be written as

$$\frac{\partial\epsilon}{\partial t}+(\boldsymbol{U}\cdot\nabla)\epsilon+\gamma\epsilon(\nabla\cdot\boldsymbol{U})=\frac{4\pi}{c^2}\eta J^2,$$

showing that the internal energy of a given volume can vary both due to compressive effects (the term proportional to $\nabla\cdot\boldsymbol{U}$) and to dissipative effects (the term proportional to $\eta$).

To obtain a conservation law we must utilize Eq. (4.39) in the MHD regime, i.e.

$$\frac{\partial}{\partial t}\left(\tfrac{1}{2}\rho U^2+\frac{P}{\gamma-1}+\frac{B^2}{8\pi}\right)+\frac{\partial}{\partial x_i}\left[U_i(\tfrac{1}{2}\rho U^2+\frac{\gamma P}{\gamma-1})+\frac{c}{4\pi}(\boldsymbol{E}\times\boldsymbol{B})_i\right]=0. \tag{5.23}$$

This equation describes the conservation of total energy density and is already written in conservative form.

Finally, we must consider the induction equation (5.16d). Generally, this equation is intrinsically non conservative, because of the presence of the dissipative terms containing $\eta$. However, it may be cast in a form that becomes conservative for ideal plasmas for which $\eta = 0$. The i-th component of $\nabla \times (\boldsymbol{U} \times \boldsymbol{B})$ is:

$$[\nabla \times (\boldsymbol{U} \times \boldsymbol{B})]_i = \epsilon_{ijk} \frac{\partial}{\partial x_j} (\epsilon_{klm} U_l B_m) = \frac{\partial}{\partial x_j} (U_i B_j - B_i U_j),$$

where $\epsilon_{ijk}$ is the antisymmetric tensor and use has been made of

$$\epsilon_{ijk}\epsilon_{klm} = \epsilon_{kij}\epsilon_{klm} = \delta_{il}\delta_{jm} - \delta_{im}\delta_{jl}.$$

The above relationship makes it possible to write the induction equation as:

$$\frac{\partial B_i}{\partial t} + \frac{\partial}{\partial x_k} (U_k B_i - U_i B_k) = \left[ \eta \nabla^2 \boldsymbol{B} - \nabla \eta \times (\nabla \times \boldsymbol{B}) \right]_i , \tag{5.24}$$

which is manifestly conservative when $\eta = 0$. In the next section we shall prove that this equation represents the magnetic flux conservation for ideal plasmas.

## 5.2 The Time Evolution of Magnetic Fields

Consider Farday's equation assuming that the resistivity $\eta$, Eq. (5.14), is a constant, independent of thermodynamical variables (e.g. density and temperature):

$$\frac{\partial \boldsymbol{B}}{\partial t} = \nabla \times (\boldsymbol{U} \times \boldsymbol{B}) + \eta \nabla^2 \boldsymbol{B}. \tag{5.25}$$

If we assume that the velocity field $\boldsymbol{U}$ is known then Eq. (5.25) does not require any other of the MHD equations for closure, and may be solved for the magnetic field once an initial field is given at an initial time. This approach to understanding the evolution of magnetic fields is known as the *kinematical approach*, and in the rest of this section we will follow it, disregarding the connection between Eq. (5.25) and the other MHD equations, in particular the momentum equation (5.16b).

From Eq. (5.25) we see that the temporal variation of $\boldsymbol{B}$ is given by two terms, each describing a different physical process. The first term, containing the fluid velocity $\boldsymbol{U}$, is a *convective* term, describing how the field changes due to convection and concentration and rarefaction from the velocity field; the second term, in which $\boldsymbol{U}$ is absent, is a *diffusive* term. The corresponding processes occur on completely different timescales. According to the usual dimensional analysis the convective term can be evaluated as ($\mathcal{U}$ being the magnitude of the velocity)

$$\frac{\mathcal{B}}{\tau_f} \quad \text{with} \quad \tau_f = \mathcal{L}/\mathcal{U},$$

while for the diffusive term one has

$$\frac{\mathcal{B}}{\tau_d} \quad \text{with} \quad \tau_d = \mathcal{L}^2/\eta,$$

with $\tau_f$ and $\tau_d$ corresponding to convective and diffusion timescales respectively. The relative importance of convection and diffusion is measured by the ratio of the magnitude of the two terms

$$\mathcal{R}_m = \frac{\tau_d}{\tau_f} = \frac{\mathcal{U}\mathcal{L}}{\eta},$$

defining the *magnetic Reynolds number*. If, as often happens in MHD, the typical convection speed is of the same order of the Alfvén speed, see Eq. (1.21), typical of magnetic phenomena, the magnetic Reynolds number is called the *Lundquist number*, defined by

$$S = \frac{c_a \mathcal{L}}{\eta}.$$

Consider now the two limiting cases of small $\mathcal{R}_m \ll 1$ and large $\mathcal{R}_m \gg 1$ Reynolds numbers.

### 5.2.1 $\mathcal{R}_m \ll 1$: *Magnetic Diffusion*

If $\mathcal{R}_m \ll 1$ the convective term can be neglected and the induction equation reduces to

$$\frac{\partial \boldsymbol{B}}{\partial t} = \eta \nabla^2 \boldsymbol{B}.$$

This being a linear differential equation for $\boldsymbol{B}$, it is possible to apply the Fourier analysis method.[1] We therefore write the magnetic field in terms of its Fourier transform

$$\boldsymbol{B}(\boldsymbol{r}, t) = \int \mathrm{d}\boldsymbol{k}\, \mathrm{d}\omega \boldsymbol{B}(\boldsymbol{k}, \omega)\, \mathrm{e}^{\mathrm{i}(\boldsymbol{k}\cdot\boldsymbol{r} - \omega t)},$$

and introduce the above representation into the equation for $\boldsymbol{B}$ to obtain

$$\int \mathrm{d}\boldsymbol{k}\, \mathrm{d}\omega (\mathrm{i}\omega - \eta k^2) \boldsymbol{B}(\boldsymbol{k}, \omega)\, \mathrm{e}^{\mathrm{i}(\boldsymbol{k}\cdot\boldsymbol{r} - \omega t)} = 0.$$

[1] A short presentation of Fourier analysis will be given in Sect. 7.1, to which we refer for details. We shall anticipate here a few fundamental concepts, which should however be already known to the reader.

Since the above equation must be identically satisfied for every $\boldsymbol{B}(\boldsymbol{k}, \omega)$, it follows that $\omega = -\mathrm{i}\eta k^2$. To take this condition into account, introduce the Dirac $\delta$ function and write

$$\boldsymbol{B}(\boldsymbol{r}, t) = \int \mathrm{d}\boldsymbol{k}\, \mathrm{d}\omega\, \delta(\omega + \mathrm{i}\eta k^2) \boldsymbol{B}(\boldsymbol{k}, \omega)\, \mathrm{e}^{\mathrm{i}(\boldsymbol{k}\cdot\boldsymbol{r} - \omega t)}.$$

Performing the $\omega$ integration we get:

$$\boldsymbol{B}(\boldsymbol{r}, t) = \int \mathrm{d}\boldsymbol{k}\, \mathrm{e}^{-\eta k^2 t} \boldsymbol{B}(\boldsymbol{k})\, \mathrm{e}^{\mathrm{i}\boldsymbol{k}\cdot\boldsymbol{r}}. \tag{5.26}$$

This is the Fourier representation of the solution of the diffusive equation: it states that the Fourier components of a generic magnetic field, given at $t = 0$ by $\boldsymbol{B}(\boldsymbol{k})$, decay exponentially with time. The magnetic energy $(B^2/8\pi)V$ (where $V$ is the volume) decreases because the presence of resistivity transforms magnetic energy into other forms of energy. A part of the original magnetic energy goes into thermal energy (Joule effect), but in general a fraction may be converted into kinetic energy from acceleration of the fluid. In fact, if $\partial B/\partial t \neq 0$ an electric field is generated which might accelerate plasma particles. Notice that the components corresponding to large values of $k$, i.e. to small wavelengths, decrease faster. Since small wavelengths describe rapid spatial variations, the field becomes smoother during its general decrease.

The diffusive time $\tau_d$ can have enormously different values for different systems. For instance, in thermonuclear fusion machines $\tau_d \simeq 10$ s, but in the liquid Earth core its value increases up to $10^4$ years, while in the Sun's interior it can reach up to$10^{10}$ years, of the same order or larger than the age of the Sun itself.

### 5.2.2 $\mathcal{R}_m \gg 1$: Alfvén's Theorem

When the electrical conductivity is very high and/or the spatial scales are very large $\mathcal{R}_m \gg 1$, so that the diffusive term in Eq. (5.25) becomes negligible. Such conditions often arise in natural plasmas (as given by the example of the Sun above), which, consequently, can be approximated by ideal plasmas, with $\eta = 0$, in many respects. In the latter case, Eq. (5.25) simplifies to

$$\frac{\partial \boldsymbol{B}}{\partial t} = \nabla \times (\boldsymbol{U} \times \boldsymbol{B}). \tag{5.27}$$

and its solutions exhibit some important features that we are now going to illustrate. The most important result is contained in

**Alfvén's theorem**: *The magnetic flux through any closed line that moves with the fluid is constant in time.*

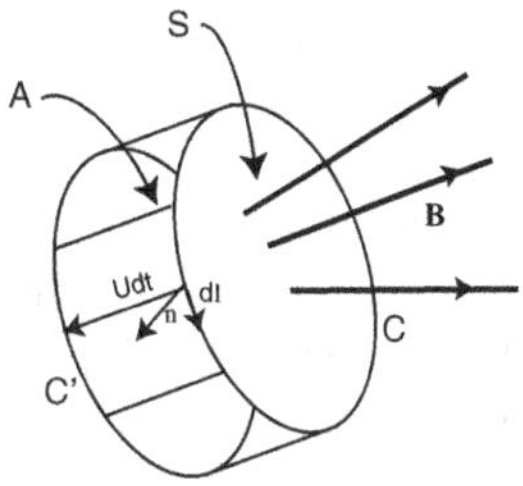


**Fig. 5.1** Surfaces employed in the proof of Alfvén's theorem

To demonstrate this theorem, consider, at time $t$, a closed curve $C$, that we identify with the particles that, in that particular moment, lie on it. Because of the motion of the plasma, the particles will be displaced and, at time $t + dt$ they will define a different curve $C'$. Since to compute the flux of $\boldsymbol{B}$ through a closed curve we can utilize any surface having that curve as a boundary, let's choose at time $t$ a generic surface $S$ and at time $t + dt$ a surface $S'$, made up of $S$ plus the surface $A$ formed by the flux lines joining $C$ and $C'$, as shown in Fig. 5.1.

The change in the flux from time $t$ to time $t + dt$ may be written as

$$d\Phi = \int_{S'} \boldsymbol{B}' \cdot \mathrm{d}\boldsymbol{S}' - \int_{S} \boldsymbol{B} \cdot \mathrm{d}\boldsymbol{S},$$

where the field $\boldsymbol{B}' = \boldsymbol{B}(t + dt, \boldsymbol{r})$. Now

$$\boldsymbol{B}' = \boldsymbol{B}(t + dt, \boldsymbol{r}) = \frac{\partial \boldsymbol{B}}{\partial t} dt + \boldsymbol{B}(t, \boldsymbol{r})$$

while

$$d\Phi = \int_{S} \left( \boldsymbol{B}' - \boldsymbol{B} \right) \cdot \mathrm{d}\boldsymbol{S} + \boldsymbol{B} \cdot \boldsymbol{n} A,$$

where we have neglected the difference between $\boldsymbol{B}'$ and $\boldsymbol{B}$ in calculating the flux through the surface $A$, which is already first order in $dt$ because $\boldsymbol{n}A = -\mathrm{d}\boldsymbol{l} \times \boldsymbol{U} dt$. Summarizing, we find that

$$d\Phi = \int_{S} \left( \boldsymbol{B}' - \boldsymbol{B} \right) \cdot \mathrm{d}\boldsymbol{S} - \int_{C} \left( \boldsymbol{U} \times \boldsymbol{B} \right) \cdot \mathrm{d}\boldsymbol{l} dt.$$

Using Stokes' theorem on the curve $C$ and dividing by $dt$ we finally obtain

$$\frac{d\Phi}{dt} = \int_{S} \left( \frac{\partial \boldsymbol{B}}{\partial t} - \nabla \times (\boldsymbol{U} \times \boldsymbol{B}) \right) \cdot \mathrm{d}\boldsymbol{S}$$

Therefore, if Eq. (5.27) is valid, the flux is constant in time and the theorem is proved.

Let $C_1$ and $C_2$ be two curves connected, at time $t$, by magnetic field lines, so as to form a *flux tube*. The total flux of $\boldsymbol{B}$ across the surface of the tube, $\Phi_B$, is obviously given by the flux across $C_1$ and $C_2$, since the flux across the lateral surface of the tube vanishes by construction. From Alfvén's theorem it follows that $\Phi_B$ will remain constant during the whole dynamical evolution of the system. Now let the area enclosed by $C_1$ and $C_2$ shrink to zero so that the flux tube becomes a single field line. At this point we may be tempted to say that *field lines move together with the fluid* or, as often stated, that *the field lines are frozen in the fluid*. Intuitive as this statement may be, a more formal proof is however required, which we are now going to present To begin write the continuity and induction equations, Eqs. (5.16a) and (5.27), respectively, in lagrangian form. Remembering that

$$\frac{\mathrm{d}}{\mathrm{d}t} = \frac{\partial}{\partial t} + \boldsymbol{U} \cdot \nabla,$$

we have

$$\frac{\mathrm{d}\rho}{\mathrm{d}t} = -\rho \nabla \cdot \boldsymbol{U}$$

and

$$\frac{\mathrm{d}\boldsymbol{B}}{\mathrm{d}t} = (\boldsymbol{B} \cdot \nabla)\boldsymbol{U} - \boldsymbol{B}(\nabla \cdot \boldsymbol{U}),$$

where use has been made of the vector identity

$$\nabla \times (\boldsymbol{F} \times \boldsymbol{G}) = \boldsymbol{F}(\nabla \cdot \boldsymbol{G}) - \boldsymbol{G}(\nabla \cdot \boldsymbol{F}) + (\boldsymbol{G} \cdot \nabla)\boldsymbol{F} - (\boldsymbol{F} \cdot \nabla)\boldsymbol{G},$$

and of the condition $\nabla \cdot \boldsymbol{B} = 0$. From the two equations above, we get:

$$\frac{\mathrm{d}}{\mathrm{d}t}\left(\frac{\boldsymbol{B}}{\rho}\right) = \frac{1}{\rho}[(\boldsymbol{B} \cdot \nabla)\boldsymbol{U} - \boldsymbol{B}(\nabla \cdot \boldsymbol{U})] + \frac{\boldsymbol{B}}{\rho}(\nabla \cdot \boldsymbol{U}) = \left(\frac{\boldsymbol{B}}{\rho} \cdot \nabla\right)\boldsymbol{U}. \tag{5.28}$$

If the fluid motion is described by a velocity field $\boldsymbol{U}(\boldsymbol{r})$, the equation that governs the motion of a generic line element $\mathrm{d}\boldsymbol{\ell}$ joining the points $\boldsymbol{r}$ and $\boldsymbol{r} + \mathrm{d}\boldsymbol{\ell}$ may be found simply as:

$$\frac{\mathrm{d}(\mathrm{d}\boldsymbol{\ell})}{\mathrm{d}t} = \frac{\mathrm{d}(\boldsymbol{r} + \mathrm{d}\boldsymbol{\ell})}{\mathrm{d}t} - \frac{\mathrm{d}\boldsymbol{r}}{\mathrm{d}t} = \boldsymbol{U}(\boldsymbol{r} + \mathrm{d}\boldsymbol{\ell}) - \boldsymbol{U}(\boldsymbol{r}) = (\mathrm{d}\boldsymbol{\ell} \cdot \nabla)\boldsymbol{U}. \tag{5.29}$$

A direct comparison of Eq. (5.28) and (5.29) shows that they are identical. Therefore the quantity $\boldsymbol{B}/\rho$ evolves precisely as $\mathrm{d}\boldsymbol{\ell}$ and it follows that, if $\mathrm{d}\boldsymbol{\ell}$ and $\boldsymbol{B}$ are parallel at a given moment, they will remain so at any later time. We thus recover, in a quantitative form, the "freezing" condition of field lines in the fluid. To describe this result in cartoon-like fashion, imagine marking all particles that, at a given moment, lie on a field line, by painting them in red. During the dynamical evolution, the line traced by the red particles will be deformed but, according to the property just shown, *it will still be a field line*. Alfvén's theorem thus allows us to identify a line of $\boldsymbol{B}$ and to follow it in time. In those cases when Alfvén's theorem *does not* apply, i.e. when $\eta \neq 0$, this is not possible. It is true that at any given moment field lines may still be drawn, but their identification with those drawn at a different time will be no longer possible. Thus, Alfvén's theorem endows field lines with a degree of reality well beyond that of a useful visualization tool. Moreover, since the fluid motion is considered to be *continuous*, the lines of $\boldsymbol{B}$ can only be deformed, but not broken, so that the field topology, i.e. the ensemble of the geometrical properties that are conserved during deformations, cannot be altered.

In an ideal plasma therefore magnetic field and matter are strongly tied together and their dynamics depends on by the dominant term in the momentum equation. If $\beta \gg 1$ the motion is determined by pressure forces and matter drags the magnetic field. On the other hand, if $\beta \ll 1$ magnetic forces are the dominant ones and matter is dragged by the magnetic field.

Having examined the extreme cases $\mathcal{R}_m \ll 1$ and $\mathcal{R}_m \gg 1$, the question arises as to which of the two is most likely to occur natural plasmas. A table of the estimated values of $S$ for a certain number of systems is given in Table. 5.1.

As shown in the Table, in most cases $S \gg 1$ and we may be tempted to neglect resistive effects all together. However, this would be wrong, as the following considerations show. First, observe that the dimensional analysis used to obtain our estimates of the relative importance of convective and diffusive terms completely neglects the vector character of the induction equation. It is true that *on average* the convective term dominates over the diffusive one by many orders of magnitude, but this is clearly untrue where the convective term vanishes or becomes very small. This may happen close to points where $\boldsymbol{U} = 0$, where $\boldsymbol{U}$ is parallel to $\boldsymbol{B}$ ($\boldsymbol{U} \times \boldsymbol{B} = 0$) or finally where $\nabla \times (\boldsymbol{U} \times \boldsymbol{B}) = 0$. In these regions the ideal plasma condition may become *locally* invalid, and the diffusive term may no longer be neglected. In these situations, Alfvén's theorem does not hold, the field topology may change and magnetic energy may be transformed into other forms of energy. This is why studies of the effects of the resistive, or other non-ideal terms in Ohm's law, are so important: in fusion machines non-ideal effects may be the cause of instabilities driving the disruption of the plasma configuration; in astrophysics heating processes and very

**Table 5.1** Spatial and temporal scales, Lundquist numbers

| System | $\mathcal{L}$(cm) | $\tau_d(s)$ | $\tau_a(s)$ | S |
|---|---|---|---|---|
| Tokamak | $10^2$ | $10^{-1}$ | $10^{-3}$ | $10^2$ |
| Earth nucleus | $10^8$ | $10^{12}$ | $10^5$ | $10^7$ |
| Sunspot | $10^9$ | $10^{14}$ | $10^5$ | $10^9$ |
| Solar corona | $10^{11}$ | $10^{18}$ | $10^6$ | $10^{12}$ |

dynamic energy releases are observed in situations where the only apparent source of energy resides in magnetic fields. But the transformation of magnetic into thermal energy and other energy forms is only possible when resistive or other non-ideal effects are at work.

## 5.3 Equilibrium States of Ideal Plasmas

A static equilibrium state is defined by the conditions $\boldsymbol{U} = 0$ and $\partial/\partial t = 0$. A system in equilibrium will remain in that state if no changes occur in the forces acting on the system or in the boundary conditions. When dealing with MHD equilibrium configurations of an ideal plasma, pressure, magnetic and all other forces of different nature acting on the system must be taken into account. In astrophysics, the most relevant force is gravity. Since angular momentum and rotation around gravitational center are common, centrifugal forces are also often important in non-static equilibria. The anisotropic nature of magnetic forces considerably contributes to the complexity of equilibrium structures, as shown by the generalized MHD *virial theorem*, a simple generalization to MHD of the well-known theorem of mechanics. One consequence of this theorem is the impossibility for an isolated plasma to be in equilibrium under the effects of the pressure and self-consistent magnetic forces only: external forces are required for confinement.

To demonstrate the theorem, start from the equation

$$\frac{\partial}{\partial x_k}\mathsf{G}_{ik} = 0,$$

that can be deduced from Eq. (4.34) by letting $\boldsymbol{U} = 0$ and $q = 0$. $\mathsf{G}_{ik}$ incorporates the entire pressure tensor $\mathsf{P}_{ik}$, including its off-diagonal part, and all the magnetic terms, namely

$$\mathsf{G}_{ik} = \mathsf{P}_{ik} + \frac{B^2}{8\pi}\delta_{ik} - \frac{B_i B_k}{4\pi} = (P + \frac{B^2}{8\pi})\delta_{ik} + \Pi_{ik} - \frac{B_i B_k}{4\pi}.$$

Multiplying by $x_i$, summing over $i$ and integrating in $\mathrm{d}V$ over the entire space leads to

$$\int_V x_i(\frac{\partial}{\partial x_k}\mathsf{G}_{ik})\mathrm{d}V = 0,$$

which, on integration by parts, becomes

$$\int_S x_i \mathsf{G}_{ik}\mathrm{d}S_k - \int_V \frac{\partial x_i}{\partial x_k}\mathsf{G}_{ik} = 0. \tag{5.30}$$

Here the first term represents the flux of $\boldsymbol{r}\mathsf{G}$ across the surface $S$ that encloses $V$, i.e. the surface at infinity. For an isolated system, with "good" convergence properties of the components of $\mathsf{G}_{ik}$, one may assume that the surface integral vanishes. Therefore the second integral, the volume integral of the trace of $\mathsf{G}$, is also equal to zero. But, since $\mathbf{Tr}\ \Pi = 0$,

$$\mathbf{Tr}\,\mathsf{G} = 3P + \frac{B^2}{8\pi} > 0.$$

Thus Eq. (5.30) cannot be satisfied, which implies that the initial assumption, namely the existence of an equilibrium under the sole action of pressure and magnetic forces, is not valid. It must be emphasized that the theorem holds only if the system is uniquely composed by the plasma and is isolated in three dimensions.

Writing the static equilibrium equation as

$$\nabla P = \frac{1}{c}\boldsymbol{J} \times \boldsymbol{B}, \tag{5.31}$$

we immediately see that

$$\boldsymbol{J} \cdot \nabla P = \boldsymbol{B} \cdot \nabla P = 0.$$

Since $\nabla P$ is perpendicular to the surfaces $P = constant$, we conclude that both vectors $\boldsymbol{B}$ and $\boldsymbol{J}$ lie on those surfaces. Moreover, Eq. (5.31) shows that even in the presence of a magnetic field, the *magnetic force* may vanish. This certainly happens when $\boldsymbol{J} = 0$, in which case the field is said to be a *potential field*. In fact, if $\boldsymbol{J}$ vanishes we must have $\nabla \times \boldsymbol{B} = 0$ and the field can be represented as the gradient of a scalar function, the magnetic potential. However, the Lorentz force can also vanish if $\boldsymbol{J} \neq 0$ when $\boldsymbol{J}$ and $\boldsymbol{B}$ are parallel to each other. These particular magnetic configurations are called *force-free fields*. We shall now examine some of their properties, before considering cases in which the magnetic force is different from zero.

### *5.3.1 Force-Free Equilibria*

Force-free fields are a common feature of rarefied plasmas whenever the pressure gradients are sufficiently small to be neglected. The equilibrium equation, $\boldsymbol{J} \times \boldsymbol{B} \propto (\nabla \times \boldsymbol{B}) \times \boldsymbol{B} = 0$, thus implies:

$$\nabla \times \boldsymbol{B} = \alpha \boldsymbol{B}, \tag{5.32}$$

where, in general, $\alpha = \alpha(\boldsymbol{r})$.

Notice that Eq. (5.32) does not come from neglecting the pressure term "with respect to" the magnetic term, and then proceeding to state that the latter is zero. Force-free equilibria arise when both the pressure gradients and the Lorentz force separately vanish. However, it is often said in the litterature that magnetic fields are force-free whenever $\beta \ll 1$, reminiscent of the statement above. What is really meant is that if the plasma pressure is everywhere much smaller than magnetic pressure, it is unlikely that gradients in the plasma pressure can supply the force required to balance any significant Lorentz force. In other words, one generally should have

$$\frac{c\,|\nabla P|}{|\boldsymbol{J} \times \boldsymbol{B}|} = \frac{4\pi|\nabla P|}{|(\nabla \times \boldsymbol{B}) \times \boldsymbol{B}|} \ll 1.$$

It is clear that if we add to $P$ or to $B$ a *constant* pressure or a *constant* magnetic field, we change the value of $\beta$, but not that of the ratio of their gradients. We thus may have a force-free field also in situations where $\beta > 1$.

Returning now to Eq. (5.32), we illustrate some of the properties of force-free fields. Taking the divergence of Eq. (5.32) we find that

$$\nabla \cdot (\nabla \times \boldsymbol{B}) = \nabla \cdot (\alpha \boldsymbol{B}) = \alpha \nabla \cdot \boldsymbol{B} + \boldsymbol{B} \cdot \nabla \alpha = 0,$$

implying

$$\boldsymbol{B} \cdot \nabla \alpha = 0.$$

Constancy along magnetic field lines is the only constraint satisfied by the function $\alpha(\boldsymbol{r})$. The preceding equation also shows that $\boldsymbol{B}$ must lie on the surfaces $\alpha = const.$.

Writing the vector Eq. (5.32) in terms of components, we immediately see that it couples different components of $\boldsymbol{B}$. If $\alpha$ is a constant independent of the coordinates, it is easy to obtain an equation for a single component of $\boldsymbol{B}$ by taking the curl of Eq. (5.32):

$$\nabla \times \nabla \times \boldsymbol{B} = -\nabla^2 \boldsymbol{B} = \alpha(\nabla \times \boldsymbol{B}) = \alpha^2 \boldsymbol{B}. \tag{5.33}$$

However, since Eq. (5.33) is a differential equation of order higher than Eq. (5.32), it is necessary to verify that the solutions of Eq. (5.33) also satisfy Eq. (5.32).

As a first example of a force-free field in plane geometry with $\alpha = const.$, consider a field given by $\boldsymbol{B} = \boldsymbol{B}(x) = [0, B_y(x), B_z(x)]$, which obviously satisfies the condition $\nabla \cdot \boldsymbol{B} = 0$. It is easy to check that, with a proper choice of the initial conditions, the solutions of Eq. (5.33) that are also solutions of Eq. (5.32) can be written as

$$B_y = B_0 \cos(\alpha\, x); \quad B_z = -B_0 \sin(\alpha\, x). \tag{5.34}$$

This is a force-free field whose lines of force are straight lines in every plane $x = const.$. However, their inclination with respect to the $y$ axis changes, making a complete turn when $x$ is increased by $2\pi/\alpha$.

A more interesting example is given by axially symmetric force-free fields, in which case the fields always have a helical structure. Axial symmetry requires that the components of $\boldsymbol{B}$ depend only on the radial coordinate $r$, the distance from the axis of symmetry, and the condition $\nabla \cdot \boldsymbol{B} = 0$ implies that $B_r = 0$ and thus $\boldsymbol{B} = \boldsymbol{B}(r) = [0, B_\theta(r), B_z(r)]$. The field lines therefore wind on coaxial cylindrical surfaces, forming a helical structure. If the *pitch* of the helix, $\kappa$, is defined as the distance covered in the $z$ direction by a representative points performing a complete revolution along a field line, it is easy to see that $\kappa(r) = 2\pi r B_z(r)/B_\theta(r)$. In cylindrical geometry Eq. (5.33) read:

$$\frac{1}{r}\frac{\mathrm{d}}{\mathrm{d}r}\left(r\frac{\mathrm{d}B_\theta}{\mathrm{d}r}\right) + \left(\alpha^2 - \frac{1}{r^2}\right)B_\theta = 0,$$

and

$$\frac{1}{r}\frac{\mathrm{d}}{\mathrm{d}r}\left(r\frac{\mathrm{d}B_z}{\mathrm{d}r}\right) + \alpha^2 B_z = 0.$$

The solutions of the above equations are:

$$B_\theta = B_0 J_1(\alpha r) \quad e \quad B_z = B_0 J_0(\alpha r), \tag{5.35}$$

where $J_n$ are the Bessel functions of order $n$. As expected, the field lines are helices whose pitch varies with $r$. On the axis ($r = 0$) the field is directed along $z$ ($J_1(0) = 0$); as $r$ increases, an azimuthal component appears and the helix pitch decreases until, in correspondence with the first zero of $J_0$, the axial component vanishes and the field is totally azimuthal. For larger values of $\alpha r$, the field is still a helix, but with the opposite sense of rotation because $J_0$ changes sign. If $r$ is further increased, the first zero of $J_1$ is encountered and the field is again purely axial, but its sense is opposite to that of the field on the axis.

An example of a force-free field with $\alpha \neq const.$ is given by

$$B_\theta = B_0\frac{kr}{1+k^2r^2}; \quad B_z = -B_0\frac{1}{1+k^2r^2}. \tag{5.36}$$

The $z$-component of $\nabla \times \boldsymbol{B}$ is given by

$$(\nabla \times \boldsymbol{B})_z = \frac{1}{r}\frac{\mathrm{d}}{\mathrm{d}r}(rB_\theta) = -B_0\frac{2k}{(1+k^2r^2)^2} = \alpha(r)B_z,$$

where

$$\alpha(r) = \frac{2k}{1+k^2r^2}.$$

Therefore this field can be considered a force-free field with $\alpha \neq const.$ at least for the $z$-component. It is however easy to verify that the equation for the $\theta$-component is also satisfied by this choice of $\alpha$. A peculiar feature of this field is to have a pitch independent of $r$.

It is also possible to find simple solutions of Eq. (5.32) in two dimensions. For instance, it is easily shown that in plane geometry the field

$$B_x = -(l/k)B_0\cos(kx)\mathrm{e}^{-lz}$$

$$B_y = -(1 - l^2/k^2) B_0 \cos(kx) \mathrm{e}^{-lz}$$
$$B_z = B_0 \sin(kx) \mathrm{e}^{-lz}$$

is a force-free field with $\alpha = (k^2 - l^2)^{1/2} = const$. This field has been used as a model for the magnetic arcades observed in the solar corona.

Force-free fields have been the object of remarkable interest. One of the reasons that justify such an interest is connected with

**Woltjer's theorem**: *For a given set of boundary conditions, the plasma state with the minimum magnetic energy is a force-free field with* $\alpha = constant$. To demonstrate this important theorem we must first introduce the concept of *magnetic helicity*, a quantity defined by:

$$\mathcal{H} = \int_V \boldsymbol{A} \cdot \boldsymbol{B} \, \mathrm{d}V, \tag{5.37}$$

where $\boldsymbol{A}$ is the *potential vector*, $\boldsymbol{B} = \nabla \times \boldsymbol{A}$. The evolution equation for $\boldsymbol{A}$ is easily obtained (see Problem 5.2) by introducing this representation of $\boldsymbol{B}$ into the induction equation:

$$\frac{\partial \boldsymbol{A}}{\partial t} = \boldsymbol{U} \times (\nabla \times \boldsymbol{A}). \tag{5.38}$$

An important property of magnetic helicity is to remain constant in time for an isolated system, provided that $\boldsymbol{A}$ does not vary on the surface $S$ that encloses the volume $V$. In fact,

$$\frac{\mathrm{d}\mathcal{H}}{\mathrm{d}t} = \int_V \left( \frac{\partial \boldsymbol{A}}{\partial t} \cdot \boldsymbol{B} + \frac{\partial \boldsymbol{B}}{\partial t} \cdot \boldsymbol{A} \right) \mathrm{d}V,$$

and, making use of the vector identity

$$\nabla \cdot (\boldsymbol{F} \times \boldsymbol{G}) = \boldsymbol{G} \cdot (\nabla \times \boldsymbol{F}) - \boldsymbol{F} \cdot (\nabla \times \boldsymbol{G}), \tag{5.39}$$

with $\boldsymbol{F} = \partial \boldsymbol{A}/\partial t$ and $\boldsymbol{G} = \boldsymbol{A}$, we obtain:

$$\frac{\mathrm{d}\mathcal{H}}{\mathrm{d}t} = \int_V \left[ \nabla \cdot \left( \frac{\partial \boldsymbol{A}}{\partial t} \times \boldsymbol{A} \right) + 2 \frac{\partial \boldsymbol{A}}{\partial t} \cdot (\nabla \times \boldsymbol{A}) \right] \mathrm{d}V.$$

The first term can be transformed into an integral over the surface $S$ where, by assumption, $\partial \boldsymbol{A}/\partial t = 0$ while the second term vanishes because of Eq. (5.38). Therefore,

$$\frac{\mathrm{d}\mathcal{H}}{\mathrm{d}t} = 0.$$

By exploiting this property of $\mathcal{H}$ we are now in the position to demonstrate Woltjer's theorem. Equilibrium states in MHD are generally found by minimizing the magnetic energy of a system, $W_B$. However, in our case we must take into account the invariance of $\mathcal{H}$, so that the minimization problem is constrained by the conservation of magnetic helicity, rather than free. As is well known, this class of problems can be solved by applying the method of Lagrangian multipliers. This consists in looking for the minimum of $(W_B - \lambda\mathcal{H})$, where $\lambda$ is a constant, and imposing the condition that $\boldsymbol{A}$ not vary on the surface $S$. Writing for convenience $\lambda = \alpha_0/8\pi$ the variational problem becomes:

$$\begin{aligned} 0 = \delta(W_B - \alpha_0/8\pi\mathcal{H}) &= \frac{1}{8\pi}\delta\int_V (B^2 - \alpha_0\boldsymbol{A}\cdot\boldsymbol{B})\,\mathrm{d}V \\ &= \frac{1}{4\pi}\int_V \left[\boldsymbol{B}\cdot\delta\boldsymbol{B} - \frac{\alpha_0}{2}(\boldsymbol{A}\cdot\delta\boldsymbol{B} + \boldsymbol{B}\cdot\delta\boldsymbol{A})\right]\mathrm{d}V \\ &= \frac{1}{4\pi}\int_V \left[(\boldsymbol{B} - \frac{\alpha_0}{2}\boldsymbol{A})\cdot\delta\boldsymbol{B} - \frac{\alpha_0}{2}\boldsymbol{B}\cdot\delta\boldsymbol{A}\right]\mathrm{d}V, \end{aligned}$$

Since

$$\delta\boldsymbol{B} = \nabla\times\delta\boldsymbol{A},$$

making use again of Eq. (5.39) with $\boldsymbol{F} = (\boldsymbol{B} - (\alpha_0/2)\boldsymbol{A})$ and $\boldsymbol{G} = \delta\boldsymbol{A}$, we obtain

$$\int_V \nabla\cdot\Big([\boldsymbol{B} - (\alpha_0/2)\boldsymbol{A}]\times\delta\boldsymbol{A}\Big)\,\mathrm{d}V + \int_V [\nabla\times\boldsymbol{B} - \alpha_0\boldsymbol{B}]\cdot\delta\boldsymbol{A}\,\mathrm{d}V = 0$$

The first integral vanishes since it can be transformed into a surface integral over $S$ where by assumption $\delta\boldsymbol{A} = 0$. The equilibrium condition thus requires the vanishing of the second integral for every $\delta\boldsymbol{A}$ or

$$\nabla\times\boldsymbol{B} = \alpha_0\,\boldsymbol{B},$$

as required by Woltjer's theorem. It is also possible to show that the configuration having the minimum magnetic energy, among all those compatible with the given boundary conditions, corresponds to a potential field $\alpha_0 = 0$.

### 5.3.2 *Equilibria in the Presence of Magnetic Forces*

If more general magnetic configurations with non-vanishing magnetic forces are considered, the equilibrium equation reads:

$$\nabla\left(P + \frac{B^2}{8\pi}\right) = \frac{1}{4\pi}(\boldsymbol{B}\cdot\nabla)\boldsymbol{B}. \tag{5.40}$$

We shall now briefly discuss some solutions of the preceding equation, restricting ourselves to cylindrical geometry.

Given a magnetic field $\boldsymbol{B} = [0, B_\theta(r), B_z(r)]$ Eq. (5.40) reduces to

$$\frac{\mathrm{d}}{\mathrm{d}r}\left(P + \frac{B_\theta^2 + B_z^2}{8\pi}\right) = -\frac{1}{4\pi}\frac{B_\theta^2}{r}. \tag{5.41}$$

We have only one differential equation for the three unknown functions $P(r)$, $B_\theta(r)$ and $B_z(r)$. If any two of them are prescribed, together with the appropriate boundary conditions, the third can be found from Eq. (5.41). It is therefore clear that an infinite number of solutions exists; we shall illustrate some examples of magnetic configurations particularly interesting for the plasma confinement in fusion machines, usually called *pinches*.

- **Theta-pinch**, $B_\theta = 0$.

Such a configuration can be produced by currents flowing in the azimuthal direction $\theta$ on the surface of a plasma column. In this case the solution of Eq. (5.40) is simply given by

$$P + \frac{B^2}{8\pi} = const.$$

The value of the constant is determined by the radial boundary conditions. If, for instance, the plasma column has a radius $a$ and is surrounded by a vacuum (or by a medium of negligible pressure) and outside of the column a constant axial magnetic field is present, $B_z = B_0$, the equilibrium condition becomes

$$P + \frac{B^2}{8\pi} = \frac{B_0^2}{8\pi}.$$

We can easily figure out why this is an equilibrium configuration: the force $\boldsymbol{J} \times \boldsymbol{B}$, directed towards the axis of the cylinder, pinches the plasma and compensates the push produced by the internal pressure. Notice that this solution is valid also in configurations that do not show an axial symmetry: it suffices that $(\boldsymbol{B} \cdot \nabla)\boldsymbol{B} = 0$, which happens, for instance, when the field lines are straight.

- **Zeta-pinch**, $B_z = 0$.

This configuration is obtained from the preceding one by switching the roles of $\boldsymbol{J}$ and $\boldsymbol{B}$. We thus assume an axial current flow in a column af radius $a$, surrounded by a vacuum. We may envisage such a current as formed by many wires in which parallel currents flow. Since, as is well known, parallel currents attract each other, the joint effect will again be the squeezing of the plasma column. The equilibrium condition now reads:

$$\frac{\mathrm{d}P}{\mathrm{d}r} = -\frac{1}{4\pi\, r} B_\theta \frac{\mathrm{d}}{\mathrm{d}r}(r\, B_\theta). \tag{5.42}$$

Multiplying both members of the equation by $r^2$ and integrating between 0 and $a$, we obtain:

$$\int_0^a r^2 \frac{\mathrm{d}P}{\mathrm{d}r}\mathrm{d}r = -\frac{1}{4\pi}\int_0^a r B_\theta \frac{\mathrm{d}}{\mathrm{d}r}(r B_\theta)\mathrm{d}r.$$

Performing the integrations and taking into account the boundary conditions we get:

$$2\int_0^a r P \mathrm{d}r = \frac{1}{8\pi}[a B_\theta(a)]^2.$$

If the plasma can be considered a perfect gas, $P = nkT$, the lhs can be written as

$$\frac{1}{\pi}\int_0^a 2\pi r(nkT)\mathrm{d}r = \frac{kT}{\pi}\int_0^a 2\pi r\, n\mathrm{d}r = \frac{N_\ell T}{\pi},$$

where the temperature $T$ is assumed to be constant inside the plasma column and we have introduced the quantity $N_\ell$, representing the number of particles per unit length of the column (*linear density*):

$$N_\ell = \int_0^a 2\pi r\, n\mathrm{d}r.$$

The integral on the rhs can be transformed by noticing that the intensity of the current flowing in the plasma column is:

$$I = \int_0^a 2\pi r J_z \mathrm{d}r,$$

and that

$$J_z = \frac{c}{4\pi}\frac{1}{r}\frac{\mathrm{d}}{\mathrm{d}r}(r B_\theta),$$

so that finally

$$\frac{2I}{c} = a B_\theta(a).$$

Making use of the above relations we obtain:

$$I^2 = 2kT N_\ell c^2,$$

known as the *Bennet's relation.* Observe that this relation is independent of the details of the pressure profile. The latter can be deduced directly from the equilibrium equation (5.42), for instance by assuming that $J_z$ is constant inside the plasma column and zero outside. In this case $B_\theta$ is equal to

$$B_\theta = \frac{2I}{a^2} r \qquad r \leq a,$$

and the pressure profile is given by

$$P(r) = \frac{1}{\pi}\left(\frac{I}{a}\right)^2 \left(1 - \frac{r^2}{a^2}\right).$$

## 5.4 Perturbed Equilibrium States

The existence of an equilibrium state is not by itself a guarantee that the equilibrium will not change with time: an equilibrium results from a precise balance of the forces acting on the system, which in turn implies a well defined set of values of the parameters characterizing the system. If one or more of those values are changed, the equilibrium cannot be maintained and the system evolves dynamically. If only small perturbations of the equilibrium parameters are considered, there are essentially two possibilities: either the resulting force is such as to push to restore the system to equilibrium or it causes a further displacement from it. In the first case, an oscillatory regime sets in and the amplitude of the perturbations remains small: the equilibrium is then said to be *stable*. In the second case, the amplitude of the perturbation increases and the equilibrium is said to be *unstable*. The intermediate case, in which perturbations do not alter the balance of forces, corresponds to a *marginal or neutral* equilibrium.

In principle, to determine the existence of an equilibrium one should check that *all* the forces acting on the system balance, i.e. the resultant of all forces should vanish. In practice, this is often not possible and to establish the equilibrium condition only the dominant forces are taken into account. Such an "equilibrium" therefore does not represent the real situation, where other, smaller, forces are present. As a consequence, equilibrium conditions are never perfectly satisfied and reality is better represented by a (theoretical) equilibrium with perturbations, a situation which makes a stability analysis of the system a requirement. In fact, if we want to set up an experimental or observational check of the actual existence of an equilibrium state, we are forced to restrict our check to stable, or moderately unstable, systems. By this we mean that the rate of change with time of the parameters that characterize the system must be such that their values will not be substantially altered during their measurement. It follows that the **physical** concept of equilibrium (not the mathematical one!) makes sense only when referred to a well defined timescale. In other words, we are not interested in finding equilibria that will last for eternity: we only want to verify the existence of states that do not undergo substantial changes over what we consider the relevant timescales.

The analysis of the dynamical evolution of an equilibrium subject to small perturbations, the so-called *linear stability analysis*, can be performed by using two different methods. The first, the *normal modes method*, basically consists in the study of the dynamics of the perturbed system. This not only determines whether a system is stable or not, but also characterizes the dynamics of small perturbations, namely the characteristic frequencies of oscillation in the case of stable systems or the growth rates of the amplitude of perturbations in the unstable case. The assumption that perturbations are small allows significant mathematical simplification coming from *linearization* of the equations. In the stable case perturbations remain small and the solutions found are valid also for long time spans. In the unstable case, perturbations grow, typicall exponentially, and the validity of solutions is limited to short time periods.

The second method, called the *energy method*, is a variational method generalizing the well known result of classical mechanics that equilibrium states correspond to energy extrema, with stable equilibria localized at energy minima and unstable ones at energy maxima. The energy method establishes if a system is stable or not, but does not yield any detailed information on the system's dynamics.

It is important to keep in mind that a system can be considered stable only if stability is achieved for **any** possible perturbation, provided that it is compatible with the constraints imposed on the system, but must be considered unstable even if a **single** growing perturbation is found.

In the following we shall utilize only the method of normal modes, that basically consists of the following steps.

- All $f$ entering the system of equations that governs the system's dynamics, are expanded as
$$f = f_0 + \epsilon f_1,$$
with $\epsilon \ll 1$. $f_0$ corresponds to the equilibrium state and $\epsilon f_1$ to the perturbation.
- This representation is introduced into the equations and terms of order higher than the first one in $\epsilon$ are discarded.
- The equation are written order by order, with terms of zeroth order separated from those of first order and the resulting equations are separately solved. The first order equations, evidently *linear* in the quantities $f_1$, determine the dynamics of perturbations. The coefficients of those linear equations are functions of the unperturbed quantities and of the other parameters that may enter in the starting set of equations.

This scheme is equally valid for any plasma model, both kinetic and fluid.

## Questions and Problems

**5.1.** Show how a dimensional estimate of the Hall term in Eq. (5.5) leads to the inequality Eq. (5.8).
**5.2.** Justify Eq. (5.38) using the *gauge* invariance properties of Maxwell's equations.
**5.3.** Introducing the (Elsasser) variables: $\boldsymbol{Z}_{\pm} = \boldsymbol{U} \pm \boldsymbol{B}/(\sqrt{4\pi\rho})$, show that the MHD equations for an incompressible ideal plasma can be written as:

$$\frac{\partial \boldsymbol{Z}_{\pm}}{\partial t} + (\boldsymbol{Z}_{\mp} \cdot \nabla)\boldsymbol{Z}_{\pm} = -\frac{1}{\rho}\nabla\left(P + \frac{B^2}{8\pi}\right).$$

**5.4.** (a) A plasma column with density $n = 10^{15}$ cm$^{-3}$ is placed inside a medium of negligible density, or "vacuum", where a constant magnetic field, $\boldsymbol{B}$, is present. Determine the value of $B$ necessary to confine a plasma whose temperature is $5 \times 10^6$ $K$. (b) If in a Zeta-pinch with a linear density of $10^{18}$ particles/cm circulates a current of $10^6$ $A$, evaluate the plasma temperature (beware of the units!).
**5.5.** Compute the helicity for the following field configurations: (a) a Theta-pinch, (b) a Zeta-pinch, (c) the field given by Eq. (5.36). Show that the answer to question (c) applies to any helical field of constant pitch.
**5.6.** Compute the helicity of a force-free field with constant $\alpha$ in plane and in cylindrical geometries.

## Solutions

**5.1.** Proceeding by dividing the Hall term with the electric field magnitude, or equivalently multiplying it by $c/(\mathcal{U}\mathcal{B})$, we obtain

$$\left[\frac{1}{en_e c}\left(\boldsymbol{J} \times \boldsymbol{B}\right)_i\right]\frac{c}{\mathcal{U}\mathcal{B}} = \omega\frac{cm_p}{e\mathcal{B}}\frac{\mathcal{B}^2}{n_e m_p}\frac{1}{\mathcal{U}^2} = \frac{\omega}{\omega_{ci}}\left(\frac{c_a}{U}\right)^2,$$

from which inequality Eq. (5.8) follows trivially.
**5.2.** If we add to the potential vector $\boldsymbol{A}$ the gradient of a scalar function the magnetic field associated with $\boldsymbol{A}$ does not change. In fact, if $\boldsymbol{A}' = \boldsymbol{A} + \nabla\psi$, we clearly have $\boldsymbol{B} = \nabla \times \boldsymbol{A} = \nabla \times \boldsymbol{A}'$. If $\boldsymbol{A}$ is replaced by $\boldsymbol{A}'$ in Eq. (5.25), taking into account that $\nabla \times \nabla\psi = 0$, we immediately obtain the requested result.
**5.3.** In an incompressible plasma the density can be considered to be uniform, in which case the continuity equation is automatically respected and $rho$ may therefore be passed into any differential operator as all its derivatives vanish. Therefore, we divide all terms in the induction equation by $\sqrt{4\pi\rho}$, while we divide the momentum equation by $\rho$. We then simply add the momentum and induction equation to obtain the equation for $\boldsymbol{Z}^{+}$, while subtracting the induction equation from the momentum

equation we obtain the equation for $\boldsymbol{Z}^{-}$:

$$\frac{\partial \boldsymbol{Z}_{\pm}}{\partial t} + (\boldsymbol{Z}_{\mp} \cdot \nabla)\boldsymbol{Z}_{\pm} = -\frac{1}{\rho}\nabla\Big(P + \frac{B^2}{8\pi}\Big).$$

**5.4.** (a) $B \simeq 4.16 \times 10^3 G$. (b) $T \simeq 36 \times 10^6 K$.
**5.5.** From $\boldsymbol{B} = \nabla \times \boldsymbol{A}$ we deduce the following relations:

$$A_z = -\int_0^r B_\theta \mathrm{d}r \quad ; \quad A_\theta = \frac{1}{r}\int_0^r r B_z \mathrm{d}r.$$

In the definition of the helicity, the integrand is: $\boldsymbol{A} \cdot \boldsymbol{B} = A_\theta B_\theta + A_z B_z$, which turns out to vanish for all cases. Moreover, for any helical field,

$$\boldsymbol{A} \cdot \boldsymbol{B} = \frac{B_\theta}{r}\int_0^r r B_z \mathrm{d}r - B_z \int_0^r B_\theta \mathrm{d}r = \frac{B_\theta}{2\pi r}\int_0^r \kappa B_\theta \mathrm{d}r - \frac{B_\theta}{2\pi r}\kappa \int_0^r B_\theta \mathrm{d}r,$$

which vanishes for $\kappa = const.$
**5.6.** In plane geometry we have $\boldsymbol{A} \cdot \boldsymbol{B} = (2B_0^2/\alpha)\cos^2(\alpha\, x/2)$. Therefore $\mathcal{H} \neq 0$ and has the same sign of $\alpha$. The helicity normalized to the volume is $\mathcal{H}/V = B_0^2/\alpha$.

In cylindrical geometry, making use of the properties of Bessel functions [see e.g. Abramowitz and Stegun, *Handbook of Mathematical Functions*], we have

$$A_\theta = \frac{B_0}{r}\int_0^r r J_0(\alpha r)\mathrm{d}r = \frac{B_0}{\alpha} J_1(\alpha r),$$

$$A_z = -B_0 \int_0^r J_1(\alpha r)\mathrm{d}r = \frac{B_0}{\alpha}\Big[J_0(\alpha r) - 1\Big].$$

Therefore,

$$\boldsymbol{A} \cdot \boldsymbol{B} = \frac{B_0^2}{\alpha}\Big[J_0^2(\alpha r) + J_1^2(\alpha r) - J_0(\alpha r)\Big].$$

In this case, it can be shown, always making use of the properties of Bessel functions, that the normalized helicity depends on the cylinder's radius, but its sign is still defined by that of $\alpha$.

# Chapter 6
# Instabilities

**Abstract** A survey of the most important linear instabilities of the fluid plasma described by MHD is presented. We start by considering instabilities in planar geometry when gravity is present, including the well-known Rayleigh-Taylor and Kelvin-Helmholtz instabilities, and discuss the stabilizing effects of magnetic fields in these cases. The instabilities stemming from the combined effects of currents, current gradients, and pressure in the ideal plasmas are then described, focusing on cylindrical geometry. Finally, the Parker or buoyancy instability of the galactic magnetic field and the magneto-rotational instability of accretion disks, topics of contemporary astrophysical relevance, are discussed in some detail.

In the preceding Chapter we have emphasized the role and importance of stability analysis for understanding possible plasma equilibrium states in the MHD limit and illustrated one of the possible approaches to the problem. This chapter is devoted to the study of the linear stability of systems described by the MHD equations using that approach, consisting in solving the eigenvalue problem and searching for the growth rates of small perturbations. As a result of our analysis, we shall be able to say if the perturbed system will show a tendency to return to the equilibrium position, giving rise to oscillatory dynamics, or to move away from it. However, nothing can be said in general about the final state of the system, which might be a different (stable) equilibrium state, but also may not exist, leading to a dynamical evolution followed by drastic changes.

An example taken from astrophysics is that of a white dwarf with a mass in excess of the Chandrasekhar limit. If the composition and the equation of state are those usually assumed for white dwarfs, a stable configuration simply does not exist. However, if the composition changes into that typical of neutron stars and the equation of state is accordingly modified, a new stable equilibrium is finally achieved. The composition change is due to the neutronization process, a process completely extraneous to the physical mechanisms determining the equilibrium of a white dwarf.

The importance of stability analysis increases with the number of degrees of freedom of the system. In a system with one degree of freedom, its energy will be a function of only one parameter and equilibria will generically correspond to energy



C. Chiuderi and M. Velli, *Basics of Plasma Astrophysics*,
UNITEXT for Physics, DOI 10.1007/978-88-470-5280-2_6

maxima or minima, so that the number of stable equilibria corresponds to a fraction $1/2$ of the total. In a system with two degrees of freedom, the energy becomes a surface in the two-parameter space and there will be four possible extrema, namely an absolute maximum (a mountain top), an absolute minimum (a valley) and two saddle points (two mountain passes). These latter states are unstable since only for a subset of perturbations along a particular direction will the system return to the saddle-equilibrium, while any noise will lead to instability. Therefore, over four possible equilibrium configurations, only one is stable. One might be tempted to generalize assuming that for a system with $n$ degrees of freedom only a fraction $1/2^n$ of the total number of equilibria is stable, a conjecture for which there is no proof. Nonetheless it is a reasonable intuition that in a system with a very large number of degrees of freedom, such as a fluid or a plasma, most equilibrium configurations might be unstable, making stability analysis a requirement.

## 6.1 Linear Stability of Ideal MHD Equilibria

Consider the ideal MHD equations:

$$\begin{aligned}
&\frac{\partial \rho}{\partial t} + \nabla \cdot (\rho \boldsymbol{U}) = 0 \\
&\rho \frac{\mathrm{d}\boldsymbol{U}}{\mathrm{d}t} = \rho\left(\frac{\partial \boldsymbol{U}}{\partial t} + (\boldsymbol{U} \cdot \nabla)\,\boldsymbol{U}\right) = -\nabla P + \frac{1}{4\pi}(\nabla \times \boldsymbol{B}) \times \boldsymbol{B} + \boldsymbol{f} \\
&\frac{\mathrm{d}}{\mathrm{d}t}(P\rho^{-\gamma}) = 0 \\
&\frac{\partial \boldsymbol{B}}{\partial t} = \nabla \times (\boldsymbol{U} \times \boldsymbol{B})
\end{aligned} \tag{6.1}$$

In agreement with the assumption of small perturbations, expand any physical variable $h$, entering the preceding equations as:

$$h = h_0 + \epsilon\, h_1 \quad \text{and} \quad \epsilon \ll 1,$$

where $h_0$ represents the equilibrium value and $\epsilon h_1$ the perturbation. In the unperturbed state for static equilibria, the velocity vanishes and consequently $\boldsymbol{U} = \epsilon \boldsymbol{U}_1$ is a first order quantity. Taking all this into account, the following equations result from an order by order expansion:

*Zeroth order*:

$$\frac{\partial \rho_0}{\partial t} = \frac{\partial \boldsymbol{B}_0}{\partial t} = 0,$$

$$P_0 \rho_0^{-\gamma} = const.$$

$$0 = -\nabla P_0 + \frac{1}{4\pi}(\nabla \times \boldsymbol{B}_0) \times \boldsymbol{B}_0 + \boldsymbol{f}_0. \tag{6.2}$$

*First order*:

$$\frac{\partial \rho_1}{\partial t} + (\boldsymbol{U} \cdot \nabla)\,\rho_0 + \rho_0(\nabla \cdot \boldsymbol{U}) = 0, \tag{6.3}$$

$$\rho_0 \frac{\partial \boldsymbol{U}}{\partial t} = -\nabla P_1 + \frac{1}{4\pi}[(\nabla \times \boldsymbol{B}_0) \times \boldsymbol{B}_1 + (\nabla \times \boldsymbol{B}_1) \times \boldsymbol{B}_0] + \boldsymbol{f}_1 \tag{6.4}$$

$$\rho_0^{-\gamma} \frac{dP_1}{dt} - \gamma P_0 \rho_0^{-\gamma-1} \frac{d\rho_1}{dt} = 0, \tag{6.5}$$

$$\frac{\partial \boldsymbol{B}_1}{\partial t} = \nabla \times (\boldsymbol{U} \times \boldsymbol{B}_0). \tag{6.6}$$

Introducing the sound speed $c_s$,

$$c_s^2 = \gamma \frac{P_0}{\rho_0},$$

and making use of Eq. (6.3), Eq. (6.5) can be writen as

$$\frac{\partial P_1}{\partial t} + (\boldsymbol{U} \cdot \nabla) P_0 + c_s^2 (\nabla \cdot \boldsymbol{U}) \rho_0 = 0, \tag{6.7}$$

The linearized MHD equations can be simplified by introducing the concept of *lagrangian displacement*, $\boldsymbol{\xi}$, and by properly choosing initial conditions. The instantaneous position of a fluid element may be characterized by:

$$\boldsymbol{r}(t) = \boldsymbol{r}_0 + \boldsymbol{\xi}(\boldsymbol{r}_0, t),$$

where, thanks to the linearization, $\boldsymbol{\xi}$ may also be taken to be a small, first order, quantity. Therefore,

$$\boldsymbol{U} = \frac{d\boldsymbol{r}}{dt} = \frac{d\boldsymbol{\xi}}{dt} = \frac{\partial \boldsymbol{\xi}}{\partial t} + (\boldsymbol{U} \cdot \nabla)\boldsymbol{\xi} \simeq \frac{\partial \boldsymbol{\xi}}{\partial t}.$$

Inserting this expression in Eq. (6.3) we obtain:

$$\frac{\partial}{\partial t}(\rho_1 + \nabla \cdot (\rho_0 \boldsymbol{\xi})) = 0,$$

or, upon integration in time

$$\rho_1 + \nabla \cdot (\rho_0 \boldsymbol{\xi}) = const.$$

The initial conditions have been chosen, without loss of generality, so that all perturbed quantities except $\dot{\xi}(\boldsymbol{r}_0, 0)$ vanish everywhere at $t = 0$.[1] With this choice, the constant of the preceding equation vanishes and therefore

$$\rho_1 = -(\boldsymbol{\xi} \cdot \nabla)\, \rho_0 - \rho_0 (\nabla \cdot \boldsymbol{\xi}). \tag{6.8}$$

Applying the same procedure to Eqs. (6.5) and (6.6) we obtain:

$$P_1 = -(\boldsymbol{\xi} \cdot \nabla) P_0 - \rho_0 c_s^2 (\nabla \cdot \boldsymbol{\xi}), \tag{6.9}$$

and

$$\boldsymbol{B}_1 = \nabla \times (\boldsymbol{\xi} \times \boldsymbol{B}_0), \tag{6.10}$$

while the momentum equation, Eq, (6.4), becomes

$$\rho_0 \frac{\partial^2 \boldsymbol{\xi}}{\partial t^2} = \boldsymbol{F}(\boldsymbol{\xi}), \tag{6.11}$$

where the "force per unit volume", $\boldsymbol{F}(\boldsymbol{\xi})$ is given by:

$$\boldsymbol{F}(\boldsymbol{\xi}) = -\nabla P_1 + \frac{1}{4\pi}[(\nabla \times \boldsymbol{B}_0) \times \boldsymbol{B}_1 + (\nabla \times \boldsymbol{B}_1) \times \boldsymbol{B}_0] + \boldsymbol{f}_1, \tag{6.12}$$

with $P_1$, and $B_1$ given by Eqs. (6.9) and (6.10).

Equation (6.11) is the starting point for the application of the method of normal modes: it describes the temporal evolution of displacements from the equilibrium position under the action of the force $\boldsymbol{F}(\boldsymbol{\xi})$. $\boldsymbol{F}(\boldsymbol{\xi})$ is a second order differential operator $[\nabla \times \boldsymbol{B}_1 = \nabla \times \nabla \times (\boldsymbol{\xi} \times \boldsymbol{B}_0)]$ involving only *spatial* coordinates, which may be written formally as an operator $\hat{\mathcal{F}}_{\boldsymbol{r}}$ acting on $[\boldsymbol{\xi}(\boldsymbol{r}, t)]$

$$\boldsymbol{F}(\boldsymbol{\xi}) = \hat{\mathcal{F}}_{\boldsymbol{r}}\, [\boldsymbol{\xi}(\boldsymbol{r}, t)].$$

Equation (6.11) is a partial differential equation in the independent variables $\boldsymbol{r}, t$ and a standard method of solution for this type of equations, in particular when the coefficients do not depend on the independent variables, is that of the *Fourier transform* (that will be discussed in more detail in Chap. 7). Presently, since the equilibrium may have spatial gradients but does not depend explicitly on time, it

[1] Think of the case of a pendulum: it is completely equivalent to move the mass from the equilibrium position and let it go or to leave it there and give it a push.

is useful to carry out the Fourier transform with respect to time of $\boldsymbol{\xi}(\boldsymbol{r},t)$, $\tilde{\boldsymbol{\xi}}(\boldsymbol{r},\omega)$, defined by

$$\tilde{\boldsymbol{\xi}}(\boldsymbol{r},\omega) = \frac{1}{2\pi}\int_{-\infty}^{\infty} \boldsymbol{\xi}(\boldsymbol{r},t)\,\mathrm{e}^{\mathrm{i}\omega t}\mathrm{d}t, \tag{6.13}$$

which implies that

$$\boldsymbol{\xi}(\boldsymbol{r},t) = \int_{-\infty}^{\infty} \tilde{\boldsymbol{\xi}}(\boldsymbol{r},\omega)\,\mathrm{e}^{-\mathrm{i}\omega t}\mathrm{d}\omega. \tag{6.14}$$

as can be easily verified multiplying Eq. (6.14) by $\mathrm{e}^{\mathrm{i}\omega' t}$, integrating in $\mathrm{d}t$ and taking into account that

$$\int_{-\infty}^{\infty} \mathrm{e}^{-i(\omega'-\omega)t}\mathrm{d}t = 2\pi\delta(\omega'-\omega).$$

Using Eq. (6.14), the equation of motion, Eq. (6.11), becomes:

$$-\int_{-\infty}^{\infty} \rho_0\,\omega^2\,\tilde{\boldsymbol{\xi}}(\boldsymbol{r},\omega)\,\mathrm{e}^{-\mathrm{i}\omega t}\mathrm{d}\omega = \int_{-\infty}^{\infty} \hat{\mathcal{F}}_{\boldsymbol{r}}\,[\tilde{\boldsymbol{\xi}}(\boldsymbol{r},\omega)]\,\mathrm{e}^{-\mathrm{i}\omega t}\mathrm{d}\omega.$$

Because the coefficients of the unknown function $\boldsymbol{\xi}$ and its derivatives are functions of the equilibrium quantities, which by definition are time-independent, the preceding expression can be written as:

$$-\rho_0\int_{-\infty}^{\infty} \omega^2\,\tilde{\boldsymbol{\xi}}(\boldsymbol{r},\omega)\,\mathrm{e}^{-\mathrm{i}\omega t}\mathrm{d}\omega = \hat{\mathcal{F}}_{\boldsymbol{r}}\left[\int_{-\infty}^{\infty} \tilde{\boldsymbol{\xi}}(\boldsymbol{r},\omega)\,\mathrm{e}^{-\mathrm{i}\omega t}\mathrm{d}\omega\right].$$

If we multiply both members by $\mathrm{e}^{\mathrm{i}\omega' t}$ and integrate over $\mathrm{d}t$ we finally get

$$-\rho_0\,\omega'^2\tilde{\boldsymbol{\xi}}(\boldsymbol{r},\omega') = \hat{\mathcal{F}}_{\boldsymbol{r}}\,[\tilde{\boldsymbol{\xi}}(\boldsymbol{r},\omega')]. \tag{6.15}$$

We have obtained an equation for the Fourier transform of $\boldsymbol{\xi}(\boldsymbol{r},t)$ which has the same form of Eq. (6.11), with the operator $\partial^2\boldsymbol{\xi}/\partial t^2$ replaced by the multiplicative factor $-\omega'^2$. In this form this becomes an eigenvalue equation, the eigenvalues being here represented by $-\rho_0\,\omega'^2$. Once this equation has been solved, the unknown function $\boldsymbol{\xi}(\boldsymbol{r},t)$ can be obtained by the simple application of the inverse Fourier transform, as defined by Eq. (6.14).

The Fourier transform can be applied also to those spatial coordinates, *ignorable coordinates*, that do not appear explicitly in the coefficients of the equations, i.e. in the unperturbed quantities. This happens when the equilibrium state exhibits some symmetry property. Let $\boldsymbol{s}$ be the set of ignorable coordinates, $\boldsymbol{r}'$ the set of the remaining coordinates and $\boldsymbol{k}$ a vector with non vanishing components only in the directions of the ignorable coordinates. The Fourier representation of $\boldsymbol{\xi}(\boldsymbol{r},t)$ then has the form

$$\boldsymbol{\xi}(\boldsymbol{r},t) = \boldsymbol{\xi}(\boldsymbol{r}',\boldsymbol{s},t) = \iint \tilde{\boldsymbol{\xi}}(\boldsymbol{r}',\boldsymbol{k},\omega)\,\mathrm{e}^{\mathrm{i}(\boldsymbol{k}\cdot\boldsymbol{s}-\omega t)}\mathrm{d}\omega\,\mathrm{d}\boldsymbol{k}. \tag{6.16}$$

Repeating the procedure adopted for the time transform, it is easy to realize that the application of the Fourier transform method is equivalent to the substitution (restricted to ignorable coordinates)

$$\frac{\partial}{\mathrm{d}t} \to -\mathrm{i}\omega, \quad \nabla \to \mathrm{i}\boldsymbol{k}, \tag{6.17}$$

and to the replacement of the unknown function with its Fourier transform.

It is possible to show that in the absence of dissipative terms the operator $\hat{\mathcal{F}}_{\boldsymbol{r}}$ is a *hermitian* or self-adjoint operator, which implies that the eigenvalues are real. Thus, in ideal MHD, $\omega^2$ is a real number. If $\omega^2$ is positive, $\omega$ must be real and $\boldsymbol{\xi}$, whose components are proportional to $\mathrm{e}^{-\mathrm{i}\omega t}$ represents an oscillation around the equilibrium position. A negative $\omega^2$ however implies a purely imaginary $\omega$, with consequent growth of perturbation amplitudes and therefore instability. Therefore, the stability of an equilibrium will be completely determined by the sign of $\omega^2$.

## 6.2 Instabilities in the Presence of Gravity

The most common force acting on astrophysical plasmas (apart from pressure gradients and magnetic field forces) is gravity. This means that the term $\boldsymbol{f}$ entering the second equation of (6.1), i.e. the momentum equation, has the form $\rho\boldsymbol{g}$ where $\boldsymbol{g}$ is the local value of the gravitational acceleration. For the sake of simplicity, we shall assume that the plasma contribution to the mass of the system is small, i.e. the plasma is not self-gravitating but is rather immersed in the gravitational potential of a more extended mass (such as for example, the plasma of a stellar corona). Therefore, $\boldsymbol{g}$ will be considered an *external* field, independent of the plasma perturbations. In the linearization process therefore one has $\boldsymbol{g}_1 = 0$ and Eq. (6.12) becomes:

$$\boldsymbol{F}(\boldsymbol{\xi}) = -\nabla P_1 + \frac{1}{4\pi}[(\nabla\times\boldsymbol{B}_0)\times\boldsymbol{B}_1 + (\nabla\times\boldsymbol{B}_1)\times\boldsymbol{B}_0] + \rho_1\boldsymbol{g}, \tag{6.18}$$

with $\rho_1$, $P_1$, and $B_1$ given by Eqs. (6.8), (6.9) and (6.10), respectively.

Moreover, we shall assume that the plasma perturbations are incompressible and that $\boldsymbol{g}$ and $\boldsymbol{B}_0$ are constant. Because incompressible perturbations can be viewed as a subset of all possible perturbations, if a system is unstable to such modes it will also be unstable more generally. We can choose our system of reference with $\boldsymbol{g}$ directed along the negative $z$-axis. In this frame, the equilibrium equation, $\nabla P_0 = \rho_0 \boldsymbol{g}$, implies that both $P_0$ and $\rho_0$ are functions of $z$ only and that

$$P_0' \equiv \frac{dP_0(z)}{dz} = -\rho_0(z) g.$$

The $x$ and $y$ coordinates are therefore ignorable. If the system is assumed to be homogeneous in the $y$-direction, we may restrict our attention to perturbations of the type $\boldsymbol{\xi} = [\xi_x, 0, \xi_z]$ and Fourier expand along $x$ all the quantities entering the equations, writing for the single Fourier component (see Eq. (6.16)):

$$\boldsymbol{\xi}(\boldsymbol{r}, \omega) = \int \tilde{\boldsymbol{\xi}}(z, k, \omega) e^{ikx} dk.$$

The vectors $\boldsymbol{k} = [k, 0, 0]$ and $\boldsymbol{g}$ define the coordinate plane $(x, z)$. In the reference system $[x, y, z]$ the vector $\boldsymbol{B}_0$ can have any orientation, but we can safely put $B_{0z} = 0$, since it is easy to realize that a (constant) $B_{0z}$ has no influence on the linear stability of the system. We shall therefore assume that $\boldsymbol{B}_0$ lies in the $(x, y)$ plane, $\boldsymbol{B}_0 = [B_{0x}, B_{0y}, 0]$. Making use of Eqs. (6.8), (6.9) and (6.10) we may cast Eq. (6.15) in the form:

$$-\omega^2 \rho_0 \boldsymbol{\xi} = \nabla(\boldsymbol{\xi} \cdot \nabla P_0) - \boldsymbol{g}(\boldsymbol{\xi} \cdot \nabla \rho_0) + \frac{1}{4\pi}\{\nabla \times [\nabla \times (\boldsymbol{\xi} \times \boldsymbol{B}_0)]\} \times \boldsymbol{B}_0, \quad (6.19)$$

where, to simplify the notation, we have written $\boldsymbol{\xi}$ instead of $\tilde{\boldsymbol{\xi}}(z, k, \omega)$ and we have used the incompressibility condition $\nabla \cdot \boldsymbol{\xi} = 0$. Observe that now the vector $\nabla$ is represented by $\nabla = [ik, 0, \partial/\partial z]$.

The magnetic term in Eq. (6.19) can be transformed into:

$$\frac{1}{4\pi}\{\nabla \times [\nabla \times (\boldsymbol{\xi} \times \boldsymbol{B}_0)]\} \times \boldsymbol{B}_0 = \frac{1}{4\pi}\{\nabla[\nabla \cdot (\boldsymbol{\xi} \times \boldsymbol{B}_0)] - \nabla^2(\boldsymbol{\xi} \times \boldsymbol{B}_0)\} \times \boldsymbol{B}_0,$$

and a long, but straightforward, calculation gives simply:

$$B_{0x}^2(-k^2 \xi_z + \xi_z'')\, \boldsymbol{e}_z,$$

where primes indicate derivatives with respect to $z$.

The final equation for $\boldsymbol{\xi} = \tilde{\boldsymbol{\xi}}(z, k, \omega)$ is thus seen to be:

$$-\omega^2 \rho_0 \boldsymbol{\xi} = \nabla(\xi_z P_0') - \boldsymbol{g}(\xi_z \rho_0') + (B_{0x}^2/4\pi)(-k^2 \xi_z + \xi_z'')\boldsymbol{e}_z, \quad (6.20)$$

whose $x$ and $z$ components are:

$$-\omega^2 \rho_0 \xi_x = ik(\xi_z P_0') \quad (6.21a)$$

$$-\omega^2 \rho_0 \xi_z = (\xi_z P_0')' + (B_{0x}^2/4\pi)(-k^2 \xi_z + \xi_z'') + g(\rho_0' \xi_z). \tag{6.21b}$$

Equations 6.21a, 6.21b must be complemented by the incompressibility condition,

$$ik\xi_x + \xi_z' = 0. \tag{6.22}$$

We now combine the latter expression with Eq. (6.21a) to eliminate $(\xi_z P_0')'$ and finally obtain:

$$\begin{aligned}\omega^2 \left[ (\rho_0 \xi_z')' - \rho_0 k^2 \xi_z \right] &= \frac{B_{0x}^2 k^2}{4\pi} (\xi_z'' - k^2 \xi_z) + k^2 g (\rho_0' \xi_z) \\ &= \frac{(\boldsymbol{k} \cdot \boldsymbol{B}_0)^2}{4\pi} (\xi_z'' - k^2 \xi_z) + k^2 g (\rho_0' \xi_z). \end{aligned} \tag{6.23}$$

We shall now turn to the illustration of some consequences of the eigenvalue equation just found.

### 6.2.1 Rayleigh-Taylor Instability

Consider first the development of instability when the equilibrium magnetic field vanishes, known as the Rayleigh-Taylor instability. An approach which quickly leads to the stability conditions consists multiplying Eq. (6.23) by $\xi_z$ and integrating in in $dz$ between $-\infty$ and $\infty$, to obtain:

$$\omega^2 \left[ \int_{-\infty}^{\infty} [\rho_0 \xi_z']' \, \xi_z \, dz - k^2 \int_{-\infty}^{\infty} \rho_0 \xi_z^2 \, dz \right] = k^2 g \int_{-\infty}^{\infty} \rho_0' \xi_z^2 \, dz.$$

Performing an integration by parts on the first integral in square brackets and assuming that $\xi_z$ and/or $\xi_z'$ vanish for $z = \pm\infty$ leads to:

$$\omega^2 = -k^2 g \frac{\int_{-\infty}^{\infty} \rho_0' \xi_z^2 \, dz}{\int_{-\infty}^{\infty} \rho_0 \, [\xi_z'^2 + k^2 \xi_z^2] \, dz}. \tag{6.24}$$

The same equation can be obtained by applying the energy method, which, as we have remarked, is a variational method. Equation (6.24) shows that all the terms appearing under the integral sign are positive definite, except $\rho_0'$. The *sufficient* condition to have stability, i.e. $\omega^2 > 0$ is therefore $\rho_0' < 0$ for every $z$, namely a density which always decreases with height. At first sight, one may think that this condition is not a *necessary* one. In fact, $\omega^2 > 0$ does not imply $\rho_0' < 0$ *everywhere*, since $\rho_0'$ is weighted in the integral by $\xi_z^2$ and if $\rho_0'$ is positive only in a certain range, the integral at the numerator of Eq. (6.24) could still be negative, Nonetheless, it is possible to show that the condition $\rho_0' < 0$ for every $z$ is also a necessary one, as we shall now see.

First observe that the expression for $\omega^2$ given by Eq. (6.24) is a presently only *formal* solution of Eq. (6.23), since it contains the quantities $\xi_z$ and $\xi_z'$ that remain unknown until Eqs. (6.21) have been explicitly solved. In the theory of calculus of variations it is proved that, if in Eq. (6.24) the functions $\xi_z$ and $\xi_z'$, solutions of the original equation, are replaced by *arbitrary* functions (satisfying the correct boundary conditions, in this case convergence properties at infinity), Eq. (6.24) gives us a value $\tilde{\omega}^2$ *larger than* the correct one (namely the one we would have obtained by using the exact solutions):

$$\tilde{\omega}^2 \geqslant \omega^2,$$

where the equal sign holds *only* when the correct solutions of the problems are inserted into the integral.

Assume now that $\rho_0' > 0$ only in a finite interval, with $\omega^2$ still positive. The variational property mentioned above tells us that we are free to choose arbitrary functions that are different from zero only within that interval. In this case, we would obtain a negative value for $\tilde{\omega}^2$. But then, according to the previous inequality, $\omega^2$ would also be negative, contrary to our assumption. It is therefore impossible to have $\omega^2 > 0$ even for a limited interval over which $\rho_0' > 0$, Therefore $\rho_0' < 0$ for every $z$ is also a necessary condition for stability.

A simple and interesting case is that of two incompressible fluids of different densities lying one atop of the other, i.e.

$$\rho = \rho_1 = const., \quad z > 0$$
$$\rho = \rho_2 = const., \quad z < 0.$$

In this model the derivative of the density is discontinuous in $z = 0$ and can be represented as: $\rho_0' = (\rho_1 - \rho_2)\,\delta(z)$. Consequently, Eq. (6.23) can be written, for $z \neq 0$,

$$\xi_z'' = k^2 \xi,$$

whose solutions are:

$$\xi_z = \xi_z(0)\mathrm{e}^{-kz}, \quad z > 0 \tag{6.25}$$
$$\xi_z = \xi_z(0)\mathrm{e}^{kz}, \quad z < 0. \tag{6.26}$$

We are now in a position to explicitly evaluate $\omega^2$ by using Eq. (6.24). The integral in the numerator gives simply

$$(\rho_1 - \rho_2)\,\xi_z^2(0).$$

The integral in the denominator can be evaluated as follows:

$$\int_{-\infty}^{\infty} [\xi_z'^2 + k^2\xi_z^2]\,\mathrm{d}z = \rho_2 \int_{-\infty}^{0-} [\xi_z'^2 + k^2\xi_z^2]\,\mathrm{d}z + \rho_1 \int_{0+}^{\infty} [\xi_z'^2 + k^2\xi_z^2]\,\mathrm{d}z$$

$$= 2(\rho_1 + \rho_2)\,k^2 \int_{0+}^{\infty} \mathrm{e}^{-2kz}\mathrm{d}z = k(\rho_1 + \rho_2)\,\xi_z^2(0),$$

where the correct expressions for $\xi_z$ and $\xi_z'$ have been used in each of the integration intervals. Therefore:

$$\omega^2 = -kg\frac{\rho_1 - \rho_2}{\rho_1 + \rho_2}. \tag{6.27}$$

An instability will set in whenever the fluid of higher density lies above that of lower density, in agreement with the general criterion discussed above. When $\rho_1 < \rho_2$ the system will be stable and the frequency of the oscillation will be given by Eq. (6.27). The maximum value of $\omega$ is reached when $\rho_1 \ll \rho_2$, in which case

$$\omega = \sqrt{kg}.$$

This is the first example of a *dispersion relation*, a relationship that expresses $\omega$ as a function of the wavevector $\boldsymbol{k}$. As we shall discuss in more detail in the next Chapter, the stable oscillations described above are an example of *dispersive waves* with a phase velocity $v_\varphi \equiv \omega/k = \frac{1}{2}\sqrt{g/k}$.

The dispersion relation derived above is also valid for surface ocean waves, that develop at the interface between the water and the atmosphere. The description of air as an incompressible fluid, may throw some legitimate doubt on the applicability of our result to sea waves. However, every fluid actually behaves as an incompressible one when the velocity of the fluid particles is negligible with respect to the sound speed. Notice that the dispersion relation we have deduced refers to the case of deep water, since we have imposed as a boundary condition the vanishing of $\xi_z$ for $z \to -\infty$. If the water has a depth $h$ the dispersion relation is modified by the fact that boundary conditions must be imposed in $z = -h$.

Even if the linear analysis does not determine the complete temporal evolution of the system, it is easy 'to visualize the initial phases of such evolution. If in the (unstable) equilibrium the dense gas lies above the light one, any perturbation will deform the originally flat surface separating the two fluids into a wavy surface with portions of lighter gas rising above of the equilibrium surface ($z = 0$) and portions of the heavier gas descending below that level. The distance between two successive regions of dense gas is of the order of $\lambda = 2\pi/k$. In the subsequent phases the vertical extension of those regions increases and the gas structure is modified until a complete overturning of the initial configuration is reached, with the lighter gas lying above

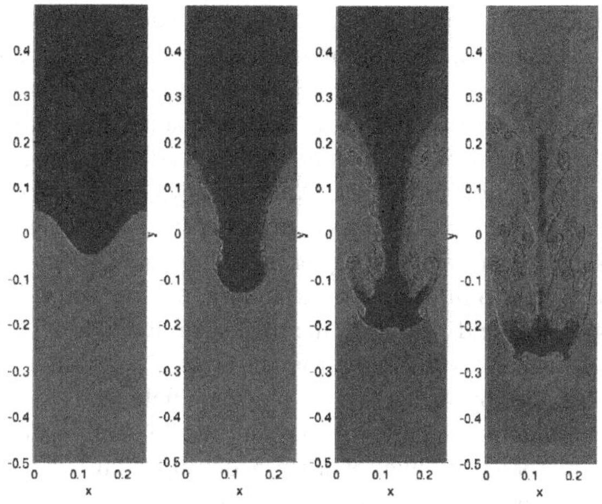


**Fig. 6.1** Numerical simulation of the development of the Rayleigh-Taylor instability

the denser one. The study of the nonlinear phase of the Rayleigh-Taylor instability and, more generally of any instability, can only be made by means of numerical simulations. An example of the results of these simulations for the Rayleigh-Taylor instability is shown in Fig. 6.1.

### *6.2.2 Kruskal-Shafranov Instability: $B_0 \neq 0$*

If gravity and magnetic fields are both present, a series of new dynamical situations arise. The basic physical mechanism underlying these new phenomena is actually simple and is due to the fact that, while plasma density and pressure are connected directly via the equation of state, the magnetic field only modifies the pressure, but not the density. The nature of the gravitational instability of the previous Section is therefore altered and, since in ideal plasmas the magnetic field is coupled to matter, the magnetic structure is altered as well. As an example, consider the case of a magnetic field occupying only a limited portion of space, with matter present everywhere. An astrophysical example of this situation is encountered when the magnetic field of a star is confined below the visible surface of the star.

Let then $\boldsymbol{B}$ be a constant horizontal field confined below the plane $z = 0$ and assume that its value is rapidly decreasing, going to zero at the top of a thin layer above that plane. Let's further assume that the gas temperature is the same everywhere. With the above assumptions, the local equilibrium condition requires that:

$$\frac{\mathrm{d}}{\mathrm{d}z}\left(P + B^2/8\pi\right) = -\rho g.$$

There will therefore be an abrupt variation of magnetic pressure crossing the surface that separates the magnetized region of the atmosphere from the field-free one. Such a variation must be compensated by a corresponding increase of the gas pressure, $\Delta P = B^2/8\pi$, in order to maintain equilibrium in the atmosphere across the discontinuity. The latter pressure variation, however, necessarily induces a density variation in the gas, $\Delta\rho = (m/k_B T)\Delta P = (m/k_B T)(B^2/8\pi)$, which will become denser with respect to that of the underlying region. The system is therefore subject to a Rayleigh-Taylor instability and regions of dense and unmagnetized gas will start to flow downward. In turn, the magnetized plasma will start rising, dragging the field with it. In the final state, the field will be concentrated in a series of vertical layers emerging from the original separation surface. This dynamics is often described by saying that the magnetic field tends to be *buoyant*.

To derive the dispersion relation, will shall again consider a situation with two incompressible fluids of different density, lying one on top of the other. Following the procedure already adopted for the case of a vanishing field, we start from Eq. (6.23) where we now keep the term proportional to $(\boldsymbol{k}\cdot\boldsymbol{B}_0)^2$.

For $z \neq 0$ Eq. (6.23) becomes:

$$[\omega^2\rho_0 - \frac{(\boldsymbol{k}\cdot\boldsymbol{B}_0)^2}{4\pi}](\xi_z'' - k^2\xi_z) = 0,$$

with $\rho_0 = \rho_1$ or $\rho_0 = \rho_2$, for $z \gtrless 0$, respectively. Therefore, no value of $\omega$ exists for which the first parenthesis is identically equal to zero, irrespective of the value of $z$. Thus, the equation for $\xi_z$ is still the one of the unmagnetized case:

$$\xi_z'' = k^2\xi,$$

whose solutions are given by Eq. (6.25). Proceeding as in the previous case we easily find

$$\omega^2 \int_{-\infty}^{\infty} \rho_0(\xi_z'^2 + k^2\xi_z^2)\,\mathrm{d}z = -k^2 g\,(\rho_1 - \rho_2)\xi_z(0)^2 + \frac{(\boldsymbol{k}\cdot\boldsymbol{B}_0)^2}{4\pi}\int_{-\infty}^{\infty}(\xi_z'^2 + k^2\xi_z^2)\,\mathrm{d}z. \tag{6.28}$$

The integrals entering Eq. (6.28) can be evaluated using Eq. (6.25), and the dispersion relation is:

$$\omega^2 = -|\boldsymbol{k}|\,g\frac{\rho_1 - \rho_2}{\rho_1 + \rho_2} + 2\frac{(\boldsymbol{k}\cdot\boldsymbol{B}_0)^2}{4\pi(\rho_1 + \rho_2)}, \tag{6.29}$$

where $k$ has been written as $|\boldsymbol{k}|$ to make the expression of $\omega$ independent of a particular frame of reference.

In the limit $\boldsymbol{B}_0 \to 0$ we find again the condition for the Rayleigh-Taylor instability, while in the case of a homogeneous plasma, $\rho_1 = \rho_2 = \rho_0$, we have

$$\omega^2 = \frac{(\boldsymbol{k} \cdot \boldsymbol{B}_0)^2}{4\pi \rho_0},$$

which, as we shall see in the following, corresponds to the propagation of Alfvén waves. Equation (6.29) shows that the presence of a magnetic field has a stabilizing effect. In fact, even if $\Delta\rho = \rho_1 - \rho_2 > 0$, $\omega^2$ can still be positive, provided that the second term dominates the rhs of Eq. (6.29), i.e. when

$$\lambda = \frac{2\pi}{|\boldsymbol{k}|} < \frac{B_0^2}{g\Delta\rho} \cos^2\theta,$$

where $\theta$ is the angle between $\boldsymbol{k}$ and $\boldsymbol{B}_0$. The most efficient stabilization is reached when $\boldsymbol{k}$ is parallel to $\boldsymbol{B}_0$, while the stabilizing effect vanishes when $\boldsymbol{k}$ and $\boldsymbol{B}_0$ are mutually orthogonal. This is easily understood since, in the first case the perturbation deforms the field lines and the magnetic tension resists the deformation, while in the second case the field lines remain straight and the tension cannot act.

A case of particular interest is one in which the plasma is confined to the region $z > 0$, while in the region $z < 0$ only a magnetic field is present. In this situation, by applying the previously adopted procedure again, we find:

$$\omega^2 = -|\boldsymbol{k}|g + \frac{(\boldsymbol{k} \cdot \boldsymbol{B}_0)^2}{4\pi \rho_1},$$

which shows how a plasma could be sustained against gravity by a magnetic field alone. This could possibly explain the existence and stability of solar prominences, i.e. structures composed by a plasma denser than that of the surrounding environment. Note however, that as long as there is a direction along which the perturbation fronts can propagate in an unstable way, the system will be unstable. In the present case, nothing can stop modes with wave-vectors orthogonal to the magnetic field from growing. Interestingly, introducing a gradient, or shear in the magnetic field with heights, which might lead to stabilization (because in this case there would be no fixed wave-vector orthogonal to the field at all heights) does not completely stabilize the instability.

### *6.2.3 Parker Instability*

A further example of an instability in the presence of gravity is provided by the so-called *Parker instability*, an interesting astrophysical application of the theory of plasma instabilities. We shall now discuss it in some detail because it permits a good appreciation of the technicalities involved in the theory, but also because it emphasizes the necessity of a close comparison with observations to assess the actual interest of the results obtained.

The problem that prompted this study, as discussed by Parker in 1966 [15], concerns the structure of the magnetic field of the Galaxy and the influence of the field on galactic configuration and dynamics. The essential observational data are the following:

- the Galaxy is permeated by a large scale magnetic field, whose indicative value is a few times $10^{-6}$ G.
- the field appears to be confined in the galactic plane.
- the magnetic field (as well as the cosmic rays trapped in it) cannot be confined by the weak pressure of the external extragalactic medium.

In this situation it is reasonable to assume that the confinement within the galactic disc is due to the weight of the *self-gravitating* interstellar gas that forms the disc itself. The freezing of the field within the plasma might then confine the field inside the disc. The gas confinement due to the gravity of the galactic nucleus itself can be ruled out since it can be shown that it would produce a number of effects that are not observed. To check the correctness of the above assumption, let's first determine the properties of the galactic disc in equilibrium.

We are going to adopt a simplified two-dimensional model, in which the magnetic field is parallel to the galactic disc. We identify this direction with the $y$-axis, $\boldsymbol{B}_0 = B_0(z)\boldsymbol{e}_y$, the $z$-axis being the direction of gravitational acceleration.
The equilibrium equation thus reads:

$$\frac{\mathrm{d}}{\mathrm{d}z}\left(P_0 + \frac{B_0^2}{8\pi}\right) = -\rho(z)g,$$

where, to keep things as simple as possible, $g$ has been assumed to be constant. To be rigorous, we should include in the equilibrium equation also the pressure generated by the cosmic rays. This contribution was included in Parker's original treatment, but here we shall neglect it, since its absence does not alter the physics of our system.

Moreover, we shall assume that the interstellar gas is isothermal, which allows us to write:

$$P_0 = \frac{k_B T_0}{\bar{m}}\rho_0 = c_0^2\rho_0,$$

and that the following relation holds

$$\frac{B_0^2}{8\pi} = \alpha P_0 = \alpha c_0^2\rho_0, \qquad \alpha = const. \tag{6.30}$$

The last assumption implies that the Alfvén speed, $c_a^2 = B_0^2/(4\pi\rho_0)$ is constant. In the above framework, the solution of the equilibrium equation is

$$\rho_0(z) = \rho_0(0)\exp(-z/H),$$

where $H$ is the scale height,

$$H = \frac{c_0^2(1+\alpha)}{g}.$$

Equation (11.17) further implies that

$$B_0(z) = B_0(0)\exp(-z/2H).$$

Summarizing, we have

$$\frac{1}{\rho_0}\frac{\mathrm{d}\rho_0}{\mathrm{d}z} = \frac{1}{P_0}\frac{\mathrm{d}P_0}{\mathrm{d}z} = \frac{2}{B_0}\frac{\mathrm{d}B_0}{\mathrm{d}z} = -\frac{1}{H}. \tag{6.31}$$

Observations suggest that $H$ is of the order of some hundreds of parsecs ($H = 1 \div 3\times 10^{20}$ cm) and that $c_0$ is of the order of a few km/sec ($c_0 = 3 \div 8\times 10^8$ cm sec$^{-1}$) corresponding to temperatures of the order of thousands of degrees, with $g = 1 \div 3\times 10^{-9}$ cm sec$^{-2}$. From the definition of $H$ we find that it is likely that $\alpha \lesssim 1$. The observed magnetic field is about $B_0 = 1 \div 5\times 10^{-6} G$ and Eq. (6.30) suggests values of $\rho$ of the order of a few hydrogen atoms per cm$^3$, close to the observed values. The proposed model thus seems to provide a reasonable representation of the equilibrium configuration of the interstellar gas-magnetic field system. The critical point, of course, is that of the stability of such equilibrium, discussed below.

It is convenient to start from the linearized equations, Eqs. (6.3)–(6.6), rather than use Eq. (6.23). In the following, we shall express $\boldsymbol{B}_1(y,z,t)$ in terms of its vector potential $\boldsymbol{A}_1(y,z,t) : \boldsymbol{B}_1 = \nabla\times\boldsymbol{A}_1$.

Since $B_{1x} = 0$ and both $y$ and $t$ are ignorable coordinates, the equilibrium quantities will be functions of $z$ only and we can write:

$$\boldsymbol{A}_1(y,z,t) = a(z)\mathrm{e}^{\mathrm{i}(ky-\omega t)}\boldsymbol{e}_x, \tag{6.32}$$

and

$$\boldsymbol{B}_1(y,z,t) = \boldsymbol{b}(z)\mathrm{e}^{\mathrm{i}(ky-\omega t)}. \tag{6.33}$$

Since

$$B_{1y} = \frac{\partial A_{1x}}{\partial z}, \qquad B_{1z} = -\frac{\partial A_{1x}}{\partial y},$$

we have:

$$b_y = \frac{\mathrm{d}a}{\mathrm{d}z} \equiv a'(z), \qquad b_z = -\mathrm{i}ka. \tag{6.34}$$

Taking into account that $U_x = i\omega\xi_x = 0$ and adopting a Fourier representation of the type given by Eq. (6.32) for all first-order quantities, from Eq. (6.6) we obtain

$$\nabla \times (\boldsymbol{A}_1 - \boldsymbol{\xi} \times \boldsymbol{B}_0) = 0.$$

The quantity in round brackets can be expressed as the gradient of a potential, which, using the gauge invariance of Maxwell's equations, can be safely taken equal to zero. We shall thus write

$$\boldsymbol{A}_1 = \boldsymbol{\xi} \times \boldsymbol{B}_0,$$

whose $z$-component provides the relation:

$$\xi_z = -\frac{a}{B_0}, \qquad \xi'(z) = -\frac{1}{B_0}\left(a' + \frac{a}{2H}\right). \tag{6.35}$$

Equation (6.35) suggests expressing all first-order quantities in terms of $a(z)$. Equation (6.8), taking into account Eqs. (6.31) and (6.35), then becomes

$$\rho_1 = -\frac{\rho_0}{H}\frac{a}{B_0} - \rho_0 ik\xi_y + \frac{\rho_0}{B_0}\left(a' + \frac{a}{2H}\right) = \frac{\rho_0}{B_0}\left(a' - \frac{a}{2H}\right) - \rho_0 ik\xi_y.$$

In a similar way, Eq. (6.9) can be cast in the form

$$P_1 = \frac{\rho_0 c_0^2}{B_0}\left[\gamma a' + (\gamma/2 - 1)\frac{a}{H}\right] - \gamma\rho_0 c_0^2(ik\xi_y).$$

To eliminate $\xi_y$, consider first the $y$-component of the momentum equation, Eq. (6.11). Writing the force term, Eq. (6.12), as

$$\begin{aligned}\boldsymbol{F}(\boldsymbol{\xi}) &= -\nabla P_1 + \frac{1}{4\pi}[(\nabla \times \boldsymbol{B}_0) \times \boldsymbol{B}_1 + (\nabla \times \boldsymbol{B}_1) \times \boldsymbol{B}_0] + \rho_1 \boldsymbol{g} \\ &= -\nabla\left(P_1 + \frac{1}{8\pi}\boldsymbol{B}_0 \cdot \boldsymbol{b}\right) + \frac{1}{4\pi}[\boldsymbol{B}_0 \cdot \nabla)\boldsymbol{b} + (\boldsymbol{b} \cdot \nabla)\boldsymbol{B}_0] + \rho_1 \boldsymbol{g},\end{aligned}$$

we find that the $y$-component of the momentum equation reduces to

$$\omega^2 \rho_0 \xi_y = ikP_1 - ik\frac{B_0}{8\pi}\frac{a}{H}.$$

Making now use of the preceding expression for $P_1$, we find:

$$\Omega^2 \xi_y = i\frac{kc_0^2}{B_0}\left[\gamma a' + (\gamma/2 - 1 - \alpha)\frac{a}{H}\right], \tag{6.36}$$

where $\Omega^2$ has been defined as

$$\Omega^2 = \omega^2 - \gamma k^2 c_0^2.$$

Equation (6.36) allows us to deduce two explicit expressions for $\rho_1$ and $P_1$ in terms of $a$ and its derivative:

$$\Omega^2 \rho_1 = \frac{\rho_0}{B_0}\left\{\omega^2 a' - [\omega^2/2 + (1-\alpha-\gamma)k^2 c_0^2]\frac{a}{H}\right\}, \tag{6.37}$$

$$\Omega^2 P_1 = \frac{\rho_0 c_0^2}{B_0}\left\{\gamma\omega^2 a' + [(\gamma/2-1)\omega^2 - \alpha\gamma k^2 c_0^2]\frac{a}{H}\right\}. \tag{6.38}$$

Finally, the $z$-component of the momentum equation, multiplied by $\Omega^2$, gives:

$$\omega^2 \rho_0(\Omega^2 \xi_z) = [(\Omega^2 P_1) + \frac{B_0}{4\pi}\Omega^2 a')]' - \frac{B_0}{4\pi}k^2\Omega^2 a + \frac{(1+\alpha)c_0^2}{H}(\Omega^2\rho_1),$$

where $g$ has been expressed in terms of $H$, according to its definition.

Inserting in the preceding equation the previously deduced expressions for $\Omega^2\rho_1$ and $\Omega^2 P_1$ and performing the derivatives, we arrive at the final form of the differential equation for $a$:

$$a''(z) = k^2 \frac{N}{D} a(z), \tag{6.39}$$

where

$$N = \left(\frac{\omega}{kc_0}\right)^4 - (\gamma + 2\alpha)\left(1 + \frac{1}{4k^2H^2}\right)\left(\frac{\omega}{kc_0}\right)^2 + 2\alpha\gamma - \frac{(1+\alpha)(1+\alpha-\gamma) - \alpha\gamma/2}{k^2H^2}, \tag{6.40}$$

and

$$D = 2\alpha\gamma - (\gamma + 2\alpha)\left(\frac{\omega}{kc_0}\right)^2. \tag{6.41}$$

The boundary conditions to be imposed on $a(z)$ are such that $a$ should vanish at the origin and remain finite for all values of $z$. This implies that $N/D$ must be a negative quantity, in which case the solution of Eq. (6.39) can be written as:

$$a(z) = (\epsilon \bar{b} H)\sin Kz, \tag{6.42}$$

where

$$K^2 = k^2\left|\frac{N}{D}\right|,$$

and the constant of integration has been written as $\epsilon\,\bar{b}\,H$. The explicit appearance of $\epsilon$ emphasizes that $a$ is a first-order quantity and $\bar{b}$ is an arbitrary constant with the dimensions of a magnetic field to be specified later on.

Since we are interested in the unstable case, $\omega^2 < 0$, let's put $\omega = \mathrm{i}/\tau$, $\tau$ representing the growth time of the instability and define

$$\nu^2 = -\left(\frac{\omega}{k^2 c_0^2}\right)^2 = \frac{1}{\tau^2 k^2 c_0^2} = \left(\frac{H}{c_0 \tau}\right)^2 \left(\frac{1}{kH}\right)^2.$$

Since $H/c_0$ represents the transit time over the scale height for a sound wave, $n = H/(c_0\tau)$ is the growth rate of the instability measured in units of the inverse of that transit time. For the unstable solutions, the term

$$D = 2\alpha\gamma + (\gamma + 2\alpha)\nu^2 > 0,$$

and the boundary conditions reduce to the requirement that $N < 0$. If $x = 1/(k^2 H^2)$, unstable solutions exist when

$$N(x, n^2) = q\,x^2 - r\,x + 2\alpha\gamma < 0, \tag{6.43}$$

with

$$q = \frac{1}{4} n^2 (4n^2 + 2\alpha + \gamma),$$

$$r = \Delta - n^2(2\alpha + \gamma),$$

$$\Delta = (1+\alpha)(1+\alpha-\gamma) - \alpha\gamma/2 = (1+\alpha)^2 - \gamma(1 + 3\alpha/2).$$

Thus, a necessary and sufficient condition to satisfy Eq. (6.43) is that $r > 0$, which implies that $\Delta > 0$. In addition, $x$ must lie between the two positive solutions of the equation $N = 0$,

$$x = \frac{1}{2q}\left(r \pm \sqrt{r^2 - 8\alpha\gamma\,q}\right), \tag{6.44}$$

which requires that

$$r^2 > 8\alpha\gamma\,q.$$

On the other hand, the definitions of $q$ and $r$ allow to write the preceding expression as:

$$n^4 - 2\lambda n^2 + \mu^2 > 0, \tag{6.45}$$

with

$$\lambda = \frac{(2\alpha + \gamma)(\Delta + \alpha\gamma)}{(\gamma - 2\alpha)^2}, \qquad \mu = \frac{\Delta}{\gamma - 2\alpha}.$$

Equation (6.45) is satisfied for $n = 0$ which means that marginal stability states exist for any value of $\alpha$ and $\gamma$. According to Eq. (6.45), the maximum value of $n^2$ corresponds to the smallest of the two positive solutions of the equation $n^4 - 2\lambda n^2 + \mu^2 = 0$,

$$n^2_{max} = \lambda - \sqrt{\lambda^2 - \mu^2}.$$

We shall verify in the following that in the cases of physical interest $\lambda^2 \gg \mu^2$; the square root can then be expanded as

$$n^2_{max} \simeq \frac{\mu^2}{2\lambda},$$

or, explicitly

$$n_{max} \simeq \frac{\Delta}{[2(\gamma + 2\alpha)(\Delta + \alpha\gamma)]^{1/2}}. \tag{6.46}$$

However, noticing that $n = n_{max}$ implies $r^2 = 8\alpha\gamma q$ from Eq. (6.44), we find $x_{max} \equiv x(n_{max}) = r/2q$. Inserting this result into the definition of $N$ (see Eq.(6.43)) we find

$$N(x_{max}, n_{max}) = 0 \quad \rightarrow \quad K = 0.$$

In this case the solution would have an infinite vertical wavelength, which makes apparent that the $n_{max}$ should be considered a limit rather than an effectively reachable value.

To summarize the preceding discussion, we have found that the gas-magnetic field system is *always* unstable, irrespective of the values of $\alpha$ e $\gamma$, provided that $\Delta > 0$. The last condition can be cast in the form

$$\gamma - 1 < \frac{\alpha(\alpha + 1/2)}{1 + 3\alpha/2}. \tag{6.47}$$

In the absence of a field, $\alpha = 0$, and, given that the $\gamma$ parameter in the gas is larger than unity, the preceding instability condition can never be satisfied and the system is stable. When a magnetic field is present Eq. (6.47) gives only an upper limit for $\gamma$. Typically in the interstellar medium $\gamma \lesssim 1$ and Eq. (6.47) is always satisfied, for all $\alpha$ values. The presence of a magnetic field thus inevitably generates an instability, known as the *Parker instability*.

To assess the real relevance of this instability for the structure and dynamics of the interstellar gas in the Galaxy, it is necessary to evaluate its growth time in the prevailing galactic conditions and compare it with the typical galactic timescales, which fall in the range of $10^8 \div 10^9$ years.

Let's choose the following set of typical values: $B_0 = 5 \times 10^{-6} G$, $\rho_0 = 5 \times 10^{-24}\,\mathrm{g\,cm^{-3}}$ (about 3 hydrogen atoms per cm$^3$), $g = 1.3 \times 10^{-9}\,\mathrm{cm\,sec^{-2}}$ and $\gamma = 1$.

The minimum growth time will then be:

$$\tau_{min} = \frac{H}{c_0\, n_{max}} \simeq \frac{c_0}{g}\frac{1+\alpha}{n_{max}}.$$

Using Eq. (11.17) to express $c_0$ as

$$c_0 = \frac{B_0}{\sqrt{8\pi\rho_0\alpha}} = \frac{c_a}{\sqrt{2\alpha}},$$

from Eqs. (6.31) and (6.46) (with $\gamma = 1$) we get:

$$\tau_{min} = \frac{c_a}{g}\frac{(1+\alpha)}{\alpha}\left[\frac{2(2\alpha+3)}{(2\alpha+1)}\right]^{1/2}. \tag{6.48}$$

With the typical values chosen above, Eq. (6.48) shows that $\tau_{min}$ is less than $10^8$ years for $\alpha > 0.5$.

It also shows that $\tau_{min}$ is a decreasing function of $\alpha$, not a surprising result since the instability is due to the presence of a magnetic field, whose importance grows with increasing $\alpha$. As an example, if $\alpha = 0.6$, Eq. (6.48) gives $\tau_{min} \simeq 8 \times 10^7$ years. Moreover, the sound speed turns out to be $c_0 \simeq 5.8\,\mathrm{km\,sec^{-1}}$, corresponding to a temperature of $T_0 \simeq 4,000\,\mathrm{K}$ and the scale height $H \simeq 130\,\mathrm{pc}$. All these values agree with the average observed properties of the galactic disc.

As already remarked, one should also include the effects connected with the presence of the cosmic rays and in particular of their pressure. It is possible to show that those effects act in the same direction of those associated with the magnetic field and therefore they contribute to the development of the instability. Therefore, to generate growth rates in the physically interesting range, values of $\alpha$ lower than those previously computed are actually sufficient.

In summary, we may conclude that the presence of a magnetic field will produce an instability whose growth times are significantly shorter the the Galaxy's lifetime.

Let's now turn to the investigation of the consequences of the Parker instability on the structure and evolution of the galactic disc. To simplify the discussion, we shall again neglect the cosmic rays and focus our attention on a "cold" gas, $c_0 \simeq 0$. This assumption seems to give rise to a problem when it is inserted into the expressions for $N$ e $D$ (Eqs. (6.40) and (6.41)), but the difficulty is only an apparent one, as can be seen by multiplying those expressions by $k^4 c_0^4$ and noticing that $\alpha c_0^2 = \frac{1}{2}c_a^2 = B_0^2/(8\pi\rho_0) \neq 0$. The solution for $a(z)$ is still given by Eq. (6.42), with

$$K^2 = \tfrac{1}{4}k^2 s^2\left[1 - \frac{4}{s^2}(1 + 4/x) - \frac{4}{s^4}x\right],$$

where

$$s = \frac{\tau}{H/c_a} \quad \text{and} \quad \mathrm{x} = 1/(\mathrm{k}^2\mathrm{H}^2).$$

The quantity $s$ is simply the growth time measured in units of the transit time of the scale height $H$ at the Alfvén speed. If it is reasonably assumed that $s \gg 1$, $K \simeq \frac{1}{2}ks$.

The complete expression for the perturbed vector potential, $\boldsymbol{A}_1(y, z, t)$, remembering that the physical parameters are real quantities, will then be given by:

$$\begin{aligned}\boldsymbol{A}_1(y, z, t) &= [\epsilon\bar{b}H \exp(t/\tau)\, \mathfrak{Re}(\exp^{iky}) \sin Kz]\, \boldsymbol{e}_x \\ &= [\epsilon\bar{b}H \exp(t/\tau) \cos ky \sin Kz]\, \boldsymbol{e}_x.\end{aligned}$$

From this, we may easily compute the magnetic field components, that turn out to be:

$$B_y(y, z, t) = B_0(z) + \epsilon\bar{b}HK \exp(t/\tau) \cos ky \cos Kz,$$

and

$$B_z(y, z, t) = \epsilon\bar{b}Hk \exp(t/\tau) \sin ky \sin Kz.$$

The field lines are defined by the equation

$$\frac{\mathrm{d}z}{\mathrm{d}y} = \frac{B_z}{B_y} = \frac{\epsilon\bar{b}Hk \exp(t/\tau) \sin ky \sin Kz}{B_0(z) + \epsilon\bar{b}HK \exp(t/\tau) \cos ky \cos Kz} \simeq \frac{\epsilon\bar{b}Hk \exp(t/\tau) \sin ky \sin Kz}{B_0(z)},$$

where in the denominator the term proportional to $\epsilon$ has been neglected with respect to $B_0$. Let's now integrate the preceding equation in the neighborhood of the point $z_0$ where the field line crosses the $z$-axis ($y = 0$). We get

$$\int_{z_0}^{z} \frac{B_0(z)}{\sin Kz} \mathrm{d}z = \epsilon\bar{b}Hk \exp(t/\tau) \int_0^y \sin ky \, \mathrm{d}y = \epsilon\bar{b}H \exp(t/\tau)(1 - \cos ky).$$

Since the rhs is of the order $\epsilon$, the lhs must be of the same order, which allows us to replace $B_0(z)$ and $\sin Kz$ with their values in $z = z_0$. In this way we obtain:

$$z - z_0 \simeq \epsilon H \exp(t/\tau)(\sin Kz_0)(1 - \cos ky),$$

where the arbitrary constant has been written as $\bar{b} = B_0(z_0)$. The instability therefore bends the initially straight field lines forming a succession of crests and troughs.

The effects of the instability are well illustrated by the behaviour of the perturbed density, $\rho_1(y, z, t)$, that we are now going to determine. Write first the physical components of the displacement in the neighborhood of the point $z = z_0$ in agreement with what was already done when computing the shape of the field lines.

From Eq. (6.35) we get

$$\xi_z = -\frac{A_1}{B_0} = -\epsilon H \exp(t/\tau) \cos ky \sin K z_0,$$

while Eq. (6.36) gives us

$$\begin{aligned}\xi_y &= \epsilon (k c_a^2 \tau^2) \exp(t/\tau) \sin K z_0 \, [\mathfrak{Re}(\mathrm{i} \exp^{iky})] \\ &= -\epsilon (k c_a \tau)^2 \exp(t/\tau) \sin K z_0 \sin ky = -\epsilon H (s^2 k H) \exp(t/\tau) \sin ky \sin K z_0.\end{aligned}$$

This shows that for $s \gg 1$ the motion of the plasma is essentially horizontal, since $\xi_y \gg \xi_z$.

Accordingly, we approximate the continuity equation (6.3) as:

$$\rho_1 \simeq -\rho_0 \frac{\partial \xi_y}{\partial y} = \epsilon H (s k H)^2 \exp(t/\tau) \cos ky \sin K z_0. \tag{6.49}$$

Figure 6.2 shows both the profile of a field line and the perturbed density in arbitrary units.

As apparent from the figure, the density increases in correspondence of the troughs of the field line and decreases in correspondence of the crests, a result that can pictorially be described by saying that the instability forces the plasma to slide along the field lines and to accumulate in the lowest points.

It is not difficult to understand the physical mechanism responsible for this behavior. In equilibrium, gravity holds matter, together with the magnetic field frozen in it. When the instability sets in, the magnetic field is distorted and portions of the field are vertically displaced. Matter attached to the field should also move up, but the action of gravity produces a downward fall of the portions of plasma along the lines of $\boldsymbol{B}$. As a result, the gravitational effects on the plasma sitting at the crests' location weaken, which favors a further distortion of the field lines. At the same time, the gravitational action becomes stronger on portions of plasma lying at the bottom

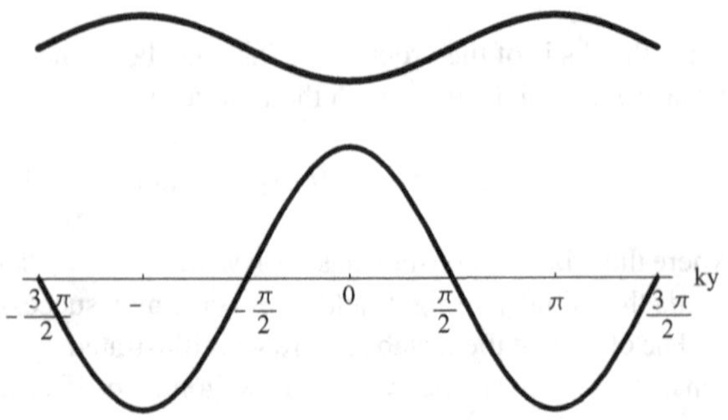


**Fig. 6.2** A magnetic field line (*upper curve*) and the perturbed density $\rho_1$ (*lower curve*) as functions of $ky$ (*arbitrary units*)

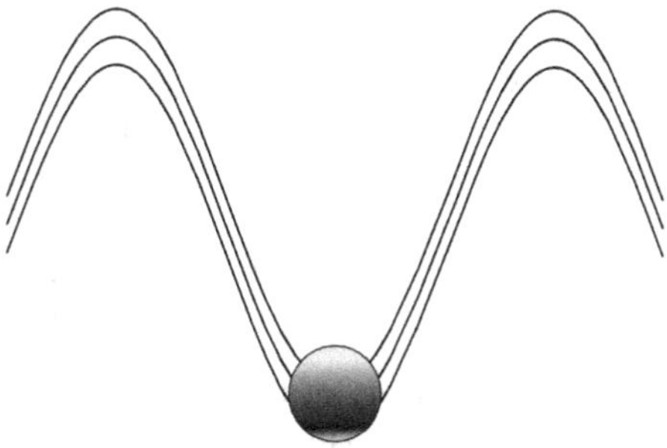

**Fig. 6.3** Sketch of a condensation at the bottom of a valley of $\boldsymbol{B}$

of the troughs of $\boldsymbol{B}$, giving a further help to the development of the instability. The final configuration thus features a series of condensations placed all along the disc, as schematically shown in Fig. 6.3.

The picture is consistent with the observations that show that the interstellar gas is not uniformly distributed, but tends to condensate into clouds, denser than the surrounding medium. If the magnetic field is absent, the formation of stable gravitationally induced condensations is only possible if their mass is larger than the so-called *Jeans mass*.[2]

$$M > M_J = \left(\frac{3}{4\pi}\right)^{1/2} \frac{c_0^3}{G^{3/2}\rho_0^{1/2}}. \tag{6.50}$$

If we use the values for the interstellar gas previously adopted, $M_J$ turns out to be of the order of $10^6 M_\odot$, which therefore should be a lower limit for the mass of interstellar clouds. On the contrary, clouds are observed with masses of the order of $10^4 M_\odot$ or less. It is therefore clear that the Parker instability cannot be the only cause of their formation: nevertheless it may contribute to the solution of the problem of cloud formation. In fact, let's look again at the configuration sketched in Fig. 6.3 and imagine that the condensations are actually a series of infinitely long cylinders, parallel to the $x$-axis, i.e. normal to the plane of the figure. The plasma they contain feels the action of a horizontal magnetic field and of a vertical gravitational field. In this configuration the charged particles inside the plasma will move with the speed given by Eq. (2.8)

$$\boldsymbol{v} = \frac{c}{e_0}\frac{m\boldsymbol{g} \times \boldsymbol{B}_0}{B_0^2},$$

where $m$ and $e_0$ are the mass and the charge of the single particle. This motion gives rise to an electric current flowing along the cylinder. The current density will be given by:

$$J = en_0(v_p - v_e) = cn_0\frac{g}{B_0}(m_p + m_e) \simeq c\frac{g}{B_0}n_0 m_p.$$

[2] The formation of a stable condensation requires that the gravitational force be dominant over the pressure (see Exercise 6.2)

Let's now consider a segment of our cylinder of length $\ell$ and cross section $S = \pi r^2$, whose volume will evidently be $S\ell$. Let $m_0$ be the mass contained in it, $n_0 m_p = m_0/S\ell$. The value of the electric current, $I$, is computed by evaluating the flux of the current density, $\boldsymbol{J}$. In our case we shall have

$$I = SJ = c\frac{g}{B_0}\frac{m_0}{\ell} \equiv c\frac{g}{B_0}\mathfrak{m},$$

where $\mathfrak{m}$ is the mass of the cylinder per unit length. The currents flowing in the cylinders are all mutually parallel and therefore two cylinders placed at a distance $d$ will attract each other with a force (per unit length) given by:

$$\mathcal{F}_B = \frac{2I^2}{c^2 d} = \frac{2}{B_0^2}\frac{\mathfrak{m}^2 g^2}{d}.$$

If we now compute the gravitational force per unit length between the same two cylinders, it is easy to show that

$$\mathcal{F}_G = \pi\frac{G\mathfrak{m}^2}{d}.$$

It follows that the ratio

$$\frac{\mathcal{F}_B}{\mathcal{F}_G} = \frac{2}{\pi}\frac{g^2}{GB_0^2},$$

can be larger than unity in the region where the galactic magnetic field is weak and the gravitational field is strong. If, for instance, we consider the situation at a height of $\simeq$100 parsec above the central galactic plane, where $g$ can be estimated to be of the order of $3 \times 10^{-9}$ cm sec$^{-2}$, and $B_0 = 5 \times 10^{-6} G$, we find

$$\frac{\mathcal{F}_B}{\mathcal{F}_G} \gtrsim 3.$$

In conclusion, we have seen how the combined action of gravity and magnetic fields, apart from giving rise to an instability, can generate an attractive force which adds to the plasma self-gravity and may justify the existence of clouds of mass substantially lower than the Jeans mass, as those that are actually observed.

### *6.2.4 Instabilities in the Presence of Plasma Flows: Kelvin-Helmholtz Instability*

In the preceding Sections we have always assumed that the unperturbed state was a *static* one, namely that the velocity was vanishing at the zeroth-order $U_0 = 0$. However, in many circumstances, both in the lab and in astrophysics, mass motions are present in the unperturbed state, a classical example being that of a fluid in laminar motion where the different layers do not move with the same speed. Therefore, the study of the possible instabilities that might arise in this or similar situations is clearly of interest. Obviously, viscosity should play a role in the system dynamics, since its action tends to reduce velocity differences between different fluid layers. In the following, however, we shall concentrate only on ideal plasmas, thus neglecting all viscous effects as well as any other effects that may modify the system's dynamics (for instance, those connected with surface tension). The most important instability that develops in a system of superposed fluids when a velocity gradient is present in the direction normal to the flow is known as the *Kelvin-Helmholtz instability*. If the velocity stratification is parallel to the direction of gravity, the Kelvin-Helmholtz instability can be thought of as a generalization of the Rayleigh-Taylor instability to a dynamical case.

To reduce the algebraic complexity of the treatment, we shall be concerned with a case analogous to that already examined for the Rayleigh-Taylor instability: two homogeneous superposed fluids of different densities moving with constant and parallel velocities, but whose speeds are different on the two sides of the separating surface. The $x$-axis will be chosen in the direction of the velocities, the $z = 0$ plane will define the surface separating the fluids and the $z$-axis coincides with the direction of the gravity assumed to be constant, $\boldsymbol{g} = -g\boldsymbol{e}_z$. We shall further assume that the two fluids are incompressible and that the system is completely homogeneous in the $y$ direction.

We shall first consider a purely hydrodynamic system ($\boldsymbol{B} = 0$), and then add the effects connected with the presence of a magnetic field.

Let the unperturbed state be characterized by:

$$\rho_0 = \rho_+ \qquad (z > 0); \qquad \rho_0 = \rho_- \qquad (z < 0),$$

and

$$\boldsymbol{U}_0 = U_+\boldsymbol{e}_x \qquad (z > 0); \qquad \boldsymbol{U}_0 = U_-\boldsymbol{e}_x \qquad (z < 0).$$

At the beginning, it turns out to be convenient to write down the equations in a general way, i.e. without considering separately the regions $z \gtrless 0$, a distinction that will be made later on. As a result derivatives (with respect to $z$) of the discontinuous quantities $\rho_0$ and $\boldsymbol{U}_0$ will appear, in spite of the fact that those derivatives actually vanish in each of the two half-spaces defined by the plane $z = 0$.

The only relevant equation at zeroth-order is the momentum equation, which in our case is identical to the one for the static case, i.e.

$$0 = -\nabla P_0 + \rho_0\boldsymbol{g}.$$

Let $\boldsymbol{u}$ e $\rho_1$ be the velocity and density perturbations, respectively. If the perturbed quantities are assumed to be proportional to $\exp[i(kx - \omega t)]$, we obtain the following first-order equations:

- *Continuity equation*

$$-\mathrm{i}\omega\rho_1 + \nabla \cdot (\rho_0 \boldsymbol{u} + \rho_1 \boldsymbol{U}_0) = \rho_0' u_z + \mathrm{i}kU_0\rho_1,$$

where the incompressibility condition, $\nabla \cdot \boldsymbol{U}_0 = \nabla \cdot \boldsymbol{u} = 0$, has been used and the prime indicates a derivation with respect to $z$. Defining $\Omega$ as

$$\Omega = \omega - kU_0,$$

from the preceding expression for $\rho_1$ we obtain

$$\rho_1 = \mathrm{i}\frac{\rho_0'}{\Omega}u_z. \tag{6.51}$$

- *Momentum equation*

$$-\mathrm{i}\rho_0\Omega\boldsymbol{u} + \rho_0 U_0' u_z \boldsymbol{e}_x = -\nabla P_1,$$

whose components are:

$$-\mathrm{i}\rho_0\Omega u_x + \rho_0 U_0' u_z = -\mathrm{i}kP_1, \tag{6.52}$$

$$-\mathrm{i}\rho_0\Omega u_z = -P_1' - \rho_1 g \tag{6.53}$$

To these equations we must add the incompressibility condition,

$$\mathrm{i}ku_x + u_z' = 0, \tag{6.54}$$

and a further condition expressing the fact the the separating surface, even if distorted, remains the border between the two fluids, which means that the particles lying on that surface will move with it. This translates into the fact that the vertical velocity must be considered as the *lagrangian* derivative of the position of the point located on the separating surface. Therefore:

$$u_z = \frac{\mathrm{d}\zeta}{\mathrm{d}t} = \frac{\partial\zeta}{\partial t} + (\boldsymbol{U}_0 \cdot \nabla)\zeta = -\mathrm{i}\Omega\zeta,$$

where $\zeta$ represents the vertical deformation of the plane $z = 0$. Since $\zeta$ is a continuous quantity, the preceding relation shows that

$$\frac{u_z}{\Omega} \text{ is a continuous function across the separating surface.}$$

It is now possible to find an equation for $u_z$ only, by eliminating $u_x$ by means of Eq. (6.54), $\rho_1$ by using Eq. (6.51) and $P_1$ by deriving Eq. (6.52) and inserting the result in Eq. (6.53). Finally, we get:

$$k^2\rho_0\Omega u_z - [\rho_0(\Omega u_z' + U_0' k u_z)]' = gk^2\rho_0' \frac{u_z}{\Omega}. \tag{6.55}$$

As already remarked, this equation is valid for all values of $z$. To impose the boundary conditions, which in this case are simply the continuity of $u_z/\Omega$, we integrate the preceding equation between $-\epsilon$ and $+\epsilon$ and notice that all terms that are not derivatives with respect to $z$ do not contribute. Consider for example the first term on the lhs of Eq. (6.55). We have to evaluate

$$\int_{-\epsilon}^{\epsilon} \rho_0\Omega u_z \mathrm{d}z = \int_{-\epsilon}^{0} \rho_0\Omega u_z \mathrm{d}z + \int_{0}^{\epsilon} \rho_0\Omega u_z \mathrm{d}z.$$

But $\rho_0\Omega$ is a constant quantity in each of the two integrals which implies

$$\rho_0\Omega \int_{-\epsilon}^{0} u_z \mathrm{d}z = [\rho_0\Omega\langle u_z\rangle]\epsilon \to 0 \quad \text{for} \quad \epsilon \to 0,$$

where $\langle u_z\rangle$ is an average value of $u_z$ in the integration interval. The same conclusions hold for the other integral. If we define $\Delta_0$ as $\Delta_0(f) = f(\epsilon) - f(-\epsilon)$, the result of the integration can be expressed as

$$\Delta_0[\rho_0(\Omega u_z' + U_0' k u_z)] = -gk^2\Delta_0(\rho_0)\frac{u_z}{\Omega},$$

where the continuity of $u_z/\Omega$ has been taken into account. Moreover, $\Delta_0(U_0' k u_z) = 0$ since $U_0' = 0$ in each of the two half-spaces. In conclusion:

$$\Delta_0[\rho_0\Omega u_z'] = -gk^2\Delta_0(\rho_0)\frac{u_z}{\Omega}. \tag{6.56}$$

Equation (6.56) must be considered as a boundary condition valid in $z = 0$, while Eq. (6.55) holds for every $z \neq 0$.

Let's now write Eq. (6.55) explicitly in the two regions $z > 0$ e $z < 0$. Since in those regions $\rho_0' = U_0' = 0$, Eq. (6.55) assumes the simple form:

$$u_z'' - k^2 u_z = 0,$$

whose solutions, convergent in $z = \pm\infty$, can be written in the form

$$u_z = A\Omega \mathrm{e}^{+kz} \qquad (z < 0),$$

$$u_z = A\Omega e^{-kz} \qquad (z > 0),$$

$A$ being an arbitrary constant that expresses the continuity of $u_z/\Omega$ in $z = 0$.

Making use of the solutions just found, we can write the condition given by Eq. (6.56) explicitly:

$$\rho_+(\omega - kU_+)^2 + (\omega - \mathrm{kU}_-)^2 = \mathrm{gk}(\rho_- - \rho_+). \tag{6.57}$$

The assumption $\rho_- > \rho_+$ ensures that in the absence of flows the system is Rayleigh-Taylor stable (see Eq. (6.27)). Defining $\alpha$ as

$$\alpha_+ = \frac{\rho_+}{\rho_+ + \rho_-} \quad \text{e} \quad \alpha_- = \frac{\rho_-}{\rho_+ + \rho_-} \quad \rightarrow \quad \alpha_+ + \alpha_- = 1,$$

and expanding the preceding equation, we finally get

$$\omega^2 - 2k(\alpha_+ U_+ + \alpha_- U_-)\omega + k^2(\alpha_+ U_+^2 + \alpha_- U_-^2) - gk(\alpha_- - \alpha_+) = 0.$$

If the discriminant of this equation is negative, $\omega$ has a non-vanishing imaginary part and the solution is therefore unstable. Evaluating explicitly the discriminant, we see that the instability condition can be written as

$$gk(\alpha_- - \alpha_+) - k^2\alpha_+\alpha_-(U_- - U_+)^2 < 0.$$

The latter expression allows us to draw a few interesting conclusions. In the first place, the discriminant vanishes for $k = 0$ and in this case $\omega$ is real and no instability arises. We also see that we always have an instability when

$$k > \frac{g(\alpha_- - \alpha_+)}{\alpha_+\alpha_-(U_- - U_+)^2}.$$

Therefore, to have an instability, there is no need of strong velocity gradients. Whatever the value of $U_- - U_+$ might be, even very small, sufficiently large values of $k$ always exist, i.e. sufficiently small wavelengths, that make the system unstable.

Figures 6.4 and 6.5 show the results of a numerical simulation of the nonlinear development of the Kelvin-Helmholtz instability and the direct observation of the same instability in the atmosphere.

Let's now examine effects that might inhibit the development of the Kelvin-Helmholtz instability. For an unmagnetized plasma or fluid, it is clear that the presence of surface tension, which contrasts the distortion of the separating surface, may contribute, within certain limits, to the stabilization of the system. In the case of a magnetized plasma, the same stabilizing role can be played by a horizontal magnetic field, coplanar with the velocity. Even without carrying out an explicit calculation, it easy to realize that a magnetic field normal to $\boldsymbol{U}$ cannot influence the development of the instability. In fact, the rippling of the separating surface can only displace the

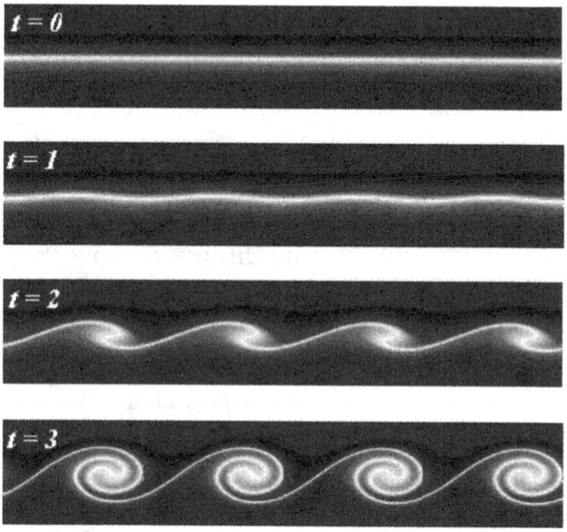


**Fig. 6.4** Numerical simulation of the development of the Kelvin-Helmholtz instability

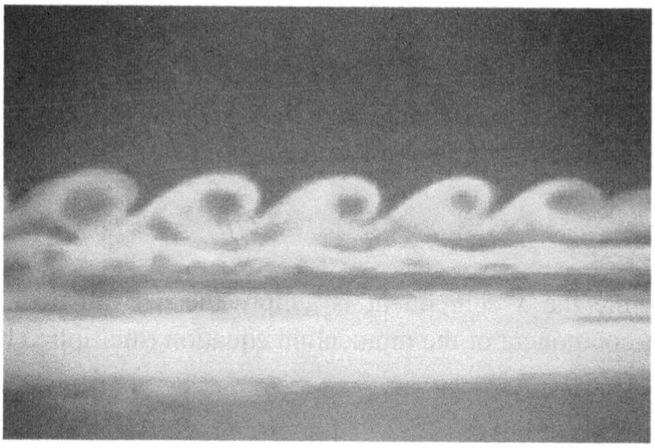

**Fig. 6.5** The Kelvin-Helmholtz observed in the atmosphere

lines of force without deforming them. A different picture arises when a magnetic field parallel to the velocity is present: the field lines are bent, with the consequent birth of a tension that opposes the distortion.

We shall therefore concentrate on a configuration where, in the unperturbed state, a magnetic field, $\boldsymbol{B_0} = B_0\boldsymbol{e}_x$, is present in both fluids.To simplify the discussion, we shall assume that the field is constant everywhere. The momentum equation now includes also the magnetic term:

$$-\mathrm{i}\rho_0\Omega\boldsymbol{u} + \rho_0 U_0' u_z \boldsymbol{e}_x = -\nabla\left[P_1 + \frac{\boldsymbol{B_0}\cdot\boldsymbol{b}}{4\pi}\right] + \frac{1}{4\pi}(\boldsymbol{B_0}\cdot\nabla)\boldsymbol{b}.$$

The discussion of the preceding case has shown that all the terms containing $U_0'$ drop out of the final expressions: we shall therefore omit them from now on. The two components of the momentum equation now read:

$$-\mathrm{i}\rho_0 \Omega u_x = -\mathrm{i}k\Pi + \mathrm{i}\frac{B_0}{4\pi}b_x,$$

$$-\mathrm{i}\rho_0 \Omega u_z = -\Pi' + \mathrm{i}\frac{B_0}{4\pi}b_z - \rho_1 g,$$

where the quantity $\Pi$ is defined as

$$\Pi = P_1 + \frac{\boldsymbol{B_0} \cdot \boldsymbol{b}}{4\pi}.$$

The induction equation can now be written as:

$$-\mathrm{i}\Omega\, \boldsymbol{b} = \mathrm{i}k\frac{B_0}{4\pi}\boldsymbol{u}.$$

The conditions $\nabla \cdot \boldsymbol{b} = 0$ and $\nabla \cdot \boldsymbol{u} = 0$, i.e. $\mathrm{i}kb_x + b_z' = 0$ and $\mathrm{i}ku_x + u_z' = 0$, can be used to eliminate from the equations the quantities $u_x$ and $b_x$ and to write the $z$-component of the induction equation as:

$$\Omega b_z = -kB_0 u_z, \tag{6.58}$$

while the $x$-component is simply the derivative of the preceding relation. The $x$-component of the momentum equation (multiplied by $k$) becomes

$$\left(\rho_0 \Omega - \frac{k^2}{\Omega}\frac{B_0^2}{4\pi}\right)u_z' = -\mathrm{i}k^2\Pi, \tag{6.59}$$

and the $z$-component reads

$$\left(\rho_0 \Omega^2 - \frac{k^2 B_0^2}{4\pi} - g\rho_0'\right)\left(\frac{u_z}{\Omega}\right) = -\mathrm{i}\Pi', \tag{6.60}$$

where Eq. (6.51) has been used. Notice that in the preceding equations explicitly contain the term $u_z/\Omega$, which, as the purely hydrodynamical case has shown, is continuous across the separating surface. Taking the derivative of Eq. (6.59) and inserting the result into Eq. (6.60), we finally obtain

$$\left\{\left(\rho_0 \Omega^2 - \frac{k^2 B_0^2}{4\pi}\right)\left(\frac{u_z'}{\Omega}\right)\right\}' = k^2\left(\rho_0 \Omega^2 - \frac{k^2 B_0^2}{4\pi} - g\rho_0'\right)\left(\frac{u_z}{\Omega}\right). \tag{6.61}$$

To impose the boundary conditions, we integrate Eq. (6.61) between $-\epsilon$ and $+\epsilon$ and obtain

$$\Delta_0\left[\left(\rho_0\Omega^2 - \frac{k^2 B_0^2}{4\pi}\right)\left(\frac{u_z'}{\Omega}\right)\right] = -k^2 g\Delta_0(\rho_0)\left(\frac{u_z}{\Omega}\right). \tag{6.62}$$

In each of the two half-spaces $z > 0$ and $z < 0$, $u_z$ satisfies the equation

$$u_z'' - k^2 u_z = 0,$$

and therefore, once again, we shall have

$$u_z = A\Omega e^{+kz} \qquad (z < 0),$$

$$u_z = A\Omega e^{-kz} \qquad (z > 0).$$

We now insert these expressions into Eq. (6.62) and obtain

$$\rho_+(\omega - kU_+)^2 + \rho_-(\omega - kU_-)^2 = gk(\rho_- - \rho_+) + k^2\frac{B_0^2}{2\pi},$$

which, compared to Eq. (6.57), clearly shows the effect of the magnetic field. Following the procedure adopted before, we find the following equation for $\omega$

$$\omega^2 - 2k(\alpha_+U_+ + \alpha_-U_-)\omega + k^2(\alpha_+U_+^2 + \alpha_-U_-^2) - gk(\alpha_- - \alpha_+) - \frac{k^2B_0^2}{2\pi(\rho_+ + \rho_-)} = 0. \tag{6.63}$$

The instability condition (i.e. the discriminant of Eq. (6.63) $< 0$) now becomes

$$gk(\alpha_- - \alpha_+) - k^2\alpha_+\alpha_-(U_- - U_+)^2 + \frac{k^2B_0^2}{2\pi(\rho_+ + \rho_-)} < 0.$$

This expression clearly shows the stabilizing effect of the magnetic field. In fact, if

$$\frac{B_0^2}{2\pi(\rho_+ + \rho_-)} \geqslant \alpha_+\alpha_-(U_- - U_+)^2$$

the Kelvin-Helmholtz instability cannot develop.

## 6.3 Instabilities in Cylindrical Geometry

All instabilities discussed so far shared the common assumption that the system could be descibed in the framework af a planar geometry. However, several situations occur where the use of a cylindrical geometry is more appropriate. For instance, in laboratory plasmas, the machines aiming at controlled thermonuclear fusion of the type known as *zeta-pinch* are substantially cylinders. The *Tokamak*, the most common fusion machine at present time, has a toroidal geometry, which in a first approximation can be treated as a cylindrical geometry (assuming a periodicity in the axial direction) if the ratio between the major and minor axes, the so-called "aspect ratio", is very large. In astrophysics it is often useful to describe the magnetic field in terms of "flux tubes", which identifies the cylindrical geometry as the most natural one to describe systems such as stellar jets or solar coronal loops.

In the following we shall discuss two examples in same detail, one of relevance to laboratory plasmas and the other to solar physics. In neither case a claim of realism is made. They are rather illustrations of the methods used to study the stability of cylindrical systems.

### 6.3.1 Instability of a Plasma Column

Our basic schematic model is a plasma column of radius $a$ and infinite length with no gravity effects. In this situation, all quantities will depend only on the coordinate $r$, the distance from the axis of the cylinder. The only non-vanishing components of magnetic field will be $B_\theta$ and $B_{0z}$. Equation $\nabla \cdot \boldsymbol{B} = 0$ implies

$$\frac{1}{r}\frac{\partial}{\partial r}(rB_r) = 0 \quad \rightarrow \quad B_r = \frac{const.}{r} = 0,$$

if we require the field be finite on the cylinder's axis. As already remarked in Chap. 5, the field is expected to show a helical structure.

The equilibrium equation (see Eq. 5.41)

$$0 = -\nabla\left(P_0 + \frac{B_0^2}{8\pi}\right) + \frac{1}{4\pi}(\boldsymbol{B}_0 \cdot \nabla)\boldsymbol{B}_0,$$

in cylindrical geometry reduces to the $r$-component only:

$$\frac{\mathrm{d}}{\mathrm{d}r}\left[P_0 + \frac{B_{0\theta}^2 + B_{oz}^2}{8\pi}\right] + \frac{1}{4\pi}\frac{B_{0\theta}^2}{r} = 0, \tag{6.64}$$

which shows that the equilibrium is completely determined by the profiles of the pressure and of the magnetic field. The density, that does not enter the equilibrium equation since gravity has been neglected, can be chosen at will. Of course, boundary conditions appropriate to the model under study must complement the preceding equation. Actually, as we shall see, the boundary conditions will fix the dispersion relation and consequently the stability conditions of the system.

The model we have chosen, represents, to a first approximation, the behaviour of a *$\theta$-pinch*: a plasma column contained inside a cylindrical, rigid and perfectly conducting wall. However, the plasma is not in direct contact with the wall, but is separated from it by a vacuum zone. The presence of a region devoid of plasma is necessary to keep the plasma off the conducting wall, where it would recombine and loose all its plasma properties. In the following we shall denote by $a$ the radius of the plasma column and by $b$ ($\gg a$) the radius of the conducting wall.

If $\tilde{\boldsymbol{B}}$ is the magnetic field in the vacuum zone external to the plasma column, the boundary conditions at the surface of the plasma column are

$$\left[\boldsymbol{n}\cdot\boldsymbol{B}\right]_{S-} = \left[\boldsymbol{n}\cdot\tilde{\boldsymbol{B}}\right]_{S+} = 0, \tag{6.65}$$

$$\left[P + \frac{B^2}{8\pi}\right]_{S-} = \left[\frac{\tilde{B}^2}{8\pi}\right]_{S+}, \tag{6.66}$$

where $\boldsymbol{n}$ is a unit vector normal to the plasma surface pointing towards the exterior[3] and the notations $S-$ and $S+$ indicate the the quantities in square brackets must be computed, respectively, inside and outside of the plasma-vacuum interface. We emphasize here that the boundary conditions must be applied to the **instantaneous** configuration of the separating surface, not to the **initial** configuration, since the instability distorts the plasma column. For instance, the external normal, $\boldsymbol{n}$, that enters Eq. (6.65), equals $\boldsymbol{e}_r$ in the equilibrium configuration, but has a different direction at later times.

These boundary conditions are easy to interpret. If the first were not satisfied, i.e. if a component of the magnetic field normal to the separating surface were be present, the vacuum region simply would not exist, since nothing could prevent the plasma from freely flowing along such a component. If the second condition were not verified, a big pressure difference would arise and the separating surface would simply be swept away.

Even if the conditions expressed by Eqs. (6.65) and (6.66) are sufficient to determine the solution of the problem, it is useful to add the condition that expresses the continuity of the tangential component of the electric field across a surface separating two different media:

$$\boldsymbol{n}\times[\boldsymbol{E} + \frac{1}{c}\boldsymbol{U}\times\boldsymbol{B}]_{S-} = \boldsymbol{n}\times[\tilde{\boldsymbol{E}} + \frac{1}{c}\boldsymbol{U}\times\tilde{\boldsymbol{B}}]_{S+}, \tag{6.67}$$

where $\tilde{\boldsymbol{E}}$ is the electric field in the vacuum region.

[3] From now on $\boldsymbol{n}$ will be referred to as **the external normal**.

The algebra of stability problems has the disappointing tendency to become very complex: we shall therefore introduce a few simplifying assumptions, that, however, do not obscure the physics involved. Let's thus assume that the unperturbed magnetic field, $\boldsymbol{B}_0$, is constant and directed along the $z$-axis, $\boldsymbol{B}_0 \equiv (0, 0, B_0)$. Equation (6.64) then implies that $P_0 = const$. Such a configuration can be thought as generated by a system of azimuthal currents circulating in the plasma column, actually in a surface layer, whose thickness tends to zero in an ideal plasma. In fact, in the interior of the column $\boldsymbol{J} \propto \nabla \times \boldsymbol{B}_0 = 0$, while on the surface currents must necessarily flow to comply with the boundary conditions. Equation (6.65) constrains only the radial component of the field and nothing forbids the presence of discontinuities in the tangential components $B_\theta$ and $B_{0z}$. To take this into account we shall write: $J_\theta = J_\theta(r)\delta(r-a)$ and $J_z = J_z(r)\delta(r-a)$. Since:

$$J_\theta = J_\theta(r)\delta(r-a) = -\frac{c}{4\pi}\frac{\mathrm{d}B_{0z}}{\mathrm{d}r},$$

we may integrate the preceding equation and obtain:

$$\int_{a-\epsilon}^{a+\epsilon} J_\theta(r)\delta(r-a)\mathrm{d}r = J_\theta(a) = -\frac{c}{4\pi}\int_{a-\epsilon}^{a+\epsilon}\frac{\mathrm{d}B_{0z}}{\mathrm{d}r}\mathrm{d}r$$
$$= -\frac{c}{4\pi}[B_{0z}(a+\epsilon) - B_{0z}(a-\epsilon)] = -\frac{c}{4\pi}[\tilde{\boldsymbol{B}}_z(a) - B_0].$$

If the $z$-component of the magnetic field is discontinuous, it follows that $J_\theta(a)$ cannot vanish. In a similar way, integrating

$$J_z = J_z(r)\delta(r-a) = \frac{c}{4\pi}\frac{1}{r}\frac{\mathrm{d}(rB_\theta)}{\mathrm{d}r},$$

we obtain

$$J_z(a) = \frac{c}{4\pi}[B_\theta(a+\epsilon) - B_\theta(a-\epsilon)] = \tilde{\boldsymbol{B}}_\theta(a),$$

and thus also $J_z(a) \neq 0$ when a discontinuity of $B_\theta$ is present.

The vacuum field will be determined by the currents flowing in the plasma column. In particular, if $I$ is value of the current globally flowing in the column's surface layer, we shall have $\tilde{\boldsymbol{B}}_\theta = (2I)/r$. If we further assume that $\tilde{\boldsymbol{B}}_z = 0$, a discontinuity of $B_z$ will be present on the surface of the plasma column and therefore $J_\theta(a) \neq 0$. The $z$-component of the vacuum field is not influenced, since only $J_z$ contributes to the current $I$.

The dynamics of the system is described by the momentum equation, Eq. (6.11),

$$\rho_0\frac{\partial^2\boldsymbol{\xi}}{\partial t^2} = \boldsymbol{F}(\boldsymbol{\xi}),$$

with

$$\boldsymbol{F}(\boldsymbol{\xi}) = -\nabla P_1 + \frac{1}{4\pi}[(\nabla \times \boldsymbol{B}_0) \times \boldsymbol{B}_1 + (\nabla \times \boldsymbol{B}_1) \times \boldsymbol{B}_0],$$

where, in our case,

$$P_1 = -\rho_0 c_s^2 (\nabla \cdot \boldsymbol{\xi}) \tag{6.68}$$

and

$$B_1 = \nabla \times (\boldsymbol{\xi} \times \boldsymbol{B}_0). \tag{6.69}$$

Equation (6.11) thus becomes:

$$\rho_0 \frac{\partial^2 \boldsymbol{\xi}}{\partial t^2} = \gamma P_0 \nabla(\nabla \cdot \boldsymbol{\xi}) + \frac{1}{4\pi}(\nabla \times \boldsymbol{B}_1) \times \boldsymbol{B}_0]. \tag{6.70}$$

Since the zeroth-order terms are constant, we can Fourier expand with respect to the variables $t$, $\theta$ and $z$. A similar expansion, however, cannot be made with respect to $r$, since in cylindrical coordinates the differential operators, besides derivatives, also contain functions of $r$. For instance, $\nabla \equiv [\partial/\partial r, (1/r)\partial/\partial\theta, \partial/\partial z]$. We shall therefore write

$$\boldsymbol{\xi} \equiv \left[\xi_r(r), \xi_\theta(r), \xi_z(r)\right] \mathrm{e}^{\mathrm{i}(m\theta + kz)} \mathrm{e}^{-\mathrm{i}\omega t},$$

where $m$ must be an integer to ensure the periodicity in the $\theta$ coordinate. To simplify the calculations we shall assume that $m = 0$, i.e. we shall restrict our study to perturbations that preserve the axial symmetry. Using the preceding expressions, it is easy to find:

$$\boldsymbol{B}_1 = \mathrm{i}kB_0\xi_r\boldsymbol{e}_r + \mathrm{i}kB_0\xi_\theta\boldsymbol{e}_\theta + B_0[\mathrm{i}k\xi_z - (\nabla \cdot \boldsymbol{\xi})]\boldsymbol{e}_z. \tag{6.71}$$

In cylindrical coordinates $\nabla \cdot \boldsymbol{\xi}$ is written as

$$\nabla \cdot \boldsymbol{\xi} = \frac{1}{r}\frac{\mathrm{d}}{\mathrm{d}r}(r\xi_r) + \mathrm{i}k\xi_z = f(r) + \mathrm{i}k\xi_z,$$

where

$$f(r) = \frac{1}{r}\frac{\mathrm{d}}{\mathrm{d}r}(r\xi_r).$$

Writing down the components of Eq. (6.70) explicitly we get:

$$-\omega^2 \rho_0 \, \xi_r = \gamma P_0 \frac{\mathrm{d}}{\mathrm{d}r}[f(r) + \mathrm{i}k\xi_z] + \frac{B_0^2}{4\pi}(\frac{\mathrm{d}f}{\mathrm{d}r} - k^2\xi_r)$$

$$-\omega^2 \rho_0 \xi_\theta = -\frac{k^2 B_0^2}{4\pi} \xi_\theta$$
$$-\omega^2 \rho_0 \xi_z = \mathrm{i}k\gamma P_0 [f(r) + \mathrm{i}k\xi_z].$$

Recalling the definitions of the characteristic speeds

$$c_s^2 = \gamma \frac{P_0}{\rho_0}, \qquad c_a^2 = \frac{B_0^2}{4\pi\rho_0},$$

the preceding equations become:

$$(k^2 c_a^2 - \omega^2)\xi_r = (c_s^2 + c_a^2)\frac{\mathrm{d}f}{\mathrm{d}r} + \mathrm{i}kc_s^2 \frac{\mathrm{d}\xi_z}{\mathrm{d}r} \tag{6.72a}$$
$$(k^2 c_a^2 - \omega^2)\xi_\theta = 0 \tag{6.72b}$$
$$(k^2 c_s^2 - \omega^2)\xi_z = \mathrm{i}kc_s^2 f. \tag{6.72c}$$

The equation for $\xi_\theta$ is uncoupled from the others and its solutions are either $\xi_\theta = 0$, or $\omega^2 = k^2 c_a^2$. Inserting the latter relation into the other two equations, it is easily seen that in this case $\xi_z = const.$ and $\xi_r$ is a linear function of $r$. If $\xi_r = \xi_z = 0$, this solution represents pure torsional oscillations.

Going back to the equations for $\xi_r$ and $\xi_z$ and putting $\xi_\theta = 0$, it is possible to eliminate $\xi_r$, and obtain an equation for $\xi_z$ only:

$$\frac{\mathrm{d}^2\xi_z}{\mathrm{d}r^2} + \frac{1}{r}\frac{\mathrm{d}\xi_z}{\mathrm{d}r} - K^2\xi_z = 0, \tag{6.73}$$

with

$$K^2 = \frac{(k^2c_a^2 - \omega^2)(k^2c_s^2 - \omega^2)}{k^2c_s^2c_a^2 - \omega^2(c_s^2 + c_a^2)} = k^2\Big[1 + \frac{(\omega/k)^4}{c_s^2c_a^2 - (\omega/k)^2(c_s^2 + c_a^2)}\Big]. \tag{6.74}$$

This is the well known equation for the Bessel functions of zero order. The choice of a specific Bessel function depends on the sign of $K^2$. If $K^2 > 0$, the solution, compatible with the condition that $\xi_z(0)$ is a finite quantity, is

$$\xi_z = I_0(Kr), \tag{6.75}$$

where $I_0$ is the modified Bessel function of the first kind. If instead, if $K^2 < 0$ the Bessel function of the first kind, $J_0(Kr)$, must be chosen. Assuming that $K^2 > 0$ and inserting the solution given by Eq. (6.75) in Eqs. (5.16a) and (5.16c) we find

$$\xi_r = \frac{K}{\mathrm{i}k}\Big[\frac{c_s^2(k^2c_a^2 - \omega^2) - \omega^2c_a^2}{c_s^2(k^2c_a^2 - \omega^2)}\Big]I_0'(Kr), \tag{6.76}$$

where the prime indicates derivation with respect to the argument of the function.

In conclusion, we have shown that $\xi_z \propto I_0(Kr)$ and $\xi_r \propto I_0'(Kr) = I_1(Kr)$, but this does not solve completely the problem since the eigenvalue $\omega^2$, entering the definition of $K$, is still unknown. To find it, it is necessary to impose the boundary conditions, obviously in their linearized form. The linearization operation requires a certain amount of care, since first-order terms may arise from the perturbations of the various quantities involved, but also as a consequence of the distortion of the surface that separates the plasma from the vacuum. In fact, a generic quantity $f(\boldsymbol{r}, t)$ entering the boundary conditions must be written in the form

$$f(\boldsymbol{r}, t) = f_0(\boldsymbol{r}, t) + f_1(\boldsymbol{r}, t) \simeq f_0(\boldsymbol{r}_0) + \boldsymbol{\xi} \cdot \nabla f_0(r_0, t) + f_1(\boldsymbol{r}_0, t).$$

Equation (6.66) thus becomes

$$P_1 + \frac{1}{4\pi}\boldsymbol{B}_0 \cdot \boldsymbol{B}_1 = \frac{1}{4\pi}\tilde{\boldsymbol{B}}_0 \cdot \tilde{\boldsymbol{B}}_1 + (\boldsymbol{\xi} \cdot \nabla)\frac{\tilde{\boldsymbol{B}}_0^2}{8\pi}.$$

To apply the boundary conditions it turns out to be convenient to use Eqs. (6.66) and (6.67). Starting from the latter, we express the vacuum electric and magnetic fields, $\tilde{\boldsymbol{E}}$ and $\tilde{\boldsymbol{B}}$, in terms of the potential vector. Since in the equilibrium state the vacuum electric field vanishes, $\tilde{\boldsymbol{E}}$ is a first-order quantity and therefore

$$\tilde{\boldsymbol{E}} = -\frac{1}{c}\frac{\partial \tilde{\boldsymbol{A}}}{\partial t},$$

and $\tilde{\boldsymbol{A}}$ is a first-order quantity as well. The total magnetic field in vacuum will be

$$\tilde{\boldsymbol{B}} = \tilde{\boldsymbol{B}}_0 + \tilde{\boldsymbol{B}}_1 = \tilde{\boldsymbol{B}}_0 + \nabla \times \tilde{\boldsymbol{A}}.$$

Taking into account the fact that $\boldsymbol{E}$, $\boldsymbol{U} = \partial\boldsymbol{\xi}/\partial t$ and $\tilde{\boldsymbol{E}}$ are first-order quantities and that for an ideal plasma the lhs of Eq. (6.67) vanishes, this condition reduces to

$$\boldsymbol{n}_0 \times \frac{\partial \boldsymbol{A}}{\partial t} = \boldsymbol{n}_0 \times \left(\partial\boldsymbol{\xi}/\partial t \times \tilde{\boldsymbol{B}}_0\right).$$

The last expression can be integrated with respect to time giving

$$\boldsymbol{n}_0 \times \boldsymbol{A} = \boldsymbol{n}_0 \times (\boldsymbol{\xi} \times \tilde{\boldsymbol{B}}_0) = (\boldsymbol{n}_0 \cdot \tilde{\boldsymbol{B}}_0)\boldsymbol{\xi} - (\boldsymbol{n}_0 \cdot \boldsymbol{\xi})\tilde{\boldsymbol{B}}_0 = -(\boldsymbol{n}_0 \cdot \boldsymbol{\xi})\tilde{\boldsymbol{B}}_0, \qquad (6.77)$$

since $\tilde{\boldsymbol{B}}_0$, whose only component is in the $\theta$ direction, is perpendicular to $\boldsymbol{n}_0 = \boldsymbol{e}_r$, On the other hand, the linearized version of this boundary condition, given by Eq. (6.66), gives

$$-\gamma P_0(\nabla \cdot \boldsymbol{\xi}) + \frac{1}{4\pi}\boldsymbol{B}_0 \cdot [\boldsymbol{B}_1 + (\boldsymbol{\xi} \cdot \nabla)\boldsymbol{B}_0] = \frac{1}{4\pi}\tilde{\boldsymbol{B}}_0 \cdot [\tilde{\boldsymbol{B}}_1 + (\boldsymbol{\xi} \cdot \nabla)\tilde{\boldsymbol{B}}_0],$$

where use has been made of the equilibrium condition

$$P_0 + \frac{B_0^2}{8\pi} = \frac{\tilde{B}_0^2}{8\pi}.$$

From Eq. (6.71) we get

$$B_{1z} = -\frac{B_0}{r}\frac{\mathrm{d}}{\mathrm{d}r}(r\xi_r),$$

which allows us to write the preceding boundary condition as:

$$-\gamma P_0(\nabla \cdot \boldsymbol{\xi}) - \frac{B_0^2}{4\pi\, r}\frac{\mathrm{d}}{\mathrm{d}r}(r\xi_r) = \frac{1}{4\pi}\tilde{\boldsymbol{B}}_0 \cdot [\nabla \times \tilde{\boldsymbol{A}} + \xi_r \frac{\mathrm{d}\tilde{\boldsymbol{B}}_0}{\mathrm{d}r}]. \tag{6.78}$$

On the other hand, Eq. (6.77) implies that $\tilde{\boldsymbol{A}}$ does not have components along $\boldsymbol{e}_\theta$ and therefore we may put

$$\tilde{\boldsymbol{A}} = q(r)\boldsymbol{e}_r + \xi_r \tilde{B}_0 \boldsymbol{e}_z,$$

where $q(r)$ is an arbitrary function. Thus,

$$\tilde{\boldsymbol{B}}_1 = \nabla \times \tilde{\boldsymbol{A}} = \left[\frac{\partial q}{\partial z} - \frac{\partial}{\partial r}(\xi_r \tilde{B}_0)\right]\boldsymbol{e}_\theta \tag{6.79}$$

Moreover, since $\nabla \times \tilde{\boldsymbol{B}}_1 = 0$ (the vacuum is current-free) we have

$$\nabla \times (\nabla \times \tilde{\boldsymbol{A}}) = \left[-\frac{\partial^2 q}{\partial z^2}\right]\boldsymbol{e}_r + \left[-\frac{\partial^2 q}{\partial r \partial z} - \frac{\partial^2}{\partial r^2}(\xi_r \tilde{B}_0)\right]\boldsymbol{e}_z = 0.$$

Equating the $r$, $z$ components of the preceding equation to zero and integrating once we get:

$$\frac{\partial q}{\partial z} = g(r),$$

and

$$\frac{\partial q}{\partial z} - \frac{\partial}{\partial r}(\xi_r \tilde{B}_0) = h(z),$$

where $g$ and $h$ are arbitrary functions of their arguments. Combining the two equations above we obtain

$$g(r) - \frac{\partial}{\partial r}(\xi_r \tilde{B}_0) = h(z).$$

But the lhs of the preceding equation is a function of $r$ only and therefore $h(z) =$ const. From Eq. (6.79) we now find that $\tilde{\boldsymbol{B}}_1 = h(z)\boldsymbol{e}_\theta$ is a constant vector. Actually it is a null vector, since initially $\tilde{\boldsymbol{B}}_1 = 0$. Equation (6.78) can now be cast in the form

$$\frac{\gamma\omega^2 P_0}{ikc_s^2}\xi_z - \frac{B_0^2(k^2c_s^2 - \omega^2)}{4\pi ikc_s^2}\xi_r = -\frac{\tilde{\boldsymbol{B}}_0^2}{4\pi\, r}\xi_r,$$

where we have taken into account that $\tilde{B}_0 \propto 1/r$ when calculating the derivative of $\tilde{\boldsymbol{B}}_0$ on the rhs At this point all we have to do is to insert in the last equation the expressions previously found for $\xi_r$ and $\xi_z$ and evaluate the result at the (initial!) plasma-vacuum interface:

$$[c_s^2\omega^2 - c_a^2(k2c_s^2 - \omega^2)]I_0(Ka) = -\frac{\tilde{B}_0^2}{B_0^2}c_a^2\frac{K}{a}\frac{I_0'(Ka)}{I_o(Ka)}.$$

The last equation allows us, finally, to deduce the dispersion relation

$$\omega^2 = k^2c_a^2 - \frac{\tilde{c_a}^2}{a^2}\left[\frac{Ka\, I_o'(Ka)}{I_0(Ka)}\right], \tag{6.80}$$

where the notation $\tilde{c_a}^2 = \tilde{B}_0^2/(4\pi\rho_0)$ has been introduced.

Obtaining the dispersion relation has required a considerable amount of work, in spite of the many simplifying assumptions adopted. Moreover, Eq. (6.80) is still an *implicit* definition of $\omega^2$, since this quantity enters also the definition of $K$, so that the effective value of $\omega^2$ can only be obtained by numerical means. An exception to this statement is provided by the incompressible limit of Eq. (6.6) which can be simply obtained by letting $\gamma$ (and therefore $c_s^2$) tend to infinity. In this limit $K = k$ and Eq. (6.80) becomes an *explicit* definition of $\omega^2$.

The dispersion relation we have found shows that a plasma column is unstable to perturbations with axial symmetry, provided that

$$B_0^2 \leqslant \frac{\tilde{B}_0^2}{(ka)^2}\left[\frac{Ka\, I_o'(Ka)}{I_0(Ka)}\right].$$

We may further observe that when $K$ is real, the quantity $[Ka\, I_o'(Ka)/I_0(Ka)] > 0$ and increases with increasing $|K|$. From the definition of the latter quantity we clearly see that when $B_0^2 = 0$ and therefore $c_a^2 = 0$, it is always possible to find solutions of Eq. (6.80) satisfying the instability condition. It follows that, without a longitudinal field inside the column, the plasma is always unstable. The introduction of a longitudinal field does stabilize the system, at least in a certain range of values of $B_0$. In fact, the equilibrium condition, $B_0^2 = \tilde{B}_0^2 - 8\pi P_0$, imposes un upper limit to $B_0$, namely

$$B_0^2 < \tilde{B}_0^2.$$

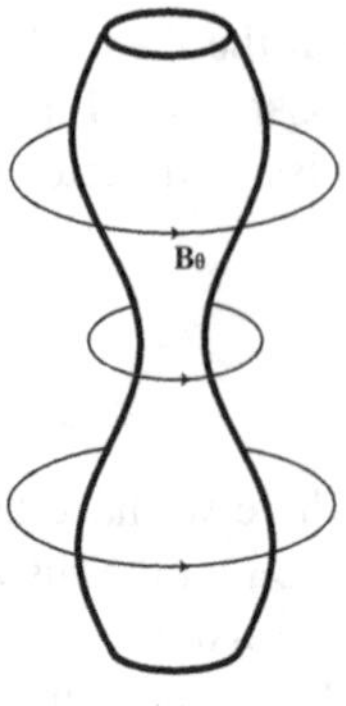


**Fig. 6.6** Instability of a plasma column with $m = 0$ (sausage instability)

If this further limitation is inserted into the *stability* condition we get

$$(ka)^2 > \left[\frac{Ka\, I_0'(Ka)}{I_0(Ka)}\right],$$

namely a limitation of the values of $k$ that allow the system to be stable. As shown in Fig. 6.6, an axially symmetric perturbation creates a series of necks and bulges along the column, which give the name to this instability, known as the *sausage instability*.

These features are easily interpreted from a physical point of view. In the neighbourhood of the necks the vacuum region is is closer to the cylinder's axis and the value of $\tilde{B}_0 \propto 1/r$ is larger than around the bulges. The force $\boldsymbol{J} \times \tilde{\boldsymbol{B}}_0$ is therefore stronger and tends to reduce even more the section of the column. The introduction of a longitudinal field in the column's interior increases the internal magnetic pressure, which tries to resist the squeeze and therefore stabilizes, at least partially, the system.

The discussions of the modes with $m \neq 0$ develops along similar lines with increased technical difficulties. Of particular interest are the $m = 1$ modes, see Fig. 6.7.

In this case the instability is due to the fact that in the regions where the plasma column is concave towards the exterior, the lines of force of $\tilde{\boldsymbol{B}}_0$ get closer increasing the value of the vacuum field, while the converse happens when the plasma column is convex. In this case too the presence of a longitudinal magnetic field opposes the instability because of the tension the deformation induces on the field lines.

### 6.3.2 The Confinement of Solar Coronal Loops

The case we have just discussed illustrates how a plasma column can be confined by the *magnetic pressure* present in a vacuum region surrounding the plasma itself. Of course, this is not the only way to obtain a confinement. A possible alternative is to use the *gas pressure* of an external medium to confine a magnetic flux tube,

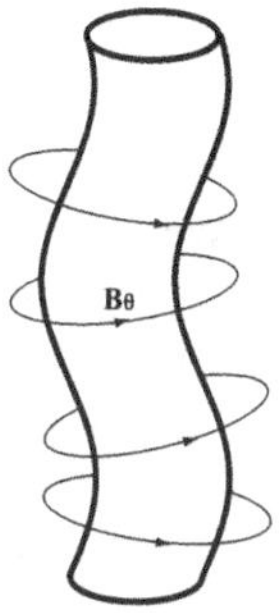


**Fig. 6.7** Instability of a plasma column with $m = 1$ (kink instability)

whose interior contains both matter and field. An astrophysical example of such a configuration is provided by solar coronal loops, shown in Fig. 6.8.

The structures we see in the figure are (unresolved) magnetic flux tubes, immersed in a plasma whose temperature is typically higher than the internal temperature of the tubes, while the external magnetic field is typically weaker than the internal one. The figure suggests that a loop should be considered as a sort of half-torus, obtained by cutting in two the entire torus and by anchoring the bases to the solar photosphere. Given that the aspect ratio of the torus is clearly very large, we may

**Fig. 6.8** Solar coronal loops observed in the ultraviolet (*photo* TRACE-NASA)

neglect the toroidal effects and model the loop as a straight cylinder of finite length, $L$ and radius $a$. If we impose some periodicity conditions in the $z$-direction, we may somehow mimic the photospheric boundary conditions.

We are now going to discuss a model of this "cylindrical" loop, without claiming it to be a realistic model. We rather consider this to be a useful example of the determination of a magnetic structure in an astrophysical context. We shall assume that the gas pressure has a minimum on the axis of the cylinder and connects smoothly to the pressure $P_c$ (assumed to be constant) of the external coronal plasma. We shall further assume that outside of the tube, i.e. for $r > a$, the magnetic field is force-free,

$$\nabla \times \boldsymbol{B}_e = \alpha \boldsymbol{B}_e, \quad \alpha = const.$$

The equilibrium of the plasma column is still defined by the equation

$$\nabla P = \frac{1}{c} \boldsymbol{J} \times \boldsymbol{B} = \frac{1}{4\pi} (\nabla \times \boldsymbol{B}) \times \boldsymbol{B}.$$

We now write the current density as $\boldsymbol{J} = \boldsymbol{J}_{\parallel} + \boldsymbol{J}_{\perp}$ where $\boldsymbol{J}_{\parallel}$ and $\boldsymbol{J}_{\perp}$ are the components parallel and perpendicular to $\boldsymbol{B}$, respectively, and observe that only the $\boldsymbol{J}_{\perp}$ component enters the equilibrium equation. We therefore need another equation to define $\boldsymbol{J}_{\parallel}$, which, obviously, can be always written in the form [16]

$$\boldsymbol{J}_{\parallel} = \frac{4\pi}{c} \lambda \boldsymbol{B}.$$

$\lambda$ is in general a function of $r$ and $t$, but in the following we shall consider it to be a constant. In cylindrical coordinates the equilibrium equation reads (see Eq. (5.41)):

$$B_{0z} B'_{0z} + B_\theta B'_\theta + \frac{B_\theta^2}{r} = -4\pi P', \tag{6.81}$$

where the prime indicates the derivative with respect to $r$. Using the definition of $J_{\parallel}$ we may write

$$\boldsymbol{B} \cdot (\nabla \times \boldsymbol{B}) = \frac{c}{4\pi} \boldsymbol{B} \cdot \boldsymbol{J}_{\parallel} = \lambda B^2,$$

and evaluating explicitly the product $\boldsymbol{B} \cdot (\nabla \times \boldsymbol{B})$ we get

$$B_{0z} B'_\theta - B_\theta B'_{0z} + \frac{B_\theta B_{0z}}{r} = \lambda (B_\theta^2 + B_{0z}^2). \tag{6.82}$$

If we divide the preceding equation by $B_{0z}^2$ and introduce the ratio of the field components

$$w(r) = \frac{B_\theta(r)}{B_{0z}(r)},$$

Equation (6.82) becomes:

$$w' + \frac{w}{r} - \lambda w^2 = \lambda. \tag{6.83}$$

Equation (6.83), the *Riccati equation*, is well known in the theory of differential equations. Its solution is obtained by defining an auxiliary function $u$ connected to $w$ by the relation

$$w = -\frac{u'}{\lambda u}.$$

In terms of $u$, Eq. (6.83) becomes

$$u'' + \frac{u}{r} + \lambda^2 u = 0,$$

namely the Bessel equation, whose general solution, not diverging in $r = 0$, is $u = C_1 J_0(\lambda r)$, $C_1$ being an arbitrary constant. The general solution for $w(r)$ is therefore

$$w(r) = \frac{J_1(\lambda r)}{J_0(\lambda r)},$$

from which we obtain

$$B_\theta(r) = B_0(r) J_1(\lambda r) \quad \text{and} \quad B_{0z}(r) = B_0(r) J_0(\lambda r), \tag{6.84}$$

$B_0(r)$ being an arbitrary function. To determine $B_0$, we insert the expressions just found for the field components in the equilibrium equation, Eq. (6.81), and get

$$\frac{\mathrm{d}}{\mathrm{d}r} B_0^2(r) = -8\pi \frac{\mathrm{d}P/\mathrm{d}r}{J_0^2(\lambda r) + J_1^2(\lambda r)} \tag{6.85}$$

This equation has to be complemented by the boundary conditions for $r = a$, which are simply the continuity of the field components and of their derivatives. When $r > a$, $\boldsymbol{B}$ is a force-free field with constant $\alpha$ and its component are a linear combination of Bessel functions of the first and second kind, $J_0$ e $Y_o$. In this case, the functions $Y$, previously discarded because of the request of regularity in $r = 0$, cannot be neglected, since the origin is no longer included in the region where $\boldsymbol{B}$ is defined. Taking into account that $J_0' = -J_1$ and that the same relation holds for the $Y$ functions, the continuity conditions for $B_{0z}$, $B_\theta$ and $B_{0z}'$ are:

$$B_0(a) J_0(\lambda a) = C_2 J_0(\alpha a) + C_3 Y_0(\lambda a) \tag{6.86a}$$

$$B_0(a) J_1(\lambda a) = C_2 J_1(\alpha a) + C_3 Y_1(\lambda a) \tag{6.86b}$$

$$\frac{\lambda}{\alpha} B_0(a) J_1(\lambda a) = C_2 J_1(\alpha a) + C_3 Y_1(\lambda a) \tag{6.86c}$$

From the direct comparison of Eqs. (6.86b) and (6.86c) we deduce that $\lambda = \alpha$ and the Eq. (6.86a) thus becomes

$$\big[B_0(a) - C_2\big] J_0(\alpha a) = C_3 Y_0(\alpha a).$$

Since $J_0$ and $Y_0$ are two independent solutions of the Bessel equation, the preceding relation can only be satisfied by

$$C_2 = B_0(a) \qquad \text{and} \qquad C_3 = 0.$$

In summary, the complete solution is

$$\begin{aligned} B_\theta(r) &= B_0(r) J_1(\alpha r) \\ B_{0z}(r) &= B_0(r) J_0(\alpha r), \end{aligned} \tag{6.87}$$

where the function $B_0^2(r)$ is the solution of Eq. (6.85)

$$\begin{aligned} B_0^2(r) &= B_0^2(a) + 8\pi \int_r^a \frac{P'(s)\,\mathrm{d}s}{J_0^2(\alpha s) + J_1^2(\alpha s)} \qquad r < a \\ B_0(r) &= B_0(a) \qquad r \geqslant a. \end{aligned} \tag{6.88}$$

Equation (6.88) shows that in our case the equilibrium is determined by the pressure profile only. An interesting feature of the field defined by Eq. (6.87) is that, in spite of the fact that it *is not* a force-free field ($P'(s) \neq 0$), its field lines perfectly coincide with those of a force-free field with the same value of $\alpha$. It follows that the knowledge of the profiles of the field lines does not allow to distinguish a force-free field from a more general one. This remark should be kept in mind when it is tried to guess the nature of a magnetic field from the reconstruction of its field lines based on observations.

We have succeeded in determining the equilibrium structure of a system that, at least to a certain extent, mimics a solar coronal loop. The computation of the stability properties of this field is rather complex and can be accomplished by the use of techniques developed for laboratory plasmas, [17] well beyond the scope of this introductory text. We shall therefore restrict ourselves to the mention of a *necessary* stability criterion, very often utilized in the studies of the stability of laboratory plasmas, the so-called *Suydam criterion.* In its version valid for a cylindrical geometry, it states that the system is *locally* stable when

$$\frac{\mathrm{d}P(r)}{\mathrm{d}r} \geqslant -r\frac{B_z^2(r)}{32\pi}\left[\frac{\mathrm{d}}{\mathrm{d}r}\ln\left(\frac{B_\theta(r)}{r B_{0z}(r)}\right)\right]^2.$$

In our case, the Suydam criterion is always satisfied for a monotonically increasing pressure profile, while it only defines a restricted stability region for profiles presenting a maximum in $r < a$. It must be emphasized that even if the Suydam criterion is satisfied, instabilities involving *global* deformations of the plasma structure cannot be excluded.

### 6.3.3 The Magnetorotational Instability (MRI)

Rotation and angular momentum and its transport play a fundamental role in astrophysical objects. Transport of angular momentum is fundamental in the formation and evolution of accretion disks, one of the natural steps in the process of gravitational collapse. Accretion disks can have widely different characteristic scales, with radii spanning dimensions from a hundred astronomical units (AU) or more for stellar disks leading to planetary systems all the way up to hundreds and thousands of kiloparsec for galaxies, and down to an astronomical unit or less around black holes. In many of these cases the rotating, accreting gas can become sufficiently hot as to become at least partially ionized (either heated by the process of accretion itself or by an outside radiation source) and must therefore be considered a plasma. In order for such a plasma to continue to fall towards the accretion center some form of viscosity must be present, which allows the removal of the angular momentum of the falling accreting gas, otherwise the accretion itself would stall due to the centrifugal barrier coming from the conservation of angular momentum. However, accreting plasma is usually so rarified that collisions may be completely neglected, so that the collisional viscosity can not be considered a viable mechanism for accretion.

A possible solution to this problem may come from instabilities naturally developing in a rotating plasma. There is a vast litterature on the instability of rotating fluids in hydrodynamics: a classic example is given by the Rayleigh instability, according to which a fluid rotating around an axis is unstable if its specific angular momentum decreases with distance from the rotation access.

A simple way to derive the Rayleigh criterion comes from perturbing the motion of a fluid element found at a distance $r$ from the center of the disk, and evaluating how the force balance between the centrifugal force and pressure gradients are altered by the time the element reaches a new distance $r'$, supposing that in the corresponding motion the fluid element preserves its angular momentum (we consider also for simplicity the density to be uniform). If only pressure and angular momentum play a role in defining the equilibrium and we assume cylindrical symmetry around the rotation axis, the following relation must hold in $r$:

$$\frac{1}{\rho}\frac{\partial P}{\partial r} = \frac{U^2(r)}{r}.$$

On the other hand in the new position $r'$ angular momentum conservation for the fluid element leads it to reach a new velocity $U'^2 = r^2 U^2 / r'^2$ around the axis, so that for $r' > r$ the fluid element will feel a restoring force towards its original radius only if

$$\left[\frac{1}{\rho}\frac{\partial P}{\partial r}\right]_{r'} = \frac{U^2(r')}{r'} > \frac{U'^2}{r'} = \frac{r^2U^2}{r'^3},$$

or in other words if the rotational velocity profile satisfies $U(r')r' > U(r)r$. In terms of the angular velocity $\Omega(r) = U(r)/r$ and the specific angular momentum $r^2\Omega$ the stability condition therefore becomes

$$\frac{\partial[r^2\Omega(r)]}{\partial r} > 0.$$

This condition is satisfied, as may be easily verified, by Keplerian orbits around a point-like center of gravity, so that a simple hydrodynamic instability leading to an effective viscosity allowing accretion is not generally found. However, as is often found in plasma dynamics, the magnetic field turns out to play a fundamental role in this case, in a way which is counter-intuitive. We have seen previously that the magnetic field tends to have a stabilizing effect on those instabilities, such as the Kelvin-Helmholtz or Rayleigh-Taylor instabilities, where the driving force is not electromagnetic in nature. This is due to the fact that deformation of magnetic lines of force by the unstable perturbing motions acts as a stabilizing energy sink for the motions themselves. However, in the case of a plasma which is orbiting around a central object, the magnetic field leads instead to a new type of instability, called the magnetorotational instability or MRI for short [21].

Consider a plasma in equilibrium rotating around a mass $M$, with an angular velocity given by $\Omega(r) = (GM/r^3)^{1/2}$, in which an equilibrium magnetic field orthogonal to the equatorial plane is present $B\boldsymbol{e}_z$. The geometry of the problem is therefore cylindrical with coordinates $(r, \theta, z)$ and we assume the equilibrium to be independent of the angular coordinate $\theta$.

Consider now small deformations of the field arising from a radial velocity fluctuation, and suppose that is such as to shift one plasma fluid element closer to the center, and another one, at a different height but very close to the previous one and belonging to the same magnetic field line, further away from the center as shown in Fig. 6.9. This kind of perturbation would be typical, for example, of an Alfvén wave propagating vertically along the field. As a result of this velocity fluctuation, the plasma element finding itself closer to the center of gravity (labelled a, in the Figure), will tend to increase its angular velocity, while the other one will tend to slow down. On the other hand the magnetic tension created in the field line will tend to brake the inner plasma parcel's (a) motion and accelerate that of the plasma element (b). Because of the loss of angular momentum, the inner parcel will tend to fall still closer in, while the second, accelerated one will tend to move further away, and the perturbation amplitude will increase. The net result will be a transfer of angular momentum, mediated by the magnetic field, from the internal regions of the disk to

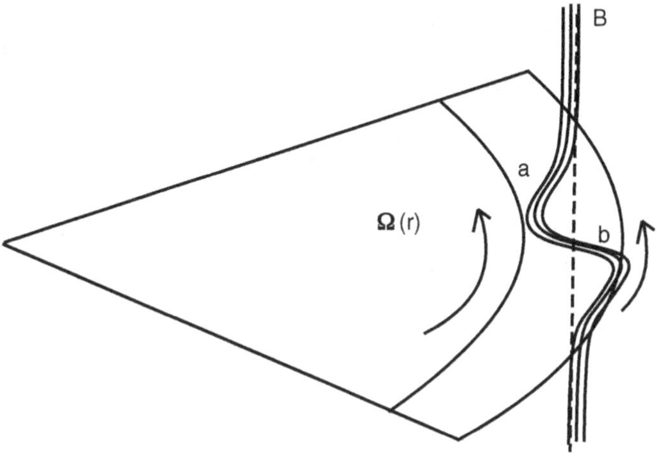


**Fig. 6.9** Deformation of a magnetic field line $B$ in the magnetorotational instability. For a weak magnetic field, the breaking of the inner plasma parcel $a$ and the acceleration of the outer plasma parcel $b$ due to the magnetic field causes $a$ to fall further and $b$ to escape further out, leading to instabiliy and growth of the perturbation and magnetic field

the external ones, together with a transfer of the kinetic energy of the rotating plasma into magnetic field energy, leading to a growth of the magnetic field intensity.

Consider now how the instability develops: take small perturbations by expanding, as usual, with the velocity field $\boldsymbol{U} = r\Omega \boldsymbol{e}_\theta + \boldsymbol{u}_1$ and the magnetic field $\boldsymbol{B} = \mathrm{B}\boldsymbol{e}_z + \boldsymbol{b}_1$. For simplicity we take incompressible plasma perturbations with axial symmetry $(\partial/\partial\theta = 0)$ and fluctuations in the form $\boldsymbol{u}_1 = \boldsymbol{u}(r)\exp(-i\omega t + ik_z z + ik_r r)$ and similarly for the field $\boldsymbol{b}_1$. If we consider fluctuations localized in space but with a large wavenumber (such as the perturbation considered in the example), the condition $|k|r \gg 1$ must hold and we may neglect radial derivatives of the envelopes u(r), $b(r)$ with respect to those in the exponential. This means assuming to lowest order that $u(r), b(r)$ are constants. On the other hand, care must be taken in the linearization to include the radial derivatives of the average Keplerian motion, i.e. of terms proportional to $\partial(\Omega r)/\partial r$. Upon linearization of the $r, \theta$ components of the momentum and magnetic induction equations (considering the ideal MHD limit) and taking account of the appropriate expressions for the differential operators in cylindrical geometry, one finds:

$$-\mathrm{i}\omega\rho_0 u_r - 2\rho_0\Omega u_\theta = -\mathrm{i}k_r\bar{P}_1 + \frac{\mathrm{i}k_z \mathrm{B} b_r}{4\pi}, \tag{6.89a}$$

$$-\mathrm{i}\omega\rho_0 u_\theta + \rho_0\big(2\Omega + r\frac{\partial\Omega}{\partial r}\big)u_r = \frac{\mathrm{i}k_z \mathrm{B} b_\theta}{4\pi}, \tag{6.89b}$$

$$-\mathrm{i}\omega\rho_0 u_z = -\mathrm{i}k_z\bar{P}_1, \tag{6.89c}$$

$$\omega b_r = -k_z \mathrm{B} u_r \tag{6.89d}$$

$$\omega b_\theta = -k_z \mathrm{B} u_\theta + \mathrm{i} b_r r\frac{\partial\Omega}{\partial r}, \tag{6.89e}$$

where $\bar{P}_1$ is the total perturbed pressure, $P_1 + \mathrm{B}b_z/4\pi$. It is possible to eliminate $\bar{P}_1$, by combining the equations along $r$ and along $z$. $u_z$ is instead eliminated using the fact that the velocity fluctations are taken to be incompressible:

$$\frac{1}{r}\frac{\partial(ru_r)}{\partial r} + \frac{\partial u_z}{\partial z} = 0.$$

If now we also ask that $k_z \gg k_r$, is is easy to see that it is also possible to neglect the fluctuations along $z$ of velocity, magnetic field and pressure. Eqs. (6.89a)–(6.89e) therefore become

$$-\mathrm{i}\omega\rho_0 u_r - 2\rho_0 \Omega u_\theta = +\frac{\mathrm{i}k_z \mathrm{B}b_r}{4\pi}, \tag{6.90a}$$

$$-\mathrm{i}\omega\rho_0 u_\theta + \rho_0 \frac{\kappa^2}{2\Omega} u_r = \frac{\mathrm{i}k_z \mathrm{B}b_\theta}{4\pi}, \tag{6.90b}$$

$$\omega b_r = -k_z \mathrm{B} u_r \tag{6.90c}$$

$$\omega b_\theta = -k_z \mathrm{B} u_\theta + \mathrm{i} b_r r \frac{\partial \Omega}{\partial r}, \tag{6.90d}$$

where we have written

$$2\Omega + r\frac{\partial \Omega}{\partial r} = \frac{\kappa^2}{2\Omega},$$

introducing the *epicyclic frequency*

$$\kappa^2 = 4\Omega^2 + \frac{\partial \Omega^2}{\partial\, lnr}.$$

The dispersion relation describing the MRI instability may therefore be found by calculating the determinant of the linear system (6.90a)–(6.90d), and it is a simple algebraic calculation that leads to:

$$\omega^4 - \left[\kappa^2 + 2(k_z c_a)^2)\right]\omega^2 + (k_z c_a)^2\left[(k_z c_a)^2 + \frac{\partial \Omega^2}{\partial\, lnr}\right] = 0. \tag{6.91}$$

In the absence of a magnetic field $c_a = 0$, Eq. (6.91) shows the existence of oscillations at the epicyclic frequency $\omega^2 = \kappa^2$, which correspond to the small oscillations arising from elliptical orbits neighboring the circular orbits of the original keplerian disk. If on the other hand we allow the rotational terms in the dispersion relation to vanish we obtain Alfvén waves propagating along the unperturbed average field B (the quartic comes from the existence of waves with independent polarizations along $r$ and along $\theta$).

The simultaneous presence of both rotational and magnetic terms leads however to the existence of unstable solutions, i.e. $\omega^2 < 0$, as long as

$$\frac{\partial \Omega(r)}{\partial r} < 0 \quad \text{and} \quad (k_z c_a)^2 < -r\frac{\partial \Omega^2(r)}{\partial r}. \tag{6.92}$$

The proof is left as a problem (Problem 6.3).

In a Keplerian disk, the gradient $\partial \Omega(r)/\partial r = -3\Omega/2r$, so that orbiting ionized plasmas tend to be generally unstable. It is important to emphasize that the MRI instability exists even for extremely weak fields, and that in this case it does not depend much on the field geometry. In addition, the MRI instability has growth rates which can reach a significant fraction of the original rotation frequency, up to a maximum value

$$|\omega_m| = \frac{1}{2}\frac{\partial \Omega^2(r)}{\partial \ln r}.$$

The nonlinear evolution of the instability can lead to developed magnetohydrodynamic turbulence and a consequent effective viscosity in accretion disks. Detailed studies of the nonlinear evolution of the MRI, the nature of the turbulence and its saturation in different accreting systems are still at the forefront of contemporary astrophysical research.

## Questions and Problems

**6.1** Consider a configuration of the theta-pinch type, but with the plasma column in direct contact with a rigid and perfectly conducting wall at a distance $a$ from the cylinder's axis. Deduce the dispersion relation for the $m = 0$ modes, assuming $c_s^2 < c_a^2$. What changes if $c_s^2 > c_a^2$?

**6.2** Deduce the condition for the gravitational collapse of a mass $M$, Eq. (6.50).

**6.3** Derive the magneto-rotational instability condition, Eq. (6.92). Show that in the keplerian case such a condition becomes $(k_z c_a)^2 < 3\Omega^2$.

## Solutions

**6.1** The equation for the perturbed quantities is still Eq. (6.73), but the boundary conditions (6.65) now require $\xi_r(0) = \xi_r(a) = 0$. To determine the explicit form of the dispersion relation we need to know the sign of $K^2$. However, the solutions corresponding to $K^2 > 0$ do not satisfy the new boundary conditions, since the function $I_0$ is a monotonically increasing function of its argument. Thus, $K^2 < 0$. Assuming that $c_s^2 < c_a^2$ from Eq. (6.74) we see that this condition can be satisfied either if $\omega^2 > k^2 c_a^2$ or if $k^2 c_s^2 c_a^2/(c_s^2 + c_a^2) < \omega^2 < k^2 c_s^2$. Therefore, the solution for $\xi_z$ is proportional to $J_0(|K|r)$ and that for $\xi_r$ is proportional to $J_0'(|K|r)$, i.e. to

$J_1(|K|r)$. The allowed frequencies are thus given by the condition

$$J_1(|K|a) = 0,$$

namely $|K|a = j_{1n}$, $n = 0, 1 \ldots$, where $j_{1n}$ is the $n-th$ zero of the function $J_1$. Writing explicitly the preceding condition we find that the frequencies are the solutions of the equation:

$$\omega^4 - \omega^2(c_s^2 + c_a^2)(k^2 + j_{1n}^2/a^2) + k^2 c_s^2 c_a^2 (k^2 + j_{1n}^2/a^2) = 0.$$

If $c_s^2 > c_a^2$, $K^2 < 0$ when either $\omega^2 > k^2 c_s^2$ or $k^2 c_s^2 c_a^2/(c_s^2 + c_a^2) < \omega^2 < k^2 c_a^2$, while the final equation for $\omega$ remains the same, as one could have predicted from the symmetry of that equation with respect to the exchange of $c_s^2$ with $c_a^2$).

**6.2** To form a gravitational condensation the gravitational force must dominate over the pressure force, namely

$$\frac{\mathrm{d}P(r)}{\mathrm{d}r} < \frac{Gm(r)\rho(r)}{r^2} \qquad \Rightarrow \qquad \frac{P_0}{\rho_0} < \frac{GM}{R}.$$

Introducing the sound speed, $c_0^2 = P_0/\rho_0$, and expressing the mass as a function of the density we obtain:

$$R > R_J = (\frac{3}{4\pi})^{1/2} \frac{c_0}{(G\rho_0)^{1/2}},$$

or, in terms of masses,

$$M > M_J = \frac{4\pi}{3}\rho_0 R_J^3 = (\frac{3}{4\pi})^{1/2} \frac{c_0^3}{G^{3/2}\rho_0^{1/2}}.$$

**6.3** Introducing the quantities $\omega_a^2 = k_z^2 c_a^2$ and $\omega_1^2 = \omega_a^2 - \omega^2$, Eq. (6.91) can be written as

$$\omega_1^4 + \kappa^2 \omega_1^2 - 4\omega_a^2 \Omega^2 = 0,$$

which has only one positive root. Unstable solutions ($\omega^2 < 0$) must have $\omega_1^2 > \omega_a^2$. If this condition is introduced into the positive root of the preceding equation, we find

$$\omega_a^2 < 4\Omega^2 - \kappa^2 = -r\frac{\partial \Omega^2}{\partial r}.$$

In the keplerian case we have $\Omega^2 = GM/r^3$, $\partial\Omega^2/\partial r = -3\Omega/r$ and the instability condition becomes $\omega_a^2 < 3\Omega^2$.

# Chapter 7
# Waves

**Abstract** The propagation of small amplitude, linear waves in plasmas is presented, starting from the MHD regime and the Alfvén and magnetoacoustic modes. High frequency effects leading to kinetic corrections are then discussed before introducing waves in completely kinetic regimes, where the phenomenon of Landau damping for ion-acoustic modes is treated in detail.

Waves are generated when a stable equilibrium is perturbed. The dynamics of such a system, as already remarked, is characterized by oscillations. This means that the value of any physical quantity at a given point oscillates around its equilibrium value. If a stone is thrown in a pond, the initially flat, horizontal water surface is warped, at first close to the point where the stone has fallen and then in increasingly larger zones. This example suggests defining a wave *a propagating disturbance*. The fluid particles within the perturbation region oscillate with a certain speed and the perturbed zone itself spreads at a different speed. It is the latter, called the *propagation speed*, that controls the dynamics of the system. Notice that in the preceding example the fluid particles oscillate essentially in the vertical direction, while the propagation speed is directed along the water surface: in this case the wave is called *transverse*. *Longitudinal* waves, in which the two speeds are parallel to each other, also exist. The best known longitudinal waves are sound waves.

In this Chapter we shall be concerned with "small amplitude" waves, or *linear waves*, meaning that the values of any oscillating quantity do not differ markedly from the equilibrium values. In the preceding example, this means that the local increase or decrease of the water level due to the oscillatory motion of the fluid particles is small with respect to the pond's depth, but **does not** mean that the propagation speed is small. It does mean however that all terms in the equations in which perturbations appear at second or higher powers may be neglected when compared to first order terms.

The starting point for the treatment of linear waves is the system of differential equations describing the dynamics of the physical system under study. This could be the system of MHD equations, but waves can exist in all regimes, from fluid to kinetic. A perturbative analysis applied to these equations reduces them to a system



C. Chiuderi and M. Velli, *Basics of Plasma Astrophysics*,
UNITEXT for Physics, DOI 10.1007/978-88-470-5280-2_7

of linear equations, to which the method of Fourier analysis can be applied. It is therefore appropriate, before delving into the discussion of waves in plasmas, to review the fundamental aspects of this method.

## 7.1 The Fourier Representation

Consider a generic function $f(x)$, where $x$ is a spatial coordinate, and its *Fourier transform*, $\tilde{f}(k)$, defined by:

$$\tilde{f}(k) = \frac{1}{2\pi} \int_{-\infty}^{\infty} f(x) e^{-ikx} \, dx. \tag{7.1}$$

In the context of the Fourier method, the original function $f(x)$ is also called the *inverse Fourier transform* of $\tilde{f}(k)$. Clearly (see (6.14)):

$$f(x) = \int_{-\infty}^{\infty} \tilde{f}(k) e^{ikx} \, dk. \tag{7.2}$$

To illustrate the relationship between a function and its Fourier transform, consider the special case where $f(x)$ is an infinitely extended, constant amplitude plane wave of wavelength $\lambda_0 = 2\pi/k_0$:

$$f(x) = A \, e^{ik_0 x}.$$

Its Fourier transform is given by:

$$\tilde{f}(k) = \frac{1}{2\pi} \int_{-\infty}^{\infty} A e^{i(k_0 - k)x} \, dx = A\delta(k_0 - k).$$

We thus see that while the function does not identify any particular region of space, its transform is localized in $k$-space with infinite precision.

A "wave packet", is a function which oscillates as above but in which the envelope $A$ is not constant, but localized in $x$-space. As an example, a Gaussian wave-packet has the form

$$f(x) = f_0 \, e^{-a^2 x^2} \, e^{ik_0 x}, \tag{7.3}$$

representing an oscillation of wavelength $\lambda = 2\pi/k_0$ whose amplitude is modulated by a gaussian function. The parameter $a$ gives us a measure of the width of the gaussian: at $x_0 = 1/a$ the value of the amplitude has dropped to $f_0/\mathrm{e}$. The Fourier transform of Eq. (7.3) is given by:

$$\tilde{f}(k) = \frac{f_0}{2a\sqrt{\pi}} \mathrm{e}^{-(k-k_0)^2/4a^2}, \tag{7.4}$$

which shows that the Fourier transform of a gaussian wave-packet is still a gaussian, whose width, defined by the range over which the amplitude remains greater than a fraction $1/\mathrm{e}$ of its maximum value, is now $k - k_0 = 2a$. A more general way to define the "localization region" of a wave-packet is to write:

$$\begin{aligned}(\Delta x)^2 &= \langle (x - \langle x \rangle)^2 \rangle = \langle x^2 \rangle - \langle 2x \langle x \rangle \rangle + \langle \langle x^2 \rangle \rangle \\ &= \langle x^2 \rangle - \langle x \rangle^2 = \langle x^2 \rangle,\end{aligned} \tag{7.5}$$

since $\langle x \rangle = 0$. Thus, for a gaussian wave-packet,

$$(\Delta x)^2 z = \langle x^2 \rangle = \frac{\displaystyle\int_{-\infty}^{\infty} x^2\, \mathrm{e}^{-a^2 x^2} \mathrm{d}x}{\displaystyle\int_{-\infty}^{\infty} \mathrm{e}^{-a^2 x^2} \mathrm{d}x}. \tag{7.6}$$

Defining the localization in $k$-space in the same way, $\langle (\Delta k)^2 \rangle$, and performing the integrals, it is easily found that for Gaussian wave-packets

$$\langle \Delta x^2 \rangle \langle \Delta k^2 \rangle = 1/4,$$

while in the more general case of a non-gaussian packet, it is possible to show that the preceding equality changes into

$$\langle \Delta x^2 \rangle \langle \Delta k^2 \rangle \geq 1/4. \tag{7.7}$$

The above results show that to a more localized function in the $x$-space, corresponds a less localized transform in $k$-space and viceversa. This property of the Fourier transforms is strongly reminiscent of the Uncertainty Principle of Heisenberg, and is, in fact, equivalent to it: if $f(x)$ is the wave-function for the position of a particle, recalling that in quantum mechanics momentum is the conjugate operator with respect to position, defined as $p = \hbar k$, one finds, mutiplying Eq. (7.7) by $\hbar$, that $\Delta x \Delta p \gtrsim \hbar$.

All preceding expressions are easily generalized to three dimensions plus time for a function $f = f(\boldsymbol{r}, t)$:

$$f(\boldsymbol{r}, t) = \int_{-\infty}^{\infty} \tilde{f}(\boldsymbol{k}, \omega) \mathrm{e}^{\mathrm{i}(\boldsymbol{k}\cdot\boldsymbol{r} - \omega t)} \mathrm{d}\boldsymbol{k}\, \mathrm{d}\omega, \tag{7.8}$$

and

$$\tilde{f}(\boldsymbol{k},\omega)=\frac{1}{(2\pi)^4}\int_{-\infty}^{\infty} f(\boldsymbol{r},t)\mathrm{e}^{-\mathrm{i}(\boldsymbol{k}\cdot\boldsymbol{r}-\omega t)}\mathrm{d}\boldsymbol{r}\,\mathrm{d}t, \tag{7.9}$$

where $\boldsymbol{k}$ and $\omega$ are real quantities.

Equation (7.8) illustrates the physical meaning of the Fourier transform: the given function is considered a superposition of "elementary waves", represented by the factor $\exp[\mathrm{i}(\boldsymbol{k}\cdot\boldsymbol{r}-\omega t)]$, whose amplitude is given by $\tilde{f}(\boldsymbol{k},\omega)$. The quantity

$$\Phi=\boldsymbol{k}\cdot\boldsymbol{r}-\omega t=k\left(\boldsymbol{r}\cdot\boldsymbol{e}_k-\frac{\omega}{k}t\right) \tag{7.10}$$

is called the "phase" of the wave.

Equation (7.10) shows that, in each elementary wave, space and time coordinates appear only in the combination $\boldsymbol{r}\cdot\boldsymbol{e}_k-(\omega/k)t$. The planes $\Phi=const.$ are therefore seen to move in the direction of $\boldsymbol{e}_k$ at the *phase velocity*:

$$\boldsymbol{v}_f=\frac{\omega}{k}\boldsymbol{e}_k. \tag{7.11}$$

This may be summarized qualitatively by saying that each elementary wave moves with its own phase velocity, given by Eq. (7.11), without changing its amplitude: at any particular time the profile of $f(\boldsymbol{r},t)$ is "reconstructed" by adding the contributions of all elementary waves. Since in general the phase speed is different for the various elementary waves, the "reconstructed" profile at time $t$ will be modified with respect to the initial profile. This phenomenon is called "dispersion". If it happens that the phase speed is the same for all elementary waves, or, equivalently, if $\omega$ is a linear function of $k$ (see Eq. (7.11)), the profile will propagate without distortion.

In describing the above procedure the dynamics of the phenomenon underlying wave propagation and the meaning of the parameter $\omega$ have been left unspecified. We have implied that $\omega$ is a function of $k$ but have not specified whence this relationship comes from. In the study of the propagation of small amplitude waves however, $f(\boldsymbol{r},t)$ is the solution of a linear, homogeneous differential equation in the variables $\boldsymbol{r}$ and $t$. This equation will, in general, contain the operators $\nabla$ and $\partial/\partial t$. If $f(\boldsymbol{r},t)$ is represented in the form of Eq. (7.8), we immediately realize that:

$$\frac{\partial f}{\partial t}=\int_{-\infty}^{\infty}(-\mathrm{i}\omega\tilde{f})\mathrm{e}^{\mathrm{i}(\boldsymbol{k}\cdot\boldsymbol{r}-\omega t)}\mathrm{d}\boldsymbol{k}\,\mathrm{d}\omega,$$

and, similarly:

$$\nabla f=\int_{-\infty}^{\infty}(\mathrm{i}\boldsymbol{k}\tilde{f})\mathrm{e}^{\mathrm{i}(\boldsymbol{k}\cdot\boldsymbol{r}-\omega t)}\mathrm{d}\boldsymbol{k}\,\mathrm{d}\omega.$$

If we assume that the equilibrium is also homogeneous in space, so that the linear system has constant coefficients, in the space of Fourier transforms the operators $\nabla$ e $\partial/\partial t$ become simply the multipliers $\mathrm{i}\boldsymbol{k}$ e $-\mathrm{i}\omega$. This result holds true also for repeated applications of the above operators, $\partial^2 f/\partial t^2 \rightarrow -\omega^2 \tilde{f}$, or for the vector calculus formulas when $f$ is a vector quantity, $\nabla \cdot \boldsymbol{f} \rightarrow \mathrm{i}\boldsymbol{k} \cdot \tilde{\boldsymbol{f}}$ and so on.

As a result, the differential homogeneous equation in $(\boldsymbol{r}, t)$ space, that we symbolically write as:

$$\tilde{D}(\nabla, \partial/\partial t)\, f = 0, \tag{7.12}$$

where $\tilde{D}$ is a given functional, in the transformed space $(\boldsymbol{k}, \omega)$ becomes:

$$\tilde{D}(\mathrm{i}\boldsymbol{k}, -\mathrm{i}\omega)\, \tilde{f} = 0. \tag{7.13}$$

Therefore, in the transformed space we have to solve an *algebraic* equation in place of a *differential* equation, and the condition for the existence of non vanishing solutions to Eq. (7.13) may simply be written as

$$\tilde{D}(\mathrm{i}\boldsymbol{k}, -\mathrm{i}\omega) = 0.$$

The preceding equation, known as the *dispersion relation*, provides the connection between $\omega$ and $\boldsymbol{k}$. In general, the dispersion relation has a finite number of discrete solutions, known as *normal modes*:

$$\omega = \omega_\alpha(\boldsymbol{k}), \qquad \alpha = 1, 2, \ldots, N. \tag{7.14}$$

The condition $\tilde{D}(\mathrm{i}\boldsymbol{k}, -\mathrm{i}\omega) = 0$ can be formally introduced into Eq. (7.8) by writing:

$$\tilde{f}(\boldsymbol{k}, \omega) = \sum_{\alpha=1}^{N} \tilde{f}_\alpha(\boldsymbol{k})\delta[\omega - \omega_\alpha(\boldsymbol{k})],$$

with the quantities $\omega_\alpha$ given by Eq. (7.14). Performing the $\omega$ integration in Eq. (7.8), we see that the general solution of our problem can be cast in the form:

$$f(\boldsymbol{r}, t) = \sum_{\alpha=1}^{N} \int_{-\infty}^{\infty} \tilde{f}_\alpha(\boldsymbol{k}) \mathrm{e}^{\mathrm{i}[\boldsymbol{k}\cdot\boldsymbol{r} - \omega_\alpha(\boldsymbol{k})\, t]} \mathrm{d}\boldsymbol{k}. \tag{7.15}$$

To simplify the notation, the quantity $\tilde{f}_\alpha(\boldsymbol{k}, \omega_\alpha(\boldsymbol{k}))$ has been written as $\tilde{f}_\alpha(\boldsymbol{k})$.

The preceding results are easily generalized to vector quantities. For instance Eq. (7.12) becomes:

$$\tilde{D}(\nabla, \partial/\partial t)\, \boldsymbol{f} = 0,$$

where $\tilde{D}$ is a functional tensor and Eq. (7.13) is now:

$$\tilde{D}(\mathrm{i}\boldsymbol{k}, -\mathrm{i}\omega)\,\tilde{\boldsymbol{f}} = 0.$$

The dispersion relation, i.e. the condition that non-vanishing solutions exist, is now given by

$$D(\boldsymbol{k}, \omega) = \mathrm{Det}[\tilde{D}(\mathrm{i}\boldsymbol{k}, -\mathrm{i}\omega)] = 0. \tag{7.16}$$

To each solution of the dispersion relation is associated an *eigenvector* $\tilde{\boldsymbol{f}}$ that characterize that particular propagation mode.

### *7.1.1 Phase Velocity and Group Velocity*

The phase velocity has been defined as the propagation speed of elementary waves, $\boldsymbol{v}_{ph} = (\omega/k)\boldsymbol{e}_k$. This velocity is not directly associated with any physical effect, and specifically can not be associated with the transmission of signals (information) or the transfer of energy (or energy flux propagation). In fact, elementary waves have a constant amplitude and are infinitely extended (not localized) in space. Their "motion" is not physically observable, since as time proceeds the wave remains identical to itself modulo a phase (which repeats periodically). Therefore, even a phase velocity larger than the speed of light, $c$, would not violate the principle of relativity stating that the speed of light provides an upper limit to the speed of any **signal**.

A completely different situation is that of a wave packet, with the wave-perturbation localized in a particular region of space. The motion of a packet is observable and this motion is associated with the propagation of information and energy. The propagation speed of a wave packet is known as the *group velocity*.

To find an expression for the group velocity, consider a wave train of finite length, composed of an exponential and a modulating amplitude, for example a gaussian packet. Further assume that the packet's "spectrum", i.e. the ensemble of wave vectors representing it in $k$-space, peaks around a particular value, $k_0$ (for the case of a gaussian wave-packet this implies $a/k_0 \ll 1$). The normal mode associated with $k_0$ will thus be "dominant" and in the integral of Eq. (7.15) only $k$ values close to $k_0$ will contribute. We may therefore expand the quantity $\omega_\alpha(\boldsymbol{k})$ in the integral in a Taylor series centered around $k_0$. Restricting ourselves to first order, we may write, dropping the index $\alpha$,

$$\omega(\boldsymbol{k}) \simeq \omega(\boldsymbol{k}_0) + \sum_i (\boldsymbol{k} - \boldsymbol{k}_0)_i \left(\frac{\partial \omega}{\partial k_i}\right)_{\boldsymbol{k}_0} = \omega_0 + (\boldsymbol{k} - \boldsymbol{k}_0) \cdot \boldsymbol{v}_g,$$

where

$$\boldsymbol{v}_g = \left(\frac{\partial \omega}{\partial \boldsymbol{k}}\right)_{\boldsymbol{k}_0}. \tag{7.17}$$

Equation (7.15) (considering only one term in the summation) can now be written as

$$f(\boldsymbol{r}, t) = \left[\int_{-\infty}^{\infty} \tilde{f}(\boldsymbol{k}) \mathrm{e}^{\mathrm{i}(\boldsymbol{k}-\boldsymbol{k}_0)\cdot(\boldsymbol{r}-\boldsymbol{v}_g t)} \mathrm{d}\boldsymbol{k}\right] \mathrm{e}^{\mathrm{i}\boldsymbol{k}_0\cdot\boldsymbol{r}-\omega_0 t}.$$

The expression in square brackets represents a generic function, $A$, of the variable $(\boldsymbol{r} - \boldsymbol{v}_g t)$, so that we finally obtain

$$f(\boldsymbol{r}, t) = A(\boldsymbol{r} - \boldsymbol{v}_g t)\, \mathrm{e}^{\mathrm{i}\boldsymbol{k}_0\cdot\boldsymbol{r}-\omega_0 t}.$$

The last equation shows that for each wave mode, namely for each value of the parameter $\alpha$ in Eq. (7.15), the solution is an infinite plane wave, whose wave vector is $\boldsymbol{k}_0$ and whose frequency is $\omega_0$, propagating with the *phase speed* $\boldsymbol{v}_{ph} = (\omega_0/k_0)\boldsymbol{e}_k$, having an amplitude modulated by the function $A(\boldsymbol{r} - \boldsymbol{v}_g t)$, which propagates with the *group velocity* $\boldsymbol{v}_g = (\partial\omega/\partial\boldsymbol{k})_{k_0}$. Because the group velocity can be associated with energy propagation, the relationship $v_g < c$ must hold quite generally.

## 7.2 Waves in the Ideal MHD Regime

The starting point for the study of ideal MHD waves is the system of linearized MHD equations (4.2), (6.9), (6.10), (6.4) and (6.12). If the only forces present are magnetic and pressure forces, we may put $\boldsymbol{f}_1 = 0$ in Eq. (6.12). If we further assume a uniform unperturbed state, we may Fourier transform with respect to both space and time coordinates. The system of equations for the transformed quantities (where for simplicity the "tilde" has been omitted since no confusion may arise) becomes:

$$\begin{aligned}
\rho_1 &= -\mathrm{i}\rho_0(\boldsymbol{k}\cdot\boldsymbol{\xi})\\
P_1 &= -\mathrm{i}\rho_0 c_s^2(\boldsymbol{k}\cdot\boldsymbol{\xi})\\
\boldsymbol{B}_1 &= \mathrm{i}\boldsymbol{k}\times(\boldsymbol{\xi}\times\boldsymbol{B}_0)\\
-\omega^2\rho_0\boldsymbol{\xi} &= -\mathrm{i}\boldsymbol{k}P_1 + \frac{\mathrm{i}}{4\pi}[(\boldsymbol{k}\times\boldsymbol{B}_1)\times\boldsymbol{B}_0]
\end{aligned} \tag{7.18}$$

Introducing the expressions for $P_1$ and $B_1$ into the last equation, an equation containing only the vector displacement $\boldsymbol{\xi}$ is obtained:

$$\omega^2 \rho_0 \boldsymbol{\xi} = \rho_0 c_s^2 \boldsymbol{k}(\boldsymbol{k} \cdot \boldsymbol{\xi}) + \frac{1}{4\pi}\{\boldsymbol{k} \times [\boldsymbol{k} \times (\boldsymbol{\xi} \times \boldsymbol{B}_0)] \times \boldsymbol{B}_0\} \tag{7.19}$$

Writing $\boldsymbol{B}_0 = B_0\, \boldsymbol{e}_b$ and introducing the Alfvén speed, $c_a^2 = B_0/(4\pi\rho_0)$, we finally get

$$\omega^2 \boldsymbol{\xi} = c_s^2 \boldsymbol{k}(\boldsymbol{k} \cdot \boldsymbol{\xi}) + c_a^2\{\boldsymbol{k} \times [\boldsymbol{k} \times (\boldsymbol{\xi} \times \boldsymbol{e}_b)] \times \boldsymbol{e}_b\} \tag{7.20}$$

### 7.2.1 Magnetic Waves

Consider first the simple case where plasma compressibility, represented here by the presence of $c_s^2$, can be neglected, i.e. $c_s \ll c_a$ Expanding the triple vector products entering Eq. (7.20) we obtain the following expression:

$$\omega^2 \boldsymbol{\xi} = c_a^2\{(\boldsymbol{k} \cdot \boldsymbol{e}_b)^2 \boldsymbol{\xi} + [(\boldsymbol{k} \cdot \boldsymbol{\xi}) - (\boldsymbol{k} \cdot \boldsymbol{e}_b)(\boldsymbol{\xi} \cdot \boldsymbol{e}_b)]\boldsymbol{k} - (\boldsymbol{k} \cdot \boldsymbol{\xi})(\boldsymbol{k} \cdot \boldsymbol{e}_b)\boldsymbol{e}_b\}. \tag{7.21}$$

Taking the scalar product of $\boldsymbol{e}_b$ with Eq. (7.21) we see that (unless $\omega^2 = 0$)

$$\boldsymbol{\xi} \cdot \boldsymbol{e}_b = 0.$$

The displacements, and therefore the velocities, of the particles for oscillations with non-vanishing frequencies are perpendicular to the direction of the unperturbed magnetic field $\boldsymbol{B}_0$. Using this condition in Eq. (7.21) and taking the scalar product with $\boldsymbol{k}$ we get

$$(\omega^2 - k^2 c_a^2)(\boldsymbol{k} \cdot \boldsymbol{\xi}) = 0. \tag{7.22}$$

The preceding equation has two possible solutions: $\boldsymbol{k} \cdot \boldsymbol{\xi} = 0$ or $\omega^2 = k^2 c_a^2$, that will now be examined in turn.

#### 7.2.1.1 $\boldsymbol{k} \cdot \boldsymbol{\xi} = 0$, Alfvén Waves

The condition $\boldsymbol{k} \cdot \boldsymbol{\xi} = 0$, which in ordinary space is equivalent to $\nabla \cdot \boldsymbol{\xi} = 0$, implies that the *perturbations* are incompressible. Clearly, this does not mean that the *plasma* is incompressible, but simply that the perturbations do not produce density variations. If the above condition is introduced into Eq. (7.21) we obtain the dispersion relation:

$$\omega^2 = (\boldsymbol{k} \cdot \boldsymbol{e}_b)^2 c_a^2 = k^2 c_a^2 \cos^2\theta, \tag{7.23}$$

where $\theta$ is the angle between the propagation vector $\boldsymbol{k}$ and the direction of the unperturbed magnetic field $\boldsymbol{B}_0$. Taking the scalar product of $\boldsymbol{e}_b$ with the third equation (7.18) we see that

$$\boldsymbol{B}_1 \cdot \boldsymbol{e}_b = 0, \tag{7.24}$$

showing that magnetic field perturbations are also perpendicular to the magnetic field $\boldsymbol{B}_0$. These waves, called *Alfvén waves*, are *transverse waves*, both for the displacement and magnetic field perturbation.

The dispersion relation for Alfvén waves shows that their phase velocity

$$\boldsymbol{v}_{ph} = (\omega/k)\boldsymbol{e}_k = \pm(c_a \cos\theta)\boldsymbol{e}_k$$

depends on the propagation angle with respect to $\boldsymbol{B}_0$. However, since the dispersion relation (7.23) contains only the component of $\boldsymbol{k}$ along $\boldsymbol{B}_0$, the group velocity is directed along $\boldsymbol{e}_b$,

$$\boldsymbol{v}_g = \pm\, c_a\, \boldsymbol{e}_b.$$

Thus, even if the Alfvén wave-front advances in a direction different from that of $\boldsymbol{B}_0$, the energy associated with the wave propagates either along or opposite to the unperturbed magnetic field $\boldsymbol{B}_0$. The waves are therefore intrinsically anisotropic in their propagation.

Using again the expression for $\boldsymbol{B}_1$ given by Eq. (7.18), the condition given by Eq. (7.24) and the dispersion relation (7.23), it is easy to derive a relationship between the velocity $\boldsymbol{U} = (\partial\boldsymbol{\xi}/\partial t) \rightarrow -i\omega\boldsymbol{\xi}$ and the magnetic field perturbation $\boldsymbol{B}_1$:

$$\frac{\boldsymbol{B}_1}{B_0} = \pm\frac{\boldsymbol{U}}{c_a}. \tag{7.25}$$

This relation characterizes Alfvén waves propagating in either direction (along or against the mean field) exactly and shows that the kinetic and the magnetic energy of the waves is the same

$$\frac{E_{kin}}{E_{mag}} = \frac{\frac{1}{2}\rho_0 U^2}{{B_1}^2/8\pi} = 1,$$

i.e. that kinetic and magnetic field energies are in equipartition. To understand the physical origin of Alfvén waves, consider the linearized form of the magnetic force, $\boldsymbol{F}_1$:

$$\boldsymbol{F}_1 \propto (\boldsymbol{k} \times \boldsymbol{B}_1) \times \boldsymbol{B}_0 = (\boldsymbol{k} \cdot \boldsymbol{B}_0)\boldsymbol{B}_1 - (\boldsymbol{B}_0 \cdot \boldsymbol{B}_1)\boldsymbol{k}.$$

The first term on the rhs derives from magnetic tension, while the second is connected with magnetic pressure (it is the linearized form of the term $\nabla(B^2/8\pi)$). Since the latter term vanishes in this case because of Eq. (7.24), we find that Alfvén waves are caused by perturbations in the magnetic tension. Therefore they bear a strong resemblance to the (transverse) oscillations of stretched elastic string, such as the vibrations of chords in a musical instrument. It is known that the phase velocity of the waves propagating on a vibrating string is given by $u_{ph} = \sqrt{T/\sigma}$, where $T$ is the tension of the string and $\sigma$ is the linear mass density, i.e. the mass per unit length. Substituting $\sigma = \rho_0 S$, where $S$ is the section of the string and the magnetic tension $T = (B_0^2/4\pi)S$, we get $u_{ph} = c_a$. This result clarifies and expands the observation made in Chap. 5 of the analogy between magnetic lines of force and elastic strings.

An important feature of Alfvén waves is that they may be a solution not only of the linearized but also of the full MHD equations. Consider a homogeneous plasma (i.e. with constant $P_0,\ \rho_0,\ \boldsymbol{B}_0$) and a perturbation $\boldsymbol{B} = \boldsymbol{B}_0 + \boldsymbol{B}_1$, without assuming that $|B_1| \ll |B_0|$. If the following conditions are satisfied

$$|\boldsymbol{B}_0 + \boldsymbol{B}_1| = const., \qquad \frac{\boldsymbol{B}_1}{B_0} = \pm\frac{\boldsymbol{U}}{c_a}, \tag{7.26}$$

it is easy to show (see Exercise 7.1) that this perturbation is an exact solution of the MHD equations. This means that if fluctuations of velocity and magnetic field obeying the preceding relation are observed, even of large amplitude, they can be identified as nonlinear Alfvén waves. Waves of this nature have been observed in the solar wind.

#### 7.2.1.2 $\boldsymbol{k} \cdot \boldsymbol{\xi} \neq 0$, Compressible Alfvén Waves

If $\boldsymbol{k} \cdot \boldsymbol{\xi} \neq 0$ the solution of Eq. (7.22) becomes simply

$$\omega^2 = k^2 c_a^2. \tag{7.27}$$

These are compressible waves with identical isotropic phase and group velocities: $\boldsymbol{v}_{ph} = \boldsymbol{v}_g = \pm c_a\, \boldsymbol{e}_k$.

Since the condition $\boldsymbol{\xi} \cdot \boldsymbol{e}_b = 0$ is still valid, but now $\boldsymbol{\xi} \cdot \boldsymbol{e}_k \neq 0$, it follows that $\boldsymbol{\xi}$ is a vector normal to $\boldsymbol{B}_0$ lying in the plane defined by $\boldsymbol{B}_0$ and $\boldsymbol{k}$. It is interesting to note that for propagation perpendicular to $\boldsymbol{B}_0$ ($\theta = \pi/2$), $\boldsymbol{\xi}$ is directed along $\boldsymbol{k}$ and the waves becomes a *longitudinal* wave. For propagation parallel to $\boldsymbol{B}_0$, ($\theta = 0, \boldsymbol{e}_k \parallel \boldsymbol{e}_b$), $\boldsymbol{\xi} \cdot \boldsymbol{e}_b = 0$ implies $\boldsymbol{\xi} \cdot \boldsymbol{e}_k = 0$ and the wave becomes *transverse* and incompressible, and thus indistinguishable from an Alfvén wave.

### 7.2.2 *Magnetosonic Waves*

Let's now restore the effects of the plasma compressibility and utilize the full form of Eq. (7.20). Expanding the triple vector products, Eq. (7.21) becomes:

$$\omega^2 \boldsymbol{\xi} = c_s^2(\boldsymbol{k}\cdot\boldsymbol{\xi})\boldsymbol{k} + c_a^2\{(\boldsymbol{k}\cdot\boldsymbol{e}_b)^2\boldsymbol{\xi} + [(\boldsymbol{k}\cdot\boldsymbol{\xi}) - (\boldsymbol{k}\cdot\boldsymbol{e}_b)(\boldsymbol{\xi}\cdot\boldsymbol{e}_b)]\boldsymbol{k} - (\boldsymbol{k}\cdot\boldsymbol{\xi})(\boldsymbol{k}\cdot\boldsymbol{e}_b)\boldsymbol{e}_b\}. \tag{7.28}$$

Following the procedure already employed for the magnetic waves, let's take the scalar product of the preceding equation with $\boldsymbol{e}_b$ and with $\boldsymbol{k}$, to obtain

$$\omega^2(\boldsymbol{\xi} \cdot \boldsymbol{e}_b) = c_s^2(\boldsymbol{k} \cdot \boldsymbol{\xi})(\boldsymbol{k} \cdot \boldsymbol{e}_b), \tag{7.29}$$

and

$$[\omega^2 - k^2(c_s^2 + c_a^2)](\boldsymbol{k} \cdot \boldsymbol{\xi}) = -k^2 c_a^2(\boldsymbol{k} \cdot \boldsymbol{e}_b)(\boldsymbol{\xi} \cdot \boldsymbol{e}_b). \tag{7.30}$$

If $(\boldsymbol{k} \cdot \boldsymbol{\xi}) = 0$, Eq. (7.29) shows that $(\boldsymbol{\xi} \cdot \boldsymbol{e}_b) = 0$ as well. Introducing these constraints in Eq. (7.28), we obtain again the dispersion relation for the Alfvén waves, Eq. (7.23), $\omega^2 = k^2 c_a^2 \cos^2\theta$. These *incompressible* waves are therefore a solution even when the medium compressibility is taken into account.

If $(\boldsymbol{k} \cdot \boldsymbol{\xi}) \neq 0$, we multiply Eq. (7.30) by $(\boldsymbol{k} \cdot \boldsymbol{e}_b)$ and use Eq. (7.29) to obtain the following equation:

$$\omega^4 - \omega^2 k^2(c_s^2 + c_a^2) + c_s^2 c_a^2 k^4 \cos^2\theta = 0. \tag{7.31}$$

The two solutions (in $\omega^2$) provide the dispersion relations of the *magnetosonic waves*:

$$\left(\frac{\omega}{k}\right)^2 = \frac{1}{2}\left[(c_s^2 + c_a^2) \pm \sqrt{c_s^4 + c_a^4 - 2c_s^2 c_a^2 \cos 2\theta}\right]. \tag{7.32}$$

The two modes corresponding, respectively, to the plus or minus sign in Eq. (7.32) are known as the fast and slow magnetosonic wave. The Alfvén wave has a phase velocity intermediate between the two.

The propagation characteristics depend on the ratio $c_s^2/c_a^2$. In fact, for propagation parallel to the magnetic field $\boldsymbol{B}_0$, $\omega/k$ tends to $c_s$ for the fast wave and to $c_a$ for the slow one, if $c_s > c_a$. The situation is just the opposite if $c_s < c_a$. To summarize, we may say that when $\theta \to 0$ the phase speed of the fast wave tends to the higher of $c_a$ and $c_a$, while the slow wave tends to the lower of the two. When $\theta \to \pi/2$, the phase speed of the slow wave tends to zero, while that of the fast wave tends to $(c_s^2 + c_a^2)^{1/2}$. It is interesting to remark, by taking the product of the squares of the phase speeds of the fast and slow modes ($v_F$ and $v_S$ respectively) from Eq. (7.32), that it is always true that $v_F^2 v_S^2 = c_s^2 c_a^2 cos^2\theta$ i.e. the product is equal to the square of the sound speed times the square of the Alfvén wave phase speed. Obtaining the group velocities of the waves requires taking the gradient of expression Eq. (7.32) with respect to the wave-vector. Since the algebra is somewhat involved the results are best summarized using polar plots of the group velocities as a function of the angle between velocity and the magnetic field, known as the Friedrichs diagram. This is shown, together with the polar plots of the phase-speeds (as a function of the angle between wave-vector and magnetic field), for the fast, slow and Alfvén waves in two cases, one for a sound speed below the Alfvén speed (plasma $\beta < 1$, Fig. 7.1) and one for the opposite case (plasma $\beta > 1$, Fig. 7.2).

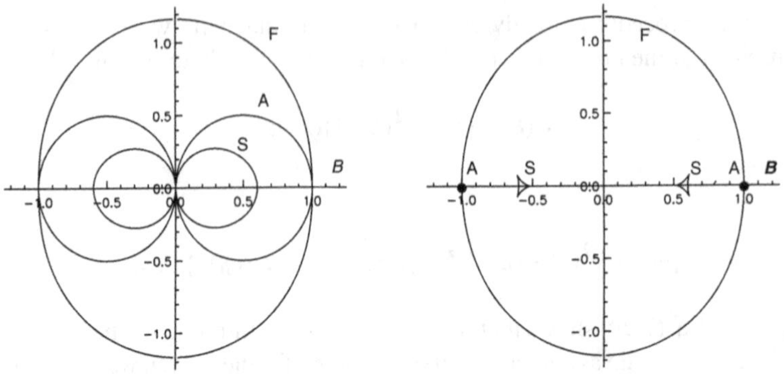


**Fig. 7.1** Phase and group velocities of MHD waves for a plasma in which $c_s/c_a = 0.6$ ($c_a$ has been normalized to 1). The *left hand panel* shows, in polar coordinates, the phase speeds of the fast ($F$), slow ($S$), and Alfvén ($A$) waves as a function of the angle between the wave-vector and the magnetic field, aligned with the $x$-axis. The *right hand panel* shows, again in polar coordinates, the magnitude of the group velocities for the fast (F), slow (S), and Alfvén waves (the *thick dot* in $x = \pm 1$), again as a function of the angle between the group velocity and the magnetic field

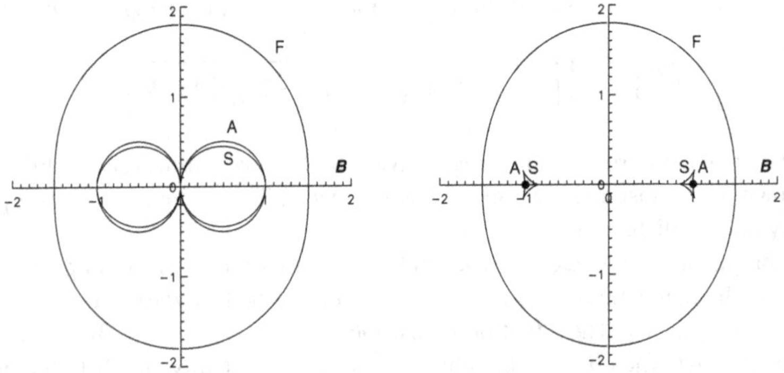


**Fig. 7.2** Phase and group velocities of MHD waves for a plasma in which $c_s/c_a = 1.5$ ($c_a$ has been normalized to 1). The *left hand panel* shows, in polar coordinates, the phase speeds of the fast ($F$), slow ($S$), and Alfvén ($A$) waves as a function of the angle between the wave-vector and the magnetic field, aligned with the $x$-axis. The *right hand panel* shows, again in polar coordinates, the magnitude of the group velocities for the fast (F), slow (S), and Alfvén waves (the *thick dot* in $x = \pm 1$), as a function of the angle between the group velocity and the magnetic field

For both cases one can see that the fast mode wave is almost isotropic, with only modest changes in speed between the perpendicular and parallel propagation as described above. In the figures, the Alfvén speed is normalized to 1 so the Alfvén wave is described by circles centered in $x = \pm 0.5$ and radius 0.5 in the polar phase speed plot. The group velocity on the other hand is given by the dark dots, identifying $v_g = c_a = \pm 1$. The main difference between $\beta < 1$ and $\beta > 1$ lies in the curves that are tangent in parallel propagation: these are the fast and Alfvén for $\beta < 1$ and the slow and Alfvén for $\beta > 1$. Also, note the strange shape of the Friedrichs diagram for the slow wave. In parallel propagation, it is equal to the sound speed, but as the angle between wave-vector and field increases, the group velocity reaches a maximum angle of propagation with respect to the magnetic field, typically a fairly small angle, and then returns to parallel propagation as the wave-vector become perpendicular to the field. The point at which the cusp forms on the x-axis, corresponding to orthogonal field and wave-vector, is called the cusp speed $v_C$. It shows that even though the slow wave has a vanishing phase speed at orthogonal propagation, like the Alfvén speed it retains a finite group velocity parallel to the magnetic field. It is left as an exercise to calculate its value

$$v_C = \frac{c_a c_s}{\sqrt{(c_s^2 + c_a^2)}}.$$

## 7.3 Fluid Waves Beyond the MHD Regime

The preceding Section was devoted to a discussion of wave modes in the MHD regime. The solutions found are exact, but, of course, their validity is limited to the range of validity of the equations from which they have been derived. Recalling the discussion in Chap. 5, where we have shown that the MHD regime is a *low frequency* regime, it is natural to ask how the wave modes change with increasing frequency, but remaining within a framework of an infinite conductivity, ideal plasma model. As a general rule, we may say that the MHD equations are valid for frequencies lower than the lowest of the characteristic frequencies of the plasma. For instance, in the physical conditions prevailing in the quiet solar corona, $n_e \simeq n_p \simeq 10^8\,\mathrm{cm}^{-3}$, $B \lesssim 10\,G$, we get $\omega_{pe} \simeq 5.6 \times 10^8\,\mathrm{s}^{-1}$, $\omega_{pi} \simeq 1.3 \times 10^7\,\mathrm{s}^{-1}$, $\omega_{ce} \simeq 1.8 \times 10^8\,\mathrm{s}^{-1}$, $\omega_{ci} \simeq 9.6 \times 10^4\,\mathrm{s}^{-1}$. Thus, in the case of magnetosonic waves for which $\omega \propto k$, $\omega$ can easily reach values taking them beyond the validity limit of MHD equations for sufficiently large $k$'s.

### 7.3.1 *Intermediate Frequencies: $\omega \lesssim \omega_{ce}$*

An intuition on the changes occurring when the frequencies becomes higher can be obtained recalling the comparison of the orders of magnitude of the terms entering the generalized Ohm's equation (5.5), restricting our attention to the "cold plasma" case, where the terms connected with pressure gradients are absent. It is easy to verify that the first term that must be kept when the frequency increases is the Hall term, namely the term proportional to $\boldsymbol{J} \times \boldsymbol{B}$. In Chap. 5 it was shown that the ratio of this term to the ideal one, $(1/c)(\boldsymbol{U} \times \boldsymbol{B})$, is $(c/\mathcal{U})^2\omega(\omega_{ce}/\omega_{pe}^2)$. Considering again the example of the solar corona and assuming $\mathcal{U} \simeq 0.01c$, we see that the two terms are comparable when $\omega \simeq \omega_{ci}$. An analytical study of the waves supported by a cold plasma for propagation at arbitrary angles with respect to the magnetic field is possible but algebraically involved. Here we shall limit ourselves to the case of parallel propagation, a simplification which still allows understanding how wave modes change when moving beyond MHD to high frequencies (close to the proton cyclotron frequency).

Consider as usual $\boldsymbol{B}_0 = B_0\boldsymbol{e}_b$, and take $\boldsymbol{k} = k\boldsymbol{e}_b$ so that the divergenceless condition $\nabla \cdot \boldsymbol{B}_1 = 0$, or equivalently $\boldsymbol{k} \cdot \boldsymbol{B}_1 = 0$, implies that $\boldsymbol{e}_b \cdot \boldsymbol{B}_1 = 0$. Fourier transforming the linearized MHD equations, but without introducing the lagrangian displacement $\boldsymbol{\xi}$, leads to:

$$\omega \boldsymbol{U} = -\frac{c_a^2}{B_0}(\boldsymbol{k} \times \boldsymbol{B}_1) \times \boldsymbol{e}_b = -\frac{kc_a^2}{B_0}[\boldsymbol{B}_1 - (\boldsymbol{e}_b \cdot \boldsymbol{B}_1)\boldsymbol{e}_b] = -\frac{kc_a^2}{B_0}\boldsymbol{B}_1.$$

The generalized Ohm equation, keeping only the dominant Hall term, reads:

$$\omega \boldsymbol{B}_1 + k\boldsymbol{e}_b \times (\boldsymbol{U} \times \boldsymbol{B}_0) = \frac{\mathrm{i}\ B_0 k^2 c}{4\pi n_0 e}\boldsymbol{e}_b \times [(\boldsymbol{e}_b \times \boldsymbol{B}_1) \times \boldsymbol{e}_b].$$

Using the preceding expression for $\boldsymbol{U}$ and expanding the vector products, we finally obtain:

$$(\omega^2 - k^2 c_a^2)\boldsymbol{B}_1 - \mathrm{i}(\omega/\omega_{ci})k^2 c_a^2(\boldsymbol{e}_b \times \boldsymbol{B}_1) = 0. \tag{7.33}$$

Choosing the $z$-axis along $\boldsymbol{B}_0$ ($\boldsymbol{e}_b = \boldsymbol{e}_z$) and writing the components of Eq. (7.33) explicitly these become:

$$\begin{aligned}
&(\omega^2 - k^2 c_a^2) B_{1x} + \mathrm{i}\,(\omega/\omega_{ci})k^2 c_a^2\, B_{1y} = 0\\
&(\omega^2 - k^2 c_a^2) B_{1y} - \mathrm{i}\,(\omega/\omega_{ci})k^2 c_a^2\, B_{1x} = 0\\
&(\omega^2 - k^2 c_a^2) B_{1z} = 0
\end{aligned} \tag{7.34}$$

If $B_{1z} \neq 0$ we must have $\omega^2 = k^2 c_a^2$, from which it follows that $B_{1x} = B_{1y} = 0$. If, on the contrary, $B_{1z} = 0$, the solution of the problem is obtained by requiring that the determinant of the coefficients of the first two equations of the system must vanish (7.34):

$$\begin{aligned}
&(\omega^2 - k^2 c_a^2)^2 - (\omega/\omega_{ci})^2 k^4 c_a^4\\
&\qquad = \left[\omega^2 - k^2 c_a^2 + (\omega/\omega_{ci})k^2 c_a^2\right]\left[\omega^2 - k^2 c_a^2 - (\omega/\omega_{ci})k^2 c_a^2\right] = 0.
\end{aligned}$$

The system then splits into two separate second degree equations, each having only one positive root. Consider first the equation:

$$\omega^2 - k^2 c_a^2 + (\omega/\omega_{ci})k^2 c_a^2 = 0,$$

whose positive root is:

$$\omega = \frac{1}{2\omega_{ci}}\left(-k^2c_a^2 + \sqrt{k^4c_a^4 + 4k^2c_a^2\omega_{ci}^2}\right).$$

The limiting values for small and large $k$'s are:

$$\omega \simeq k\,c_a \quad (k \to 0), \quad \text{and} \quad \omega \simeq \omega_{ci} \quad (k \to \infty).$$

Turning now to the solution of equation:

$$\omega^2 - k^2c_a^2 - (\omega/\omega_{ci})k^2c_a^2 = 0,$$

we have

$$\omega = \frac{1}{2\omega_{ci}}\left(k^2c_a^2 + \sqrt{k^4c_a^4 + 4k^2c_a^2\omega_{ci}^2}\right),$$

with limiting values,

$$\omega \simeq k\,c_a \quad (k \to 0) \quad \text{but} \quad \omega \simeq \frac{k^2c_a^2}{\omega_{ci}} \quad (k \to \infty).$$

We thus see the wave modes obey the same dispersion relation for $k \to 0$, namely for low frequencies, in agreement with the results found for parallel propagating MHD waves [see Eqs. (7.23) and (7.27)]. However, for large values of $k$, the two dispersion relations are considerably different: in the first case, the frequency remains finite and tends to $\omega_{ci}$, while in the second case it diverges quadratically with $k$. However, our treatment cannot be extended to exceedingly large values of $k$, and consequently of $\omega$, since it only takes into account the Hall term in the equation for $\boldsymbol{B}$ and it is easy to see that this approximation ceases to be valid for values of $\omega$ considerably higher than $\omega_{ci}$.

Before moving to still higher frequencies, it is interesting to determine the characteristic features of the two wave modes just found. If we consider the first solution when $k \to \infty$ and replace $\omega$ with $\omega_{ci}$ in Eq. (7.34), we get

$$-B_{1x} + \mathrm{i}\,B_{1y} = 0 \quad \text{namely} \quad \frac{\mathrm{i}\,B_{1y}}{B_{1x}} = 1.$$

This relation implies that we are in the presence of circularly polarized waves that appear to rotate clockwise when we look in a direction antiparallel to $\boldsymbol{B}_0$. To verify this statement, remember that the quantities $B_{1x}$ and $B_{1y}$, entering the preceding equations, are actually the Fourier components of the respective physical quantities. Introducing again the notation $\tilde{f}$ for the Fourier transform of a generic quantity $f$, we may write the condition found above as

$$\tilde{B}_{1y} = -\mathrm{i}\tilde{B}_{1x}.$$

But,

$$B_{1x} \equiv \Re\mathrm{e}(\tilde{B}_{1x}) = |B_1| \cos \omega t$$

and

$$B_{1y} \equiv \Re\mathrm{e}(\tilde{B}_{1y}) = -\Re\mathrm{e}(\mathrm{i}\tilde{B}_{1x}) = -|B_1| \sin \omega t,$$

where $|B_1|$ is the magnitude of the vector $\boldsymbol{B}_1$. Recalling Maxwell's equation for $\nabla \times \boldsymbol{E}, (\boldsymbol{e}_z \times \tilde{\boldsymbol{E}} \propto \boldsymbol{B}_1)$, we see that the same relationship holds among the components of $\tilde{E}$. Therefore

$$E_x = |E| \cos \omega t\,; \quad E_y = -|E| \sin \omega t.$$

At time $t = 0$ the field is directed along $x$. As time increases, $E_x$ decreases but remains positive, while $E_y$ takes up negative values. This shows that the vector $\boldsymbol{E}$ rotates in the sense indicated above. If we recall the discussion of Chap. 2, we see that this is the sense of rotation of the *positive* particles under the action of the magnetic field $B_0$. Therefore, the sense of rotation of the wave's electric field coincides with that of positive ions, thus justifying the name of *ion cyclotron waves* given to this mode. It is clear that this situation enhances the interaction between the wave and the positive ions, thus favoring the birth of resonance phenomena.

Repeating the procedure for the second solution, we get:

$$B_{1x} + \mathrm{i}\, B_{1y} = 0 \quad \text{or} \quad \frac{\mathrm{i}\, B_{1y}}{B_{1x}} = -1,$$

and in this case the sense of rotation agrees with that of the electrons. It is not surprising therefore that the exact theory for this type of waves predicts that when $k \to \infty$, the frequency tends to $\omega_{ce}$ and that a resonance might take place between the wave and the electrons. For intermediate values of $\omega$, $\omega_{ci} \ll \omega \ll \omega_{ce}$, it is possible to show that this so-called *whistler* wave obeys the dispersion relation

$$\omega \simeq \frac{k^2 c^2 \omega_{ce}}{\omega_{pe}^2}$$

#### 7.3.1.1 High Frequencies: $\omega \simeq \omega_{pe}$

Once again we shall restrict ourselves to a simple case, that of a cold unmagnetized plasma. At sufficiently high frequencies, we can no longer neglect the displacement current in Maxwell's equation for $\nabla \times \boldsymbol{B}$. On the other hand ions, because of their larger mass, are unable to follow the oscillations over the fast timescales $\simeq \omega^{-1}$. Therefore they can be approximated by a fixed background, their only function being to provide the charge required to ensure plasma quasi-neutrality. Our description will thus refer to the electronic component only. The relevant equations are:

$$\begin{aligned} -\mathrm{i}\omega \boldsymbol{u} &= -\frac{e}{m_e}\boldsymbol{E}, \\ \boldsymbol{k} \times \boldsymbol{E} &= \frac{\omega}{c}\boldsymbol{B}_1, \\ \mathrm{i}\boldsymbol{k} \times \boldsymbol{B}_1 &= \frac{4\pi}{c}\boldsymbol{J}_1 - \mathrm{i}\frac{\omega}{c}\boldsymbol{E}, \end{aligned} \tag{7.35}$$

where $\boldsymbol{u}$ is the electrons' velocity and $\boldsymbol{J}_1 = -en_0\boldsymbol{u}$. Notice that all quantities entering the preceding equations are Fourier transforms. By combining the above equations, we arrive without difficulty at the system:

$$(\omega_{pe}^2 - \omega^2 + k^2c^2)\boldsymbol{E} = c^2(\boldsymbol{k} \cdot \boldsymbol{E})\boldsymbol{k}. \tag{7.36}$$

Taking the scalar product of the preceding expression with $\boldsymbol{k}$, we get:

$$(\omega_{pe}^2 - \omega^2)(\boldsymbol{k} \cdot \boldsymbol{E}) = 0, \tag{7.37}$$

that shows that two types of waves are possible, depending on whether $(\boldsymbol{k} \cdot \boldsymbol{E})$ is different from zero or not.

In the first case

$$\omega = \omega_{pe},$$

and, introducing the above equality in Eq. (7.36), we see that $\boldsymbol{E}$ is directed along $\boldsymbol{k}$. On the other hand, the first of Eq. (7.35) tells us that also $\boldsymbol{u}$ is parallel to $\boldsymbol{k}$. We therefore have a *longitudinal wave* with a vanishing magnetic field, $\boldsymbol{B}_1 = 0$, as shown by the second of Eq. (7.35). These *electrostatic* waves are known as *plasma waves* or *Langmuir waves*. In the cold plasma limit so far adopted, these waves do not propagate and are therefore stationary oscillations, caused by the reaction of the plasma to local violations of charge neutrality: since $(\boldsymbol{k} \cdot \boldsymbol{E}) \neq 0$, the equation for $\nabla \cdot \boldsymbol{E}$, $\mathrm{i}\boldsymbol{k} \cdot \boldsymbol{E} = 4\pi q = -4\pi\, e\, n_1$ confirms that a non-vanishing charge density is present.

The introduction of thermal effects modifies the picture, as shown by the introduction of the term proportional to the pressure gradient in the equation of motion of the electron fluid. In this case:

$$m_e n_0 \frac{\partial \boldsymbol{u}}{\partial t} = -\nabla P_1 - en_0\boldsymbol{E} = \gamma c_s^2 \nabla \rho_1 - en_0\boldsymbol{E},$$

where use has been made of the definition

$$c_s^2 = \frac{k_B T}{m_e}.$$

The Boltzmann constant has been written as $k_B$ to avoid confusion with the magnitude of the wave vector $\boldsymbol{k}$. Using the continuity equation to eliminate $\rho_1$ and Fourier transforming, we get:

$$\omega^2 \boldsymbol{u} = \gamma k^2 c_s^2 \boldsymbol{u} - \mathrm{i}\, e\omega \boldsymbol{E}.$$

Since we are dealing with longitudinal waves, the only non-vanishing component of the preceding equation is along $\boldsymbol{e}_k$. Therefore:

$$(\omega^2 - \gamma k^2 c_s^2)u = -\mathrm{i}\frac{e}{m_e}\omega E = -\omega_{pe}^2 u,$$

the third of Eq. (7.35) (with $\boldsymbol{B}_1 = 0$) has been used together with the definition of $\boldsymbol{J}_1 = -en_0\boldsymbol{u}$. We thus arrive at the dispersion relation:

$$\omega^2 = \omega_{pe}^2 + \gamma k^2 c_s^2.$$

Therefore, in a "warm" plasma, the wave propagates, but only for frequencies higher than the plasma frequency, $\omega > \omega_{pe}$. When $\omega \gg \omega_{pe}$, the phase speed tends to $\gamma c_s$. The value of $\gamma$ is connected with the number of degrees of freedom, $s$, of the plasma particles, $\gamma = 1 + 2/s$. In our case the particle motion is strictly one-dimensional, since the oscillations take place in the $k$-direction and we have neglected the effects of collisions that may scatter the electrons in other directions. Therefore, $s = 1$ and $\gamma = 3$. The final form of the dispersion relation thus reads:

$$\omega^2 = \omega_{pe}^2 + 3k^2 c_s^2 = \omega_{pe}^2 + 3k^2(\frac{k_B T}{m_e}). \tag{7.38}$$

Consider now the second solution of Eq. (7.37),

$$\boldsymbol{k} \cdot \boldsymbol{E} = 0.$$

In this case Eq. (7.36) immediately yields the dispersion relation

$$\omega^2 = \omega_{pe}^2 + k^2 c^2. \tag{7.39}$$

Now the wave's magnetic field is given by $\boldsymbol{B}_1 = (\omega/c)(\boldsymbol{k} \times \boldsymbol{E}) \neq 0$ and is orthogonal to both $\boldsymbol{E}$ e $\boldsymbol{k}$. This solution thus represents *transverse electromagnetic waves*, the generalization of the well known electromagnetic waves that propagate in vacuum.

Their phase velocity is

$$v_{ph} = \frac{\omega}{k} = c\sqrt{1 + (\omega_{pe}^2/k^2c^2)},$$

greater than the speed of light $c$. As already remarked, this does not constitute a problem. However, we may verify that the group velocity $v_g < c$. In fact,

$$v_g = \frac{\partial \omega}{\partial k} = c^2\frac{k}{\omega} = \frac{c^2}{v_{ph}} < c.$$

These waves also propagate only for frequencies $\omega > \omega_{pe}$, a circumstance of great relevance for radio broadcasting. In fact, low-frequency waves propagating in the terrestrial atmosphere eventually reach its farthermost layer, formed by an *ionized* gas (the so-called ionosphere) and may find themselves in a situation where their frequency is lower than the local plasma frequency, $\omega < \omega_{pe}$. The propagation is then stopped and the waves undergo total reflection back towards the Earth's surface. This phenomenon allows the connection by radio of points on the terrestrial surface that are not "in sight" of each other as first demonstrated by Marconi.

## 7.4 Waves in Kinetic Regimes: The Case of Landau Damping

We shall now examine in some detail a phenomenon of great interest that opens the way to a better understanding of the physics of plasmas at microscopic level and of the connection between kinetic and fluid models.

So far, we have been concerned with the description of waves starting from fluid models in different regimes: MHD, non-MHD one-fluid. When fluid models cease to be adequate, we must go back to the more general kinetic description and analyze wave phenomena in this context. In 1946 Landau attacked the problem of the propagation of electrostatic waves in a *collisionless* plasma and showed that the amplitude of such waves must eventually decrease with time. This result is very surprising, since we are used to thinking that wave damping is caused by the action of dissipative processes that, in turn, depend on the presence of collisions.

As we have already seen in Chap. 3, a non-collisional plasma is described by the Vlasov equation, that therefore will be chosen as the basis of our treatment. Assume that the plasma is homogeneous, with a vanishing magnetic field at equilibrium, and that the ions form as before a fixed background providing charge neutrality. Given these assumptions, consider again only the electron component, whose distribution is denoted by $f(\boldsymbol{r}, \boldsymbol{v}, t)$. In agreement with the procedure already adopted for the fluid case, consider an equilibrium state, represented by $f_0$ and a perturbation $f_1$, with $|f_1| \ll |f_0|$, around which the Vlasov equation for the electrons will be linearized. In the unperturbed situation, the electric field must vanish, otherwise the plasma particles would be permanently accelerated without the possibility of reaching an equilibrium state. It is then clear that any arbitrary function of the velocity, $f_0 = f_0(\boldsymbol{v})$, would be a solution of Vlasov's equation at the zeroth order:

$$\frac{\partial f_0}{\partial t} + \boldsymbol{v} \cdot \frac{\partial f_0}{\partial \boldsymbol{r}} = 0.$$

In the following we shall assume that $f_0$ is the maxwellian distribution function:

$$f_0 = n_0 \left( \frac{m}{2\pi k_B T} \right)^{3/2} \exp\left( -\frac{m v^2}{2 k_B T} \right).$$

The total electron distribution function is written as:

$$f(\boldsymbol{r}, \boldsymbol{v}, t) = f_0(\boldsymbol{v}) + f_1(\boldsymbol{r}, \boldsymbol{v}, t),$$

and the linearized form of the Vlasov equation will then be:

$$\frac{\partial f_1}{\partial t} + \boldsymbol{v} \cdot \frac{\partial f_1}{\partial \boldsymbol{r}} - \frac{e E}{m} \cdot \frac{\partial f_0}{\partial \boldsymbol{v}} = 0, \tag{7.40}$$

where $m$ is the electron mass, the electric field being a first order quantity.

The preceding equation must be coupled with that for the electric field:

$$\nabla \cdot \boldsymbol{E}(\boldsymbol{r}, t) = 4\pi q = -4\pi e \int f_1(\boldsymbol{r}, \boldsymbol{v}, t) \mathrm{d}\boldsymbol{v}.$$

Finally, because of the assumption of homogeneity, we can Fourier transform with respect to the spatial coordinates and write:

$$f_1(\boldsymbol{r}, \boldsymbol{v}, t) = \int \tilde{f}_1(\boldsymbol{k}, \boldsymbol{v}, t) \, \mathrm{e}^{\mathrm{i}\boldsymbol{k} \cdot \cdot \boldsymbol{r}} \mathrm{d}\boldsymbol{k}.$$

$$\boldsymbol{E}(\boldsymbol{r}, t) = \int \tilde{\boldsymbol{E}}(\boldsymbol{k}, \boldsymbol{v}, t) \, \mathrm{e}^{\mathrm{i}\boldsymbol{k} \cdot \cdot \boldsymbol{r}} \mathrm{d}\boldsymbol{k}.$$

The Fourier transform of Eq. (7.40) now reads:

$$\frac{\partial \tilde{f}_1}{\partial t} + \mathrm{i}\boldsymbol{k} \cdot \boldsymbol{v} \tilde{f}_1 - \frac{e \tilde{\boldsymbol{E}}}{m} \cdot \frac{\partial f_0}{\partial \boldsymbol{v}} = 0. \tag{7.41}$$

Since we are restricting to the case of longitudinal waves,

$$\nabla \times \boldsymbol{E} = 0, \quad \Rightarrow \quad \mathrm{i}\boldsymbol{k} \times \tilde{\boldsymbol{E}} = 0,$$

we are allowed to choose a reference system where

$$\boldsymbol{k} = [k, 0, 0], \quad \tilde{\boldsymbol{E}} = [\,\tilde{E}, 0, 0\,], \quad \boldsymbol{v} = [\,u, v_y, v_z\,].$$

Unless explicitly indicated, in the following we shall always assume that $k > 0$.

The formal solution of the problem is found by solving first the equations for he Fourier transforms

$$\frac{\partial \tilde{f}_1(k, \boldsymbol{v}, t)}{\partial t} + \mathrm{i}\, k u\, \tilde{f}_1(k, \boldsymbol{v}, t) - \frac{e\,\tilde{E}}{m}\frac{\partial f_0(\boldsymbol{v})}{\partial u} = 0 \tag{7.42}$$

and

$$\mathrm{i}\, k\tilde{E}(k, t) = -4\pi\, e \int \tilde{f}_1(k, \boldsymbol{v}, t)\mathrm{d}\boldsymbol{v}, \tag{7.43}$$

and then transforming back to the coordinates in physical space.

As far as time dependence is concerned, one might Fourier transform with respect to this variable as well, since the unperturbed state is stationary. However, if we assume that our perturbation is applied at time $t = 0$ it is preferable, following the original treatment by Landau, to perform a Laplace transform with respect to time. Let's begin by defining and reviewing some aspects of the Laplace transform.

Given a generic function of the time $g(t)$, we define its Laplace transform as:

$$\hat{g}(p) = \int_0^\infty g(t)\,\mathrm{e}^{-pt}\mathrm{d}t,$$

where $p$ is a complex number with $\Re\mathrm{e}\, p > 0$. It may be shown that the inverse Laplace transform is obtained by an integration along a line parallel to the imaginary axis, lying *to the right* of all the singularities of $\hat{g}(p)$:

$$g(t) = \frac{1}{2\pi\mathrm{i}}\int_{\sigma-\mathrm{i}\infty}^{\sigma+\mathrm{i}\infty} \hat{g}(p)\,\mathrm{e}^{pt}\mathrm{d}p.$$

Also notice that the Laplace transform is formally equivalent to a Fourier transform in which the substitution $-\mathrm{i}\omega \to p$ has been made. The use of the Laplace transform turns out to be especially useful when the solution of an initial value differential equation is sought. Consider for instance the first order equation:

$$\frac{dg}{dt} + a = b(t)\,; \quad g(t = 0) = g(0)$$

with $a = const.$ and $b(t)$ a known function of $t$. Laplace transforming the preceding equation we get:

$$\int_0^\infty \frac{dg}{dt} \mathrm{e}^{-pt}\mathrm{d}t + a\int_0^\infty \mathrm{e}^{-pt}\mathrm{d}t = \int_0^\infty b(t)\mathrm{e}^{-pt}\mathrm{d}t.$$

An integration by parts of the first term gives:

$$g(t)\mathrm{e}^{-pt}\Big|_0^\infty + p\,\hat{g}(p) + a\,\hat{g}(p) = \hat{b}(p),$$

and finally

$$-g(0) + (a+p)\,\hat{g}(p) = \hat{b}(p).$$

Notice that the initial condition for $g(t)$ is automatically included in the solution for $\hat{g}(p)$. The function $g(t)$, solution of the original equation and satisfying the given initial condition, is obtained by performing the inverse Laplace transform of $\hat{g}(p)$.

Returning to our problem, take the Laplace transform of Eq. (7.42):

$$p\,\hat{f}_1(k, \boldsymbol{v}, p) - \tilde{f}_1(k, \boldsymbol{v}, t=0) + \mathrm{i}k\,u\,\hat{f}_1(k, \boldsymbol{v}, p) - \frac{e}{m}\hat{E}(k, p)\frac{\partial f_0}{\partial u} = 0,$$

To simplify the notations we shall omit the explicit indication of the variables $k$ and $p$ in $\hat{f}_1$ and $\hat{E}$ and drop the index 1 in $f_1$. The preceding equation now reads:

$$(p + \mathrm{i}k\,u)\,\hat{f} = \tilde{f}(0) + \frac{e}{m}\hat{E}\frac{\partial f_0}{\partial u}. \tag{7.44}$$

Repeating the procedure with Eq. (7.43) and using of Eq. (7.44) we obtain:

$$\mathrm{i}\,k\hat{E} = -4\pi\,e\int\frac{\tilde{f}(0)}{p+\mathrm{i}ku}\mathrm{d}\boldsymbol{v} - 4\pi\frac{e^2}{m}\,\hat{E}\int\frac{(\partial f_0/\partial u)}{p+\mathrm{i}ku}\mathrm{d}\boldsymbol{v},$$

that we write in the form

$$\left[1 - \mathrm{i}\frac{4\pi\,e^2}{m\,k}\int\frac{(\partial f_0/\partial u)}{p+\mathrm{i}ku}\mathrm{d}\boldsymbol{v}\right]\hat{E} = \mathrm{i}\frac{4\pi\,e}{k}\int\frac{\tilde{f}(0)}{p+\mathrm{i}ku}\mathrm{d}\boldsymbol{v}. \tag{7.45}$$

Introduce the notation:

$$F_0(u) = \frac{1}{n_0}\int f_0(\boldsymbol{v})\,\mathrm{d}v_y\,\mathrm{d}v_z = \left(\frac{m}{2\pi k_B t}\right)^{1/2}\exp\left(-\frac{mu^2}{2k_B T}\right), \tag{7.46}$$

if $f_0$ is a maxwellian. If we further define:

$$D(k, p) = 1 - \mathrm{i}\frac{4\pi e^2 n_0}{m k}\int_{-\infty}^{\infty}\frac{(\mathrm{d}F_0/\mathrm{d}u)}{p + \mathrm{i}ku}\mathrm{d}u$$

$$= 1 - \mathrm{i}\frac{\omega_{pe}^2}{k}\int_{-\infty}^{\infty}\frac{(\mathrm{d}F_0/\mathrm{d}u)}{p + \mathrm{i}ku}\mathrm{d}u, \tag{7.47}$$

where $n_0$ is the electron density in the unperturbed state, we finally arrive at an explicit expression for $\hat{E}(k, p)$:

$$\hat{E}(k, p) = \mathrm{i}\frac{4\pi e}{k}\frac{1}{D(k, p)}\int\frac{\tilde{f}(k, \boldsymbol{v}, t = 0)}{p + \mathrm{i}ku}\mathrm{d}\boldsymbol{v}. \tag{7.48}$$

To obtain $E(\boldsymbol{r}, t)$ the inverse Laplace and Fourier transform of Eq. (7.48) must be taken and this introduces a few difficulties, starting with the inverse Laplace transform of $\hat{E}(k, p)$, namely

$$\tilde{E}(k, t) = \frac{1}{2\pi\mathrm{i}}\int_{\sigma-\mathrm{i}\infty}^{\sigma+\mathrm{i}\infty}\hat{E}(k, p)\,\mathrm{e}^{pt}\mathrm{d}p. \tag{7.49}$$

To find the value of the parameter $\sigma$, we must know the position of all the singularities of $\hat{E}(k, p)$ in the complex $p$-plane. However, if $\tilde{f}(k, \boldsymbol{v}, t = 0)$ and $(\mathrm{d}F_0/\mathrm{d}u)$ are analytic functions of $u$, it is possible to show that the only singularities of $\hat{E}(k, p)$ are the poles corresponding to the zeroes of $D(k, p)$, which we shall assume to be all distinct, for simplicity. Some of these poles could have a positive real part, which would imply that $\sigma > 0$. In this case the function under the integral in Eq. (7.49) would contain a factor $\mathrm{e}^{\sigma t}$, diverging when $t \to \infty$. However, the integration path in the complex $p$-plane can be modified in the way shown in Fig. 7.3. Notice that the vertical segments of the integration path (where $\mathfrak{Re}\, p = -\alpha$) do not contribute to the integral when $t \to +\infty$, because of the factor $\mathrm{e}^{-\alpha t}$. The contributions from the oppositely directed horizontal segments cancel out as well. Thus the only part left is the contribution from the paths (of infinitesimal radius) circling the poles:

$$\tilde{E}(k, t) = \frac{1}{2\pi\mathrm{i}}\oint\hat{E}(k, p)\,\mathrm{e}^{pt}\mathrm{d}p.$$

If the solutions of the equation

$$D(k, p_j) = 0, \tag{7.50}$$

**Fig. 7.3** **a** Original integration path; **b** Modified integration path

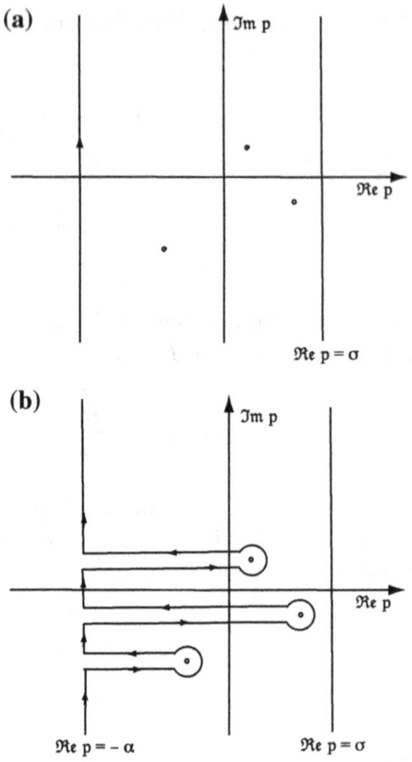


are indicated by $p_j$, ($j = 1, 2 \ldots N$), writing

$$p_j(k) = \gamma_j(k) - \mathrm{i}\,\omega_j(k),$$

the following expression for $\tilde{E}(k,t)$ is obtained:

$$\tilde{E}(k,t) = \sum_{j=1}^{N} R_j e^{\gamma_j t}\, e^{-\mathrm{i}\omega_j t}, \tag{7.51}$$

where $R_j$ is a quantity known in the theory of functions of complex variable as the *residue* of $\hat{E}(k,p)$ in $p = p_j$.

The electric field is therefore given by a superposition of waves, whose amplitude increases or decreases exponentially with time, according to the sign of $\gamma$. All terms with $\gamma_j < 0$ represent damped oscillations, the more so the highest the value of $|\gamma_j|$. When the time elapsed from the initial excitation is sufficiently long, the dominant term will be the one corresponding to the pole closest to the imaginary axis of $p$. The terms with $\gamma_j > 0$, on the other hand, represent amplified oscillations that very quickly get out of the linear regime. To stay within the framework of the present theory of small amplitude waves, we must assume that, even if poles with $\Re e\, p > 0$ do exist, they lie close to the imaginary axis of $p$. In both cases, we are concerned only with poles of $D(k, p)$ whose $\gamma$ is sufficiently small, i.e. $|\gamma/\omega| \ll 1$.

Going back to the dispersion relation of our low-amplitude electrostatic waves, Eq. (7.50), if we examine the expression for $D(k, p)$, Eq. (7.47), we see that the integrand has a pole for $u = ip/k$. Originally, the Laplace transform had been defined for $\Re e\, p > 0$, which implies that the pole lies in the half-space $\Im m\, u > 0$, external to the integration path that runs along the real axis of $u$. However, when we deformed the integration path of the inverse Laplace transform, we assumed that $\Re e\, p = \alpha < 0$. Therefore, we must find an analytical continuation of the integral entering Eq. (7.47) when $\Re e\, p < 0$. Such a continuation is obtained, once again, by deforming the integration path so that the pole remains located *above* the new path, as shown in Fig. 7.4.

floatbelowskip-3pc

Making use of this prescription, known as Landau prescription, we can define $D(k, p)$ for any value of $p$, in particular for $\Re e\, p = \gamma \to 0$, which is the most interesting case for as, as remarked above. By applying the above procedure and recalling the definition of *principal part* of an integral in the complex plane:

$$P \int_{-\infty}^{\infty} \frac{f(z)}{z - z_0} dz = \lim_{\epsilon \to 0} \left[ \int_{-\infty}^{-\epsilon} \frac{f(z)}{z - z_0} dz + \int_{\epsilon}^{\infty} \frac{f(z)}{z - z_0} dz \right],$$

we may write the integral appearing in Eq. (7.47) as

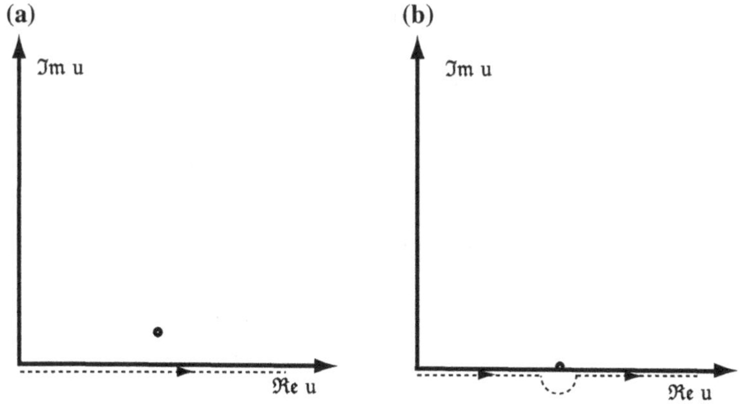


**Fig. 7.4** The integration path according to the Landau prescription. **a** Original path, **b** Modified path

$$\int_{-\infty}^{\infty} \frac{(\mathrm{d}F_0/\mathrm{d}u)}{p+\mathrm{i}ku}\mathrm{d}u = \frac{1}{\mathrm{i}k}\int_{-\infty}^{\infty} \frac{F_0'}{u-\mathrm{i}\,p/k}\mathrm{d}u$$

$$= \frac{1}{\mathrm{i}k}\left[P\int_{-\infty}^{\infty} \frac{F_0'}{u-\mathrm{i}\,p/k}\mathrm{d}u + \mathrm{i}\pi\, F_0'(\mathrm{i}\,p/k)\right]$$

$$= P\int_{-\infty}^{\infty} \frac{F_0'}{p+\mathrm{i}ku}\mathrm{d}u + \frac{\pi}{k}F_0'(\mathrm{i}\,p/k). \tag{7.52}$$

The dispersion relation will be given by the solution of the equation:

$$D(k,p) = 1 - \mathrm{i}\frac{\omega_{pe}^2}{k}\left[P\int_{-\infty}^{\infty} \frac{F_0'}{p+\mathrm{i}ku}\mathrm{d}u + \frac{\pi}{k}F_0'(\mathrm{i}\,p/k)\right]$$

$$= 1 + \omega_{pe}^2\left[P\int_{-\infty}^{\infty} \frac{F_0}{(p+\mathrm{i}ku)^2}\mathrm{d}u - \mathrm{i}\frac{\pi}{k^2}F_0'(\mathrm{i}\,p/k)\right] = 0, \tag{7.53}$$

where an integration by parts has been carried out. For small $k$-values we take a series expansion of $(p+\mathrm{i}ku)^{-2}$ to get

$$1 + \frac{\omega_{pe}^2}{p^2}\int_{-\infty}^{\infty} F_0\left(1 - \frac{2\mathrm{i}ku}{p} - \frac{3k^2u^2}{p^2} + \cdots\right)\mathrm{d}u - \mathrm{i}\frac{\pi\omega_{pe}^2}{k^2}F_0'(\mathrm{i}\,p/k) = 0,$$

Note that the imaginary terms in the integrand are odd functions of $u$ and therefore the corresponding integrals vanish, since $F_0$ is an even function of the same variable. The preceding equation therefore reduces to:

$$-p^2 \simeq \omega_{pe}^2\left(1 - 3\frac{k^2c_s^2}{p^2}\right) - \mathrm{i}\frac{\pi\omega_{pe}^2}{k^2}p^2\,F_0'(\mathrm{i}\,p/k), \tag{7.54}$$

where use has been made of the integrals:

$$\int_{-\infty}^{\infty} F_0\,\mathrm{d}u = 1, \qquad \int_{-\infty}^{\infty} u^2\,F_0\,\mathrm{d}u = \frac{k_B T}{m} = c_s^2.$$

In the limit $k \to 0$, the dispersion relation reduces to

$$-p^2 = \omega^2 - \gamma^2 + 2\mathrm{i}\gamma\omega = \omega_{pe}^2,$$

since $F_0$ and all its derivatives converge so fast when the argument tends to infinity that the last term of Eq. (7.54) tends to zero. In this limit, $\gamma = 0$ and as a consequence the wave's electric field, given by Eq. (7.51), is constant in time. The wave frequency coincides with that of a Langmuir wave, $\omega = \pm\omega_{pe}$.

Waves whose amplitude is time-dependent appear when the corrections connected with finite values of $k$ are taken into account. At the lowest order, we may keep the term in $k^2$ at the rhs of Eq. (7.54), but replace $p^2$ with $-\omega_{pe}^2$, namely with the solution of zeroth-order in $k^2$. From the real part of Eq. (7.54) we get:

$$\omega^2 = \omega_{pe}^2 + 3k^2 c_s^2, \tag{7.55}$$

which is the dispersion relation of Langmuir waves corrected for thermal effects [see Eq. (7.38)].

Finally, the imaginary part of Eq. (7.54) gives us:

$$\gamma = \frac{\pi}{2}\frac{\omega_{pe}^3}{k^2}\, F_0'(\omega_{pe}/k). \tag{7.56}$$

Equation (7.56) is one of the most fundamental results of the theory of collisionless plasmas. Before illustrating the physical mechanism underlying the above result, it is important to notice that the sign of $\gamma$ depends upon the sign of the first derivative of the equilibrium distribution function, taken at the position where *the speed of the particles equals the phase velocity of the wave*. The damping is thus related to the *resonant interaction* of a particular group of particles with the electrostatic wave. It is therefore not surprising that this phenomenon does not show up in a fluid description, where the "anomalous" behaviour of a group of peculiar particles is cancelled by the process of averaging over the velocities that lies at the basis of those models. Since the (maxwellian) equilibrium function is a decreasing function of the velocity for $u > 0$, $\gamma$ will have negative values and the wave will be subject to the so-called *Landau damping*.

It could be objected that $F_0(u)$ has a positive derivative for $u < 0$, which might suggest that an amplification of the wave amplitude is also a possibility. However, when $u < 0$ the resonant particles have $k < 0$, contrary to what we have so far assumed. We must then repeat the whole procedure for this new case and it's easy to realize that now the pole of $D(k, p)$ is located in half-plane $\mathfrak{Im}\, u < 0$ and that when $\gamma \to 0$ the integration path must be modified to maintain the pole *under* the path. The half circle around the pole is now travelled in a *clockwise* sense and this implies a sign change in the imaginary terms of Eqs. (7.53) and (7.54). Equation (7.56) now reads:

$$\gamma = -\frac{\pi}{2}\frac{\omega_{pe}^3}{k^2}\, F_0'(-\omega_{pe}/|k|),$$

and the sign of $\gamma$ is still negative. We thus conclude that if the particles obey a maxwellian distribution function the electrostatic waves are always damped. The explicit form of Eq. (7.56) for a maxwellian distribution function can be written in the following form:

$$\gamma = -\sqrt{\frac{\pi}{8}} \frac{\omega_{pe}}{|k\lambda_D|^3} \exp\left[ -\frac{1}{2(k\,\lambda_D)^2} - \frac{3}{2} \right], \tag{7.57}$$

where use has been made of

$$\omega_{pe}\lambda_D = c_s,$$

and the term proportional to $k^2$ in Eq. (7.55) has been retained, because of the strong dependence of $\gamma$ on $(k\lambda_D)$.

The damping appears to be very modest for small values of $(k\lambda_D)$, namely for wavelengths large with respect to $\lambda_D$, but it increases rapidly with the decrease of $\lambda$, reaching a maximum for $k\lambda_D \simeq 0.6$. Notice that Eq. (7.57) has been obtained by a series expansion in powers of the parameter $k\lambda_D \ll 1$ which loses its validity close to one. If Eq. (7.53) is exactly solved (by a numerical procedure), the maximum disappears and $\gamma$ is found to be a monotonically increasing function of $k\lambda_D$. The conclusion is that when the latter quantity is of the order or larger than unity, the waves are so strongly damped that in practice they do not exist.

The preceding discussion sheds some light on the physical process underlying Landau damping. Imagine observing the phenomenon from a reference system moving at the wave's phase speed. In that system both the electric field and its associated electrostatic potential would appear time-independent. A particle whose total energy is less than the maximum of the electrostatic potential energy of the wave will be reflected at the points marked by $A$ and $B$ in Fig. 7.5 and will therefore be trapped between two successive maxima of the potential.

In our system of reference, a particle moving at a speed slightly lower than the wave's phase speed will be seen to slowly drift towards the left to be eventually

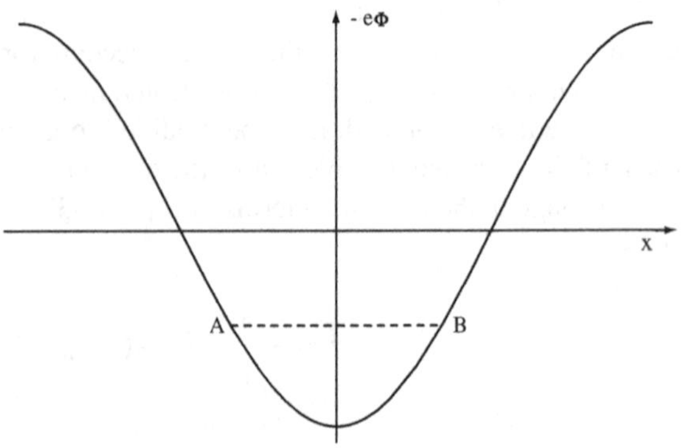


**Fig. 7.5** Potential energy, $-e\Phi$, as a function of $x$ in the wave's reference system

reflected at the point $A$, where it will acquire energy from the wave. Conversely, a particle moving at a speed slightly higher than the wave phase speed will drift to the right and will reflect at the point $B$, transferring part of its energy to the wave. If the distribution function is a *decreasing* function of the velocity close to the phase speed of the wave, the number of particles with speeds $u < \omega/k$ will be larger then that of the particles having $u > \omega/k$. The absorption of energy from the wave will thus be larger than the transmission of energy to the wave and the final result will be a damping of the wave.

If the distribution function presents a secondary maximum, for instance for $k > 0$, and if the phase speed of the wave falls in the region where $F_0'(u) > 0$, $\gamma$ becomes positive and the wave amplitude grows feeding on the energy of the particles. Such a situation may arise when in a thermal plasma (maxwellian distribution) a beam of particles with speeds substantially higher than the thermal speed of the plasma is also present. As already remarked, an exponential growth of the wave amplitude forces the system out of the linear regime, thus invalidating our treatment. However, it easy to guess what will be the state of the system in the non linear phase. The growth of the wave amplitude will slow down the particles: particles with $u > \omega/k$ will be moved to regions where $u < \omega/k$, and this will eventually "cut" the secondary bump of the distribution function, resulting in a saturation of the wave amplitude.

Finally, notice that our physical explanation of the linear phase of Landau damping only applies for times shorter than the "bouncing time" of the particle in the wave's potential well. In fact, a particle reflected at the point $A$ will eventually reach point $B$ where it will lose energy to the wave, thus partially compensating the damping. The bouncing frequency $\omega_b$ can be evaluated to a first approximation as follows.

In the wave reference system, let the electric field be given by $E = E_0 \sin(kx)$. The corresponding potential will then be $\Phi = (E_0/k)\cos(kx) = \Phi_0 \cos(kx)$ and the potential energy of a test electron will be $V = -e\Phi = -(eE_0/k)\cos(kx)$, as shown in Fig. 7.5.

The equation of motion of the test electron, with energy slightly higher than the minimum of $V$, will then be:

$$m\ddot{x} = -eE = -eE_0 \sin(kx) \simeq -eE_0 k\, x,$$

from which it follows that the particles trapped close to the region of minimum potential energy will harmonically oscillate with a frequency

$$\omega_b = \sqrt{eE_0 k/m_e} = k\sqrt{e\Phi_0/m_e}.$$

The ratio between the bouncing time, $\tau_b$ and the wave period $\tau_{pe}$ for $\omega \simeq \omega_{pe}$, is therefore

$$\frac{\tau_b}{\tau_{pe}} \simeq \frac{\omega_{pe}}{kc_s}\sqrt{\frac{k_B T}{e\Phi_0}} \gg 1,$$

where we have taken into account both of Eq. (7.55) and of the basic condition for plasmas $e\Phi_0/kT \ll 1$.

We thus see that the linear phase of the Landau damping might last for many wave periods.

## Questions and Problems

**7.1.** Show that in a homogeneous plasma an incompressible perturbation with $\boldsymbol{B} = \boldsymbol{B}_0 + \boldsymbol{B}_1$, with an arbitrary $\boldsymbol{B}_1$, satisfying the conditions given by Eq. (7.26) is an exact solution of the full MHD equations. Show also that the solution corresponds to a transverse wave that propagates with speed $c_a$.

**7.2.** Calculate the polarization of slow and fast magnetosonic waves, $\boldsymbol{\xi}_{s,f}$, propagating at an arbitrary angle $\theta$ with respect to the magnetic field and show that $\boldsymbol{\xi}_f \cdot \boldsymbol{\xi}_s = 0$ always holds. The eigenmodes of the slow, fast and Alfvén waves are therefore mutually orthogonal.

**7.3.** Calculate the limit of the group velocity of the slow mode wave when the angle of propagation is close to 90°. First, expand the expression for the square of the frequency when $\theta \simeq \pi/2 - \epsilon$ with small $\epsilon$, then calculate the components of the group velocity, and then take the limit $\epsilon -> 0$.

## Solutions

**7.1.** Let $\boldsymbol{B}_0 = B_0\boldsymbol{e}_z$, $c_a = B_0/\sqrt{4\pi\rho_0}$ and $\boldsymbol{b} = \boldsymbol{B}_1/\sqrt{4\pi\rho_0}$. The momentum and induction equations are:

$$\frac{\partial \boldsymbol{U}}{\partial t} - c_a\frac{\partial \boldsymbol{b}}{\partial z} = (\boldsymbol{b}\cdot\nabla)\boldsymbol{b} - (\boldsymbol{U}\cdot\nabla)\boldsymbol{U},$$

$$\frac{\partial \boldsymbol{b}}{\partial t} - c_a\frac{\partial \boldsymbol{U}}{\partial z} = (\boldsymbol{b}\cdot\nabla)\boldsymbol{U} - (\boldsymbol{U}\cdot\nabla)\boldsymbol{b},$$

where the following conditions have been used: $\nabla\cdot\boldsymbol{U} = 0$, $P_0 = const.$ and $|\boldsymbol{B}_0 + \boldsymbol{B}_1| = const.$

If the characteristic relations for the Alfvén waves hold, namely $\boldsymbol{b} = \pm\boldsymbol{U}$, the rhs of the preceding equations vanish identically. Deriving the first equation with respect to $t$ and making use of the second we get

$$\frac{\partial^2 \boldsymbol{U}}{\partial t^2} - c_a^2\frac{\partial^2 \boldsymbol{U}}{\partial z^2} = 0.$$

**7.2.** The polarizations of the slow and fast waves are obtained solving the equation for the eigenvectors obtained by inserting the value of the frequency, Eq. (7.32), into Eqs. (7.29) and (7.30). If $y$ is the coordinate orthogonal to $\boldsymbol{B}$ in the $\boldsymbol{k}$, $\boldsymbol{B}$ plane, we have

$$\left(\frac{\xi_y}{\xi_z}\right)_{s,f} = \frac{\omega_{s,f}^2 - c_s^2 k^2 \cos^2\theta}{c_s^2 k^2 \sin\theta\cos\theta} \tag{7.58}$$

The quantity $\boldsymbol{\xi}_f \cdot \boldsymbol{\xi}_s$ can be written as

$$\boldsymbol{\xi}_f \cdot \boldsymbol{\xi}_s = (\xi_z)_s(\xi_z)_f\left[\left(\frac{\xi_y}{\xi_z}\right)_s\left(\frac{\xi_y}{\xi_z}\right)_f + 1\right].$$

Using the previously found expressions for $\left(\frac{\xi_y}{\xi_z}\right)_{s,f}$ and the properties of the solutions of Eq. (7.31) we find the required result.

**7.3.** Expanding Eq. 7.32 for $\theta = \pi/2 - \epsilon$ and taking the minus sign for the slow wave leads immediately to

$$\omega_s^2/k^2 \simeq \frac{1}{2}\left[(c_s^2 + c_a^2)\left(1 - \sqrt{1 - 4\frac{\epsilon^2 c_s^2 c_a^2}{(c_s^2 + c_a^2)^2}}\right)\right].$$

Expanding the square root, we find that

$$\omega_s^2/k^2 \simeq \epsilon^2 \frac{c_s^2 c_a^2}{(c_s^2 + c_a^2)},$$

however $k^2\epsilon^2$ is nothing but the square of the parallel wave-vector $k_x^2$. So the group velocity in the $y$ direction vanishes, while in the $x$-direction taking the derivative with respect to $k_x$ yields

$$\frac{\partial\omega}{\partial k_x} \simeq \frac{k_x c_s^2 c_a^2}{\omega(c_s^2 + c_a^2)} \simeq \frac{c_s c_a}{\sqrt{(c_s^2 + c_a^2)}}$$

which is the desired result.

# Chapter 8
# Shocks

**Abstract** Large amplitude waves, or shocks, are the subject of this chapter. Shocks in neutral gases are briefly reviewed. MHD shocks are then treated in detail while a brief discussion of collisionless shocks is presented in the final section.

So far we have restricted our attention to small amplitude waves. However, situations may arise where the equilibrium state undergoes substantial changes that can no longer be considered as small perturbations. The whole procedure of linearization therefore becomes meaningless and a new approach is required. The most important qualitative change when the amplitude of perturbations, or "waves", in a fluid becomes large is the appearance of spatial discontinuities in the physical quantities characterizing the fluid. These discontinuities are usually referred to as *shock waves* or simply *shocks*.

Shocks are generated both in plasmas and neutral gases, where they are produced by supersonic motions. This can be understood by considering the effect of the motion of a body through a gas, that we shall assume to be a perfect gas. The body generates a disturbance that propagates in the gas at the sound speed $c_s = (\gamma P/\rho)^{1/2}$. If the velocity of the object is less than $c_s$ (*subsonic* motion), the perturbation, namely the compressive sound wave, will always move ahead of the body which caused it. However, if the motion of the object is *supersonic* the perturbation will lag behind and the unperturbed gas will not be "alerted" of the arrival of the body by the sound wave, as was the case for subsonic motions. The physical parameters of the gas will then be subject to abrupt variations, whose amplitude may became extremely large, thus forbidding a treatment in terms of the small perturbations of the linear theory. The region where such variations take place is called *shock front*. The structure of a shock front is determined by a delicate balance among steepening, arising from the nonlinear terms (convective derivative in a fluid, and convective derivatives and field aligned gradients in MHD), and the smoothing effect associated with viscosities and resistivities, which depend on collisions or other irreversible processes which allow heating of the gas across the shock.

As remarked several times, the dissipative effects connected to thermal conductivity, viscosity (and resistivity) are generally negligible in gases and plasmas. However, they may become important in regions of strong gradients, as can be seen by



C. Chiuderi and M. Velli, *Basics of Plasma Astrophysics*,
UNITEXT for Physics, DOI 10.1007/978-88-470-5280-2_8

examining the mathematical structure of the dissipative terms in the fluid equations. These are given by the product of a coefficient (usually depending on the thermodynamical parameters) times the *gradient* of a fluid variable. In the case of thermal effects we have the coefficient of thermal conductivity multiplied by the temperature gradient, viscous effects are described by the viscosity coefficient multiplied by the velocity gradient, and so on.

It is therefore clear that if the gradients of the fluid variables become large, the importance of the dissipative terms will accordingly increase, at least locally. On the other hand, the formation of strong gradients is a typical nonlinear effect, coming into play when the amplitude of perturbations increases. This statement can be easily verified by considering the acceleration of a fluid particle, appearing on the lhs of the one-dimensional equation of motion:

$$\frac{dU}{dt} = \frac{\partial U}{\partial t} + U\frac{\partial U}{\partial x}.$$

The second term is manifestly nonlinear and, to illustrate its effects, assume for example to have a sinusoidal velocity perturbation so that $U \propto \sin(kx)$, and $dU/dx \propto \cos(kx)$. Thus:

$$U\frac{\partial U}{\partial x} \propto \sin(kx)\cos(kx) \propto \sin(2kx)$$

and this shows that the nonlinear term generates quantities that vary on a spatial scale $1/2k$, smaller than the scale of the gradients of $U$, namely $1/k$. If $U$ is given by a Fourier expansion containing a finite number of terms of non negligible amplitude, we easily realize that the product of that series by the corresponding series of derivatives will eventually create a disturbance with increasingly strong spatial gradients until their further growth is stopped by the counteracting effect of the dissipative terms. The balance, in the case of small dissipative coefficients, requires large values of the gradients across the shock front, which in turn demands a small shock thickness. Since dissipative coefficients are generally very small in gases and plasmas, the shocks can be practically considered as discontinuities.

This qualitative introductory discussion suggests that shocks may be described as a transition zone (of vanishingly small thickness) between a region of undisturbed flow, that will be identified by the index 1, and a region that has been subject to the action of the shock, identified by the index 2. These regions are often indicated as the zone *ahead of* and *behind* the shock or *upstream* and *downstream* of the shock. It is usually assumed that both regions are *homogeneous*, which is not true in practice, since the region behind the shock relaxes in time to a state different from that prevailing immediately after the passage of the shock. The homogeneity assumption makes sense only in the framework of a *local* analysis, namely for times shorter than the typical relaxation times and for distances small with respect to the macroscopic scale of the system.

The equations describing a discontinuity are simply a series of relationships that connect the values of the fluid (and electromagnetic) variables in the two homogeneous regions upstream and downstream of the shock, where it is reasonable to assume that the ideal fluid (or MHD) equations are valid. These relationships are known as the *Rankine-Hugoniot relations* or *jump conditions*. In the spirit of the local analysis we just mentioned, we may consider the situation as stationary and adopt a frame of reference attached to the shock front. In this system, the fluid enters the shock with a speed $\boldsymbol{U}_1$ and leaves with a speed $\boldsymbol{U}_2$. In the system that better corresponds to the physical situation, the so-called "laboratory system", the plasma velocity in Region 1 is zero and the shock front moves into that region with a velocity $\boldsymbol{U}_{shock} = -\boldsymbol{U}_1$, while in Region 2 the plasma moves with the speed $\boldsymbol{U} + \boldsymbol{U}_{shock} = \boldsymbol{U}_2 - \boldsymbol{U}_1$.

The treatment of shocks in plasmas is considerably more complex than in neutral gases, because of the larger number of effective degrees of freedom of a plasma. In fact, let's consider the motion of a piston in a neutral gas producing a compressive wave in the direction of motion. If transverse motions are also present, they do not alter the situation and the system has therefore only one effective degree of freedom. But if we consider the motion of a conducting piston in an ideal magnetized plasma, any transverse motion of the piston would drag the magnetic field, generating a magnetic component of the perturbation. A magnetized plasma described by MHD has three effective degrees of freedom, one associated with each of the propagating modes. The situation is reminiscent of that encountered when studying the linear MHD waves, where we had an incompressible transverse mode (the Alfvén waves) and two "mixed" modes (the magnetosonic waves) whose nature is intermediate between a sound wave and a magnetic wave. In the nonlinear case as well, we have three waves with different propagation speeds, generally called *fast*, *intermediate* and *slow*. The waves propagating at the intermediate speed are purely transverse and do not produce shocks, while the other two exhibit both longitudinal and transverse components and give rise to shocks.

Before discussing the formation of discontinuities within magnetohydrodynamics, it is important to mention that the mechanisms leading to the formation of such discontinuities, typically nonlinear steepening, may also lead to the formation of gradients that are too strong for the fluid description of the plasma to remain valid. In fluid shocks the thickness of the discontinuity, as we shall see, is typically set by the collisional particle mean free path, or taking into account magnetic fields, the particle gyration radius. If however the scale of gradients turns out to be smaller than such length-scales, then phenomena other than those described by MHD, for example wave-particle interactions, must mediate the dissipation of energy required across the discontinuity. In fact, in most space and astrophysical plasmas, the shocks that form are collisionless, i.e. the shock structure is not well described by MHD. A detailed discussion of the physics of collisionless shocks is beyond the scope of this work: we will however return to them briefly at the end of the chapter.

## 8.1 The Jump Conditions

Consider a plane shock front of vanishing thickness (i.e. a discontinuity occurring along only one direction with all quantities homogeneous in planes orthogonal to this direction) moving in a direction normal to the the front itself and identify such a direction with the unit vector $\boldsymbol{e}_x$. The planes $x = const.$ will thus be parallel to the shock front. With the assumed symmetry, all quantities will depend on the variable $x$ only.

Although there may be discontinuities in some of the fluid quantities across the front, the dynamics must not violate the fundamental conservation laws of mass momentum and energy, which maintain their validity. To derive relationships between the quantities downstream and upstream of the shock we therefore take the MHD equations in conservative form and integrate them in a right cylindrical volume $V$ containing the shock front (the classic pill box) defined as follows. The bases of the volume, $A_1$ and $A_2$ straddle the shock, while the side, of height, $h$, is parallel to the shock normal direction and will eventually tend to zero. From the continuity equation, we get:

$$\int \frac{\partial \rho}{\partial t} \mathrm{d}V + \int_V \nabla \cdot (\rho\, \boldsymbol{U})\, \mathrm{d}V = \int \frac{\partial \rho}{\partial t} \mathrm{d}V + \int_S (\rho\, \boldsymbol{U}) \cdot \boldsymbol{n}_s\, \mathrm{d}S = 0,$$

where $\boldsymbol{n}_s$ is the unit vector normal to the surface $S$ pointing outwards from the volume itself. In our reference frame therefore $\boldsymbol{n}_s = \pm \boldsymbol{e}_x$. The volume integral tends to zero when $h \to 0$ (and $V \to 0$ as well). The surface integral splits into the two integrals on the base surfaces, because the contribution from the sides vanishes as the cylinder shrinks to zero,

$$\int_{A_1} (\rho \boldsymbol{U}) \cdot \boldsymbol{e}_x\, \mathrm{d}S - \int_{A_2} (\rho \boldsymbol{U}) \cdot \boldsymbol{e}_x\, \mathrm{d}S = 0.$$

Since this relation has to remain valid for any area of the base surfaces, in the limit $h \to 0$ we obtain:

$$(\rho U_x)_1 = (\rho U_x)_2,$$

where the indices 1 and 2 indicate that the quantities in brackets are evaluated at the upstream and downstream bases, respectively. In the theory of shocks, it is customary to introduce the notation:

$$\left[ G \right] \equiv G_2 - G_1,$$

to denote the jump in any quantity $G$ (scalar or vector). The first jump condition, deduced from the continuity equation, thus reads:

$$\left[\rho\, U_x\right] = \left[\rho\, \boldsymbol{U} \cdot \boldsymbol{e}_x\right] = 0.$$

More generally, without making reference to any particular coordinate system, if $\boldsymbol{n}$ is the unit vector normal to the shock front, we write

$$\left[\rho(\boldsymbol{U} \cdot \boldsymbol{n})\right] = 0. \tag{8.1}$$

Notice that this jump condition can be obtained directly from the (stationary) continuity equation by applying the formal substitution $(\boldsymbol{\nabla} \cdot) \to (\boldsymbol{n} \cdot)$ and setting the jump of the quantity obtained in this way to zero. Following the same procedure, the jump conditions deriving from the momentum and energy equations (5.22 and 5.23) may be found and read:

$$\left[\rho\, \boldsymbol{U}(\boldsymbol{U} \cdot \boldsymbol{n}) + \left(P + \frac{B^2}{8\pi}\right)\boldsymbol{n} - \frac{\boldsymbol{B}}{4\pi}(\boldsymbol{B} \cdot \boldsymbol{n})\right] = 0, \tag{8.2}$$

and

$$\left[\left(\tfrac{1}{2}\rho U^2 + \frac{\gamma P}{\gamma - 1}\right)(\boldsymbol{U} \cdot \boldsymbol{n}) + \frac{c}{4\pi}(\boldsymbol{E} \times \boldsymbol{B}) \cdot \boldsymbol{n}\right] = 0 \tag{8.3}$$

If we now adopt the following representation for the magnetic field

$$\boldsymbol{B} = \boldsymbol{B}_n + \boldsymbol{B}_t = B_n \boldsymbol{n} + \boldsymbol{B}_t,$$

thus separating the component normal to the plane containing the discontinuity from the "tangential" component lying on that plane, and analogous representations for the other vectors, Maxwell's equations $\boldsymbol{\nabla} \cdot \boldsymbol{B} = 0$ and $\boldsymbol{\nabla} \times \boldsymbol{E} = -\partial \boldsymbol{B}/\partial t = 0$ (with the replacement $\boldsymbol{\nabla} \to \boldsymbol{n}$) provide the conditions

$$\left[B_n\right] = 0,$$

and

$$\left[\boldsymbol{E}_t\right] = 0,$$

namely the conservation relations already known from electrodynamics. Using the equation $\boldsymbol{E} = -(1/c)(\boldsymbol{U} \times \boldsymbol{B})$ to eliminate the electric field, the jump conditions can be cast in the following form:

$$\left[\rho U_n\right] = 0, \tag{8.4}$$

$$\left[\rho U_n^2 + P + \frac{B_t^2}{8\pi}\right] = 0, \tag{8.5}$$

$$\left[\rho U_n \boldsymbol{U}_t - \frac{B_n}{4\pi}\boldsymbol{B}_t\right] = 0, \tag{8.6}$$

$$\left[\left(\tfrac{1}{2}\rho U^2 + \frac{\gamma P}{\gamma - 1} + \frac{B_n^2 + B_t^2}{4\pi}\right)U_n - \frac{B_n}{4\pi}(\boldsymbol{U}\cdot\boldsymbol{B})\right] = 0 \tag{8.7}$$

$$\left[B_n\right] = 0, \tag{8.8}$$

$$\left[\boldsymbol{E}_t\right] = 0, \tag{8.9}$$

$$\left[\boldsymbol{U}_n \times \boldsymbol{B}_t + \boldsymbol{U}_t \times \boldsymbol{B}_n\right] = 0, \tag{8.10}$$

The presence of a discontinuity in the flow does not necessarily imply that a shock wave is present. Discontinuities may be classified on the basis of the jump conditions described above, and their nature depends on whether $U_n = 0$ or $U_n \neq 0$. In the first case, if there is no jump in density, $[\rho] = 0$, no discontinuity is present, while if $[\rho] \neq 0$ and there is a jump in density, but no flow across it, we have a so-called *contact discontinuity*. In the second case in which there is flow upstream and ($U_n \neq 0$), if there is no jump in density $[\rho] = 0$ we are in the presence of a *rotational discontinuity*. Only if there is a flow upstream, $U_n \neq 0$, and a finite jump in density across the discontinuity, $[\rho] \neq 0$, the discontinuity is classified as a *shock*.

### *8.1.1 Contact Discontinuities*

In a contact discontinuity ($U_n = 0$) no matter flows across the discontinuity, but a density jump is present. If $B_n \neq 0$, Eq. (8.10) implies that $[\boldsymbol{U}_t] = 0$, while from Eq. (8.6) it follows that $[\boldsymbol{B}_t] = 0$. Finally, Eq. (8.5) shows that $[P] = 0$. As a consequence, the pressure and all components of $\boldsymbol{B}$ and $\boldsymbol{U}$ are continuous across the discontinuity and the density is the only quantity with a discontinuity. From the equation of state it follows that the temperature must also abruptly change across the discontinuity. This discontinuity is therefore nothing but the passage between two plasma states in pressure equilibrium but different entropies (a hotter, lower density plasma in contact with a cooler, higher density plasma).

If $B_n = 0$ the jump conditions show that it is also possible to have $[\boldsymbol{U}_t] \neq 0$, $[B_t] \neq 0$ and

$$\left[P + \frac{B_t^2}{8\pi}\right] = 0.$$

In this circumstance we speak of a *tangential discontinuity*. In this type of flow, the velocity and the magnetic field are parallel to the surface of discontinuity, but their value and/or direction change across that surface, while the total pressure remains unaltered. Generally a current flows along the interface between the two plasma states, and if the tangential velocity doesn't vanish and has a finite discontinuity, not only is the interface a current sheet, but also a vorticity sheet. The entropy may or may not be different, depending on whether the density jumps across the discontinuity as well.

An astrophysical example of a tangential discontinuity is provided by the *magnetopause*, i.e. the surface separating the terrestrial magnetic field, the *magnetosphere*, from the solar wind. In the absence of processes violating the Alfvén theorem, namely of phenomena of magnetic reconnection (which will be discussed in the next Chapter), the normal components of the velocity and of the magnetic field turn out to be negligible ($U_n \simeq 0$ and $B_n \simeq 0$) and the magnetosphere is said to be "closed", because the solar wind and the magnetic field associated with it cannot penetrate inside. In this situation the magnetopause behaves practically as a tangential discontinuity and has the same pressure in the Sunward and Earthward directions. Similar configurations have been observed also in the magnetospheres of other planets.

### 8.1.2 Rotational Discontinuities

Rotational discontinuities are characterized by a non-vanishing normal velocity and no density jump across the discontinuity: $U_n \neq 0$, but $\left[\rho\right] = 0$. Eqs. (8.4), (8.5) and (8.6) now imply that

$$\left[U_n\right] = 0, \tag{8.11}$$

$$\left[P + \frac{B_t^2}{8\pi}\right] = 0, \tag{8.12}$$

$$\left[\boldsymbol{U}_t - \boldsymbol{Q}\right] = 0, \tag{8.13}$$

where we have introduced the vector

$$\boldsymbol{Q} = \frac{B_n}{4\pi\rho\, U_n}\boldsymbol{B}_t.$$

In terms of $\boldsymbol{Q}$ Eq. (8.7) may now be written as:

$$\left[\tfrac{1}{2}U_t^2 + \frac{\gamma}{\gamma - 1}\frac{P}{\rho} + \frac{B_t^2}{4\pi\rho} - \boldsymbol{Q}\cdot\boldsymbol{U}_t\right] = 0. \tag{8.14}$$

The absence of a jump for a generic vector $\boldsymbol{A}$, $\left[\boldsymbol{A}\right] = 0$, implies also that $\left[(\boldsymbol{A})^2\right] = 0$. Applying this to Eq. (8.13), it follows that :

$$0 = \left[(\boldsymbol{U}_t - \boldsymbol{Q})^2\right] = \left[U_t^2 + Q^2 - 2\boldsymbol{Q}\cdot\boldsymbol{U}_t\right].$$

Now eliminate $\boldsymbol{Q} \cdot \boldsymbol{U}_t$ between the preceding equation and Eq. (8.14) to obtain:

$$\left[\frac{\gamma}{\gamma - 1}\frac{P}{\rho} + \frac{B_t^2}{4\pi\rho}\left(1 - \frac{1}{2}\frac{B_n^2/4\pi\rho}{U_n^2}\right)\right] = 0. \tag{8.15}$$

From the continuity of both $B_n$ and $U_n$ Eq. (8.10) may be rewritten as:

$$U_n\boldsymbol{n} \times \left[\boldsymbol{B}_t\right] = B_n\boldsymbol{n} \times \left[\boldsymbol{U}_t\right] = B_n\boldsymbol{n} \times \left[\boldsymbol{Q}\right] = \frac{B_n^2}{4\pi\rho\, U_n}\boldsymbol{n} \times \left[\boldsymbol{B}_t\right],$$

or

$$\left(\boldsymbol{n} \times \left[\boldsymbol{B}_t\right]\right)\left(U_n - \frac{B_n^2}{4\pi\rho\, U_n}\right) = \frac{\boldsymbol{n} \times \left[\boldsymbol{B}_t\right]}{U_n}(U_n^2 - B_n^2/4\pi\rho) = 0.$$

Since $U_n \neq 0$ and $\boldsymbol{n} \times \left[B_t\right] \neq 0$ it follows that

$$U_n^2 = \frac{B_n^2}{4\pi\rho} \equiv c_{an}^2, \tag{8.16}$$

where $c_{an}^2 = B_n^2/(4\pi\rho)$ is the Alfvén speed based on the normal magnetic field. Therefore the rotational discontinuity advances with the Alfvén speed corresponding to the magnetic field along the normal direction. Introducing Eq. (8.16) into Eq. (8.15) we get

$$\left[\frac{\gamma}{\gamma - 1}\frac{P}{\rho} + \frac{B_t^2}{8\pi\rho}\right] = 0.$$

By comparing the preceding expression with Eq. (8.12), we conclude that they are compatible only if

$$\left[P\right] = 0 \quad \text{and} \quad \left[B_t^2\right] = 0.$$

This type of discontinuity is characterized by the fact that no density or pressure variations are associated with the discontinuity, the transverse component of the magnetic field does change, $\left[\boldsymbol{B}_t\right] \neq 0$, but its magnitude remains unaltered, $\left[B_t^2\right] = 0$. Thus, both the magnetic field and the velocity rotate when the discontinuity is crossed, thus justifying the name given to it. The discontinuity itself moves with respect to the unperturbed fluid at the speed $U_n = \pm c_{an}$, and depending on the sign the correlation between the tangential velocity and magnetic field discontinuities changes accordingly. We therefore recover the result of Chap. 7 (Exercise 7.1) concerning large amplitude, circularly polarized, Alfvén waves. For small amplitudes, $\left[U_t\right]$ and $\left[B_t\right]$ can be identified with the perturbations $U_1$ and $B_1$ of the linear theory and the rotational discontinuity simply becomes an incompressible Alfvén wave. In fact the rotational discontinuity may be viewed as nothing but a propagating, large amplitude Alfvén wave steepened to a discontinuous wave front.

Discontinuities of this type are observed when the magnetic field associated with the solar wind has a direction opposite to the terrestrial one. In this case, magnetic reconnection becomes important, with the effect of generating non negligible components of the velocity and of the magnetic field along the normal to the surface of discontinuity ($U_n \neq 0$, $B_n \neq 0$). The wind and the field that goes along with it penetrate the magnetosphere, which is then said to be "open" and the magnetopause becomes a rotational discontinuity.

## 8.2 MHD Shocks

According to the classification introduced above, MHD shocks correspond to jumps in both density and the velocity normal to the shock front, $\left[U_n\right] \neq 0$ and $\left[\rho\right] \neq 0$. In this case the jump conditions can not be easily simplified, but the treatment can be made more tractable by a clever choice of an appropriate reference frame, as we shall show in the following. To begin, let's examine the two simple cases of:

- *Perpendicular shocks*, characterized by the vanishing of the magnetic field normal to the plane of the shock, $B_{1n} = 0$. As we shall see, in this case it is possible to choose a reference frame where $U_{1t} = 0$ and consequently $\boldsymbol{U}_1 \cdot \boldsymbol{B}_1 = 0$, which justifies the name given to these discontinuities.
- *Parallel shocks*, for which both $\boldsymbol{U}_1$ and $\boldsymbol{B}_1$ are parallel to the shock normal $\boldsymbol{n}$.

The general case of an arbitrary angle between $\boldsymbol{U}_1$ and $\boldsymbol{B}_1$ is known as an *oblique shock*.

### 8.2.1 Perpendicular Shocks

So far we have only defined the axis normal to the shock plane but not a coordinate system within the plane itself. In the case of a perpendicular shock when the magnetic field lies completely in the plane, it is natural to choose one of the axes in the direction of $\boldsymbol{B}_1$. Identify $\boldsymbol{n}$ with the $x$-axis, $\boldsymbol{n} = \boldsymbol{e}_x$, and choose the $z$-axis in the direction of $\boldsymbol{B}_1$. In this frame, therefore, $\boldsymbol{B}_1 = \left[0, 0, B\right]$. Eqs. (8.4) and (8.6) then show that $\left[\boldsymbol{U}_t\right] = 0$, which means that the transverse speed remains unaltered when the discontinuity is crossed. We can therefore move to a new frame moving at a constant speed parallel to $\boldsymbol{U}_t$, where obviously the transverse speed will vanish, $\boldsymbol{U}_1 = \left[U, 0, 0\right]$. The jump conditions now read:

$$\left[\rho U\right] = 0. \tag{8.17}$$

$$\left[\rho U^2 + P + \frac{B^2}{8\pi}\right] = 0, \tag{8.18}$$

$$\left[\tfrac{1}{2}U^2 + \frac{\gamma}{\gamma-1}\frac{P}{\rho} + \frac{B^2}{4\pi\rho}\right] = \left[\tfrac{1}{2}U^2 + \frac{c_s^2}{\gamma-1} + c_a^2\right] = 0, \tag{8.19}$$

$$\left[\frac{B}{\rho}\right] = 0. \tag{8.20}$$

Le condition expressed by Eq. (8.20) simply reflects the "freezing in" condition of the magnetic field in the plasma [see Eq. (5.28)].

If we let $\boldsymbol{B} = 0$ in the above relations we recover the Rankine-Hugoniot relations for a hydrodynamic shock. In this case it is useful, making use of the jump conditions, to express the ratios of the values of the various physical parameters upstream and downstream of the shock in terms of the upstream Mach number,

$$M = \frac{U_1}{c_{s1}},$$

where $c_{s1}$ is the upstream sound speed. A long algebraic calculation gives:

$$\frac{\rho_2}{\rho_1} = \frac{U_1}{U_2} = \frac{(\gamma+1)M^2}{2+(\gamma-1)M^2}, \tag{8.21}$$

$$\frac{P_2}{P_1} = \frac{2\gamma M^2 - (\gamma-1)}{\gamma+1}. \tag{8.22}$$

It is also possible to show that, as a consequence of the Second Principle of Thermodynamics, which requires that $S_2 \geqslant S_1$ a hydrodynamic shock must always be compressive, $\rho_2 \geqslant \rho_1$, $P_2 \geqslant P_1$ and that $M \geqslant 1$. Also, $M_2 = U_2/c_{s2} \leqslant 1$. Thus, in the frame of the shock, the upstream fluid motion is supersonic, while downstream of the shock the gas motion is subsonic . From the preceding equations it follows that in the *strong shock* limit, $M \gg 1$, the pressure ratio $P_2/P_1 \propto M^2$ becomes arbitrarily large,[1] while the density ratio tends to the finite value $(\gamma+1)/(\gamma-1)$. Consequently, the ratio of temperatures can also assume very large values for strong shocks. The presence of very high temperatures behind the shock front can initiate ionization processes, so that a plasma can be generated as a consequence of the passage of a gas through a shock front, even if the gas ahead of the shock is neutral.

In a hydrodynamic shock part of the kinetic energy of the incoming flux is transformed into thermal energy. In fact, from Eq. (8.18) with $B = 0$ we get:

$$\frac{\gamma P_2}{\gamma-1}(1 - P_1/P_2) = \tfrac{1}{2}\rho_1 U_1^2 (1 - U_1/U_2),$$

[1] It is customary to use $P_2/P_1$ as a measure of the *shock strength*.

and finally, making use of Eqs. (8.21) and (8.22) (with $M \neq 1$):

$$\frac{\gamma P_2}{\gamma-1} = \tfrac{1}{2}\rho_1 U_1^2 \left(\frac{2}{\gamma+1} - \frac{\gamma-1}{\gamma(\gamma+1)}\frac{1}{M^2}\right).$$

The expression in brackets on the rhs is a decreasing function of $M^2$ that approaches $2/(\gamma+1)$ when $M \gg 1$. Thus for a strong shock,

$$P_2 \simeq \frac{2(\gamma-1)}{\gamma(\gamma+1)}\left(\tfrac{1}{2}\rho_1 U_1^2\right),$$

which shows that a strong shock transforms a fraction $\frac{2}{\gamma(\gamma+1)}$ of the initial kinetic energy into thermal energy. For $\gamma = 5/3$, 9/20, almost one half, of the kinetic energy is converted into heat.

When a magnetic field is present, the relations equivalent to Eqs. (8.17)–(8.20) become considerably more complicated. Introducing, in addition to $M$, the parameters:

$$X = \frac{\rho_2}{\rho_1},$$

and

$$\beta = \frac{P_1}{B_1^2/8\pi} = \frac{2}{\gamma}\left(\frac{c_{s1}}{c_{a1}}\right)^2,$$

a long calculation leads to the following expressions:

$$\frac{B_2}{B_1} = \frac{U_1}{U_2} = X_0, \tag{8.23}$$

$$\frac{P_2}{P_1} = 1 + \frac{\gamma M^2 (X_0 - 1)}{X_0} - \frac{X_0^2 - 1}{\beta}, \tag{8.24}$$

where $X_0$ is the positive solution of the equation for the density jump:

$$f(X) \equiv aX^2 + bX - c = 0,$$

with

$$a = 2(2-\gamma), \quad b = \gamma[2\beta + (\gamma-1)\beta M^2 + 2], \quad c = \gamma(\gamma+1)\beta M^2$$

Since $1 \leqslant \gamma \leqslant 2$, it is easy to verify that the only positive root of the preceding equation is

$$X_0 = \frac{1}{2a}\left[-b + \sqrt{b^2 + 4ac}\,\right].$$

Using the expressions for the coefficients, it is easily verified that when $\beta \gg 1$, $X_0$ can be written as:

$$X_0 \approx \frac{c}{b} - \frac{ac^2}{b^3} = \frac{(\gamma+1)M^2}{2+(\gamma-1)M^2} - \mathcal{O}(1/\beta).$$

The hydrodynamic case corresponds to the limit $\beta \to \infty$ and the preceding expression shows that the effect of a magnetic field is that of *reducing* the value of $X_0$ with respect to the hydrodynamic case. Since it is possible to show that in any case the shock is compressive, $X_0 \geqslant 1$, Eq. (8.24) shows that the shock strength is reduced by the presence of a magnetic field. This is a consequence of the fact that part of the kinetic energy of the incoming flux can be converted not only into heat, but also into magnetic energy.

The function $f(X)$ exhibits a single minimum for $X = -b/(2a) < 0$. Thus $f(X) < 0$ for $0 \leqslant X \leqslant X_0$ and, since $X_0 \geqslant 1$, it follows that $f(X = 1) \leqslant 0$. Explicitly evaluating this expression, we find:

$$M^2 \geqslant 1 + \frac{2}{\beta\gamma} = 1 + \frac{c_a^2}{c_s^2}.$$

Thus, $U_1^2 \geqslant c_s^2 + c_a^2$, namely the upwind flow, or the shock's speed in the lab system, must be greater than the propagation speed of a fast magnetosonic wave.

### *8.2.2 Parallel Shocks*

In a completely parallel shock, with $\boldsymbol{B}_1 = B_{1n}\boldsymbol{n}$ and $B_{1t} = 0$, a possible solution of the jump conditions is given by $[B_t] = 0$, i.e. $\boldsymbol{B}_2 = \boldsymbol{B}_1$. Since in addition $U_{1t} = 0$ and Eq. (8.10) guarantees that $[U_t] = 0$, it also follows that $\boldsymbol{U}_2$ is parallel to $\boldsymbol{n}$. In this situation we have both $[B_n] = 0$ and $[B_t] = 0$ and all magnetic terms drop out of the jump conditions, which thus turn out to be identical to those of a hydrodynamic shock.

However, the preceding solution is not unique. In fact,as already remarked, in an MHD shock part of the upstream kinetic energy can be converted into magnetic energy. So there are possible solutions with $|\boldsymbol{B}_2| > |\boldsymbol{B}_1|$. But, the normal component of $\boldsymbol{B}$ does not change crossing the shock front and this means that, during the crossing, a tangential component of $\boldsymbol{B}$ that was absent in the upstream region may be generated. In other words, the shock "switches on" a magnetic field (or better a field component), which justifies the name of *switch-on shocks* given to this kind of waves. We shall return to this point below.

### 8.2.3 Oblique Shocks

In the general case, $\boldsymbol{B}_1$ will have both normal and tangential components to the shock plane. Adopting again the reference frame used for perpendicular shocks, i.e. the $x$-axis in the direction of $\boldsymbol{n}$ and the $z$-axis along the tangential component of $\boldsymbol{B}_1$, we can write:

$$\boldsymbol{B}_1 \equiv \left[B_x, 0, B_{1z}\right]. \tag{8.25}$$

Because the normal component of the field is continuous, the component $B_x$ refers indifferently to the upstream or downstream region [see Eq. (8.8)]. As for the velocity vector, $\boldsymbol{U}_1$ will in general have all components different from zero,

$$\boldsymbol{U}_1 \equiv \left[U_{1x}, U_{1y}, U_{1z}\right].$$

The discussion of oblique shocks is greatly simplified by the choice of an appropriate frame of reference, called the *de Hoffmann-Teller* frame (hereafter referred to as dHT frame), where the vectors $\boldsymbol{B}$ and $\boldsymbol{U}$ are parallel to each other both upstream and downstream of the shock. To show that such a frame exists, write Eqs. (8.6) and (8.10) as:

$$(\rho U_n)\left[\boldsymbol{U}_t\right] = \frac{1}{4\pi} B_n \left[\boldsymbol{B}_t\right], \tag{8.26}$$

and

$$(\rho U_n)\left[\frac{\boldsymbol{B}_t}{\rho}\right] = B_n \left[\boldsymbol{U}_t\right], \tag{8.27}$$

where $\rho U_n$ and $B_n$ are conserved quantities. Eliminating $\left[\boldsymbol{U}_t\right]$ between the above equations, we get

$$(\rho U_n)^2 \left[\frac{\boldsymbol{B}_t}{\rho}\right] = \frac{B_n^2}{4\pi} \left[\boldsymbol{B}\right]_t.$$

Recalling the definition of the symbol $[\,]$, wee see that the preceding equation implies

$$\left(\frac{(\rho U_n)^2}{\rho_1} - \frac{B_n^2}{4\pi}\right)\boldsymbol{B}_{1t} = \left(\frac{(\rho U_n)^2}{\rho_2} - \frac{B_n^2}{4\pi}\right)\boldsymbol{B}_{2t}.$$

Since $\rho_1 \neq \rho_2$, we conclude that the vectors $\boldsymbol{B}_{1t}$ and $\boldsymbol{B}_{2t}$ are parallel to each other and to the vector $\left[\boldsymbol{B}_t\right]$, which, in turn, is parallel to $\left[\boldsymbol{U}_t\right]$, as shown by Eq. (8.26). The parallelism of $\boldsymbol{B}_{1t}$ and $\left[\boldsymbol{U}_t\right]$ translates, in our reference frame, into the fact that the $y$-component of $\boldsymbol{U}$ is continuous across the shock $U_{1y} = U_{2y} \equiv U_y$. It is therefore clear that in a reference frame moving at a speed $\boldsymbol{V} = U_y \boldsymbol{e}_y$ with respect to the original frame, the $y$-components of the velocities will vanish. In the new system therefore:

$$\boldsymbol{U}_1 \equiv \left[U_{1x}, 0, U_{1z}\right] \quad \text{and} \quad \boldsymbol{U}_2 \equiv \left[U_{2x}, 0, U_{2z}\right].$$

The situation is illustrated in Fig. 8.1.

For the sake of simplicity, we have continued to use the same notation for the components of the vectors $\boldsymbol{B}$ and $\boldsymbol{U}$ adopted in the original frame, but we must remember that they now refer to the new frame. The vectors $\boldsymbol{B}$ and $\boldsymbol{U}$ are now coplanar and the problem has been reduced to a two-dimensional one in the coordinate plane $(x, z)$. Consider now a further reference frame moving along the $z$-axis with speed $\boldsymbol{W} = W\boldsymbol{e}_z$, where $W$ is to be determined. In this system, calling the velocity components $U'_{1i}$, $(i = x, z)$, we have:

$$U'_{1x} = U_{1x}, \quad U'_{1z} = U_{1z} - W,$$

while the components of $\boldsymbol{B}$ remain unaltered. An appropriate choice for $W$ then leads to $\boldsymbol{U}$ and $\boldsymbol{B}$ becoming parallel to each other. This happens when:

$$\frac{U'_{1z}}{U'_{1x}} = \frac{U_{1z} - W}{U_{1x}} = \frac{B_{1z}}{B_{1x}},$$

namely when

$$W = U_{1z} - U_{1x}\frac{B_{1z}}{B_{1x}}.$$

We have thus shown that with the appropriate choice of two kinematic (Galilean) transformations, it is possible to find a reference frame where $\boldsymbol{B}_1$ and $\boldsymbol{U}_1$ are parallel to each other. This is the dHT frame. Note that the method used here would not work for perpendicular shocks, since for them $B_{1x} = 0$.

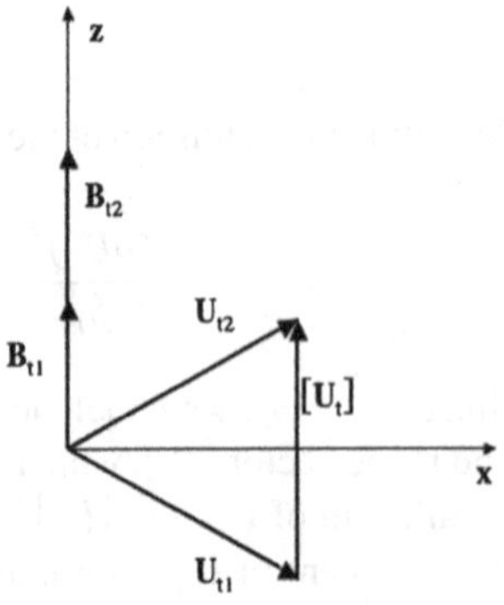


**Fig. 8.1** The vectors $\boldsymbol{B}_{1t}$, $\boldsymbol{B}_{2t}$, $\boldsymbol{U}_{1t}$, $\boldsymbol{U}_{2t}$ and $\left[\boldsymbol{U}_t\right]$ in the new frame

The advantage of the dHT frame is clear: in fact, not only are the vectors $\boldsymbol{U}$ and $\boldsymbol{B}$ parallel upstream of the shock, but they maintain that property also in the downstream region, as shown by Eq. (8.10), that now reads:

$$U'_{2x}B_{2z} - B_{2x}U'_{2z} = U'_{1x}B_{1z} - B_{1x}U'_{1z}.$$

The rhs of this equation vanishes, due to the choice of the reference system: the lhs therefore vanishes as well and the vectors $\boldsymbol{U}$ and $\boldsymbol{B}$ stay parallel when crossing the shock. Moreover, since both upstream and downstream of the shock the plasma can be considered ideal, in the dHT frame the electric field, $\boldsymbol{E} = -(1/c)(\boldsymbol{U} \times \boldsymbol{B})$, is equal to zero and the condition (8.3) reduces to

$$\left[\tfrac{1}{2}U^2 + \frac{\gamma P/\rho}{\gamma - 1}\right](\rho U_x) = 0 \Longrightarrow \left[\tfrac{1}{2}U^2 + \frac{c_s^2}{\gamma - 1}\right] = 0. \tag{8.28}$$

The terms containing the magnetic field drop out from the energy jump condition, that now is identical to that of a hydrodynamic shock. The other jump conditions are still those given by Eqs. (8.4)–(8.10), that however turn out to be significantly simplified in the new reference frame:

$$\left[\rho U_x\right] = 0, \tag{8.29}$$

$$\left[\rho U_x^2 + P + \frac{B_z^2}{8\pi}\right] = 0, \tag{8.30}$$

$$\left[\rho U_x U_z - \frac{B_x B_z}{4\pi}\right] = 0, \tag{8.31}$$

$$\left[\tfrac{1}{2}U^2 + \frac{c_s^2}{\gamma - 1}\right] = 0, \tag{8.32}$$

$$\left[B_x\right] = 0, \tag{8.33}$$

$$U_{2x}B_{2z} - B_{2x}U_{2z} = U_{1x}B_{1z} - B_{1x}U_{1z}. \tag{8.34}$$

Once more, though we have used the same notation for the components of $\boldsymbol{B}$ and $\boldsymbol{U}$, we have to keep in mind that they are now evaluated in the de Hoffman-Teller frame. To deduce the value of the same quantities in the original frame of reference, we must apply the two Galileian transformations in the opposite sense. Eqs. (8.29)–(8.34) are seven relations among the unknown quantities ($\rho$, $c_s$, $U_x$, $U_z$, $B_x$, $B_z$) evaluated upstream and downstream of the shock, a total of 12 quantities. The usual way of proceeding is to fix $\rho_1$, $c_{s1}$, $B_{1x}$, $B_{1z}$ and the ratio of the densities in order to obtain a system of seven relations connecting the seven remaining unknowns, which will be therefore expressed in terms of the five quantities whose values have been fixed. Introducing the Alfvén speed,

$$c_a = c_{a1} \equiv \frac{B_1}{\sqrt{4\pi\rho_1}},$$

a long algebraic calculation allows the preceding relations to be cast in the form:

$$\frac{U_{2x}}{U_{1x}} = \frac{\rho_1}{\rho_2} = X_0^{-1}, \tag{8.35a}$$

$$\frac{U_{2z}}{U_{1z}} = \frac{U_1^2 - c_a^2}{U_1^2 - X_0 c_a^2}, \tag{8.35b}$$

$$\frac{B_{2x}}{B_{1x}} = 1, \tag{8.35c}$$

$$\frac{B_{2z}}{B_{1z}} = \frac{(U_1^2 - c_a^2)X_0}{U_1^2 - X_0 c_a^2}, \tag{8.35d}$$

$$\frac{P_2}{P_1} = X_0 \frac{c_{s2}}{c_{s1}} = X_0 \left(1 + \frac{\gamma - 1}{2} \frac{U_1^2 - U_2^2}{c_{s1}^2}\right), \tag{8.35e}$$

where $X_0$ is a positive solution of the third degree equation:

$$\begin{aligned} \left(U_1^2 - X\, c_a^2\right)^2 \left[X\, c_{s1}^2 + \tfrac{1}{2} U_1^2 \cos^2\theta \left(X\,(\gamma - 1) - (\gamma + 1)\right)\right] \\ + \tfrac{1}{2} c_a^2 U_1^2 X \sin^2\theta \left[\left(\gamma + X\,(2 - \gamma)\right) U_1^2 \right. \\ \left. + X \left(X\,(\gamma - 1) - (\gamma + 1)\right)\right] = 0, \end{aligned} \tag{8.36}$$

and $\theta$ is the angle between $\boldsymbol{B}$ (or $\boldsymbol{U}$) and the normal to the shock plane, $\boldsymbol{e}_x$.

To the three solutions of Eq. (8.36) correspond three types of waves, that are characterized by the value of the propagation speed along the normal to the shock plane, i.e. by the value of $U_{1x} = U_1 \cos\theta$. Ordering the solutions for increasing values of $U_{1x}$, we will have a *slow shock*, an *intermediate (or alfvenic) shock* and a *fast shock*. These shocks correspond to the three linear MHD waves discussed in Chap. 7. In fact, in the limit $X \to 1$, when the discontinuity becomes infinitesimal, Eq. (8.36) factorizes as follows:

$$\left(U_{1x}^2 - c_a^2 \cos^2\theta\right)\left(U_{1x}^4 - \left(c_{s1}^2 + c_a^2\right) U_{1x}^2 + c_{s1}^2 c_a^2 \cos^2\theta\right) = 0,$$

and the Alfvén wave and the slow and fast magnetosonic waves of the linear theory are recovered.

The intermediate shock is not really a shock at all, but a rotational discontinuity. If $U_1 = c_a$, one of the solutions of Eq. (8.36) is simply $X_0 = 1$, or, equivalently, $[\rho] = 0$. But $U_{1x} \neq 0$ so we are back to the already discussed situation characterizing rotational discontinuities. In the dHT frame of reference the $x$-components of $\boldsymbol{U}$ and $\boldsymbol{B}$ are continuous, while their $z$-components change sign when crossing the discontinuity.

The study of the general properties of oblique shocks involves cumbersome algebra: therefore we shall only mention that it is possible to show that oblique shocks are always *compressive*, $X_0 > 1$ and that the sign of the tangential component of the magnetic field is conserved, so that the ratio $B_{2z}/B_{1z}$ is always positive. Eq. (8.35d) then clearly shows that we have two possibilities:

$$U_1^2 \leqslant c_a^2 < X_0 c_a^2, \qquad \text{or} \qquad U_1^2 \geqslant X_0 c_a^2 > c_a^2.$$

In the first case Eq. (8.35d) implies that $B_{2z} < B_{1z}$. This is the fundamental property of *slow shocks*: the magnetic field lines are refracted towards the shock normal and thus the magnitude of $\boldsymbol{B}$ *decreases* when crossing the shock front.

In the second case (*fast shocks*), $B_{2z} > B_{1z}$ and therefore the magnetic lines are refracted away from the shock normal and the magnitude of $\boldsymbol{B}$ *increases*.

Two interesting cases occur when the equal sign holds in the preceding relations. In the case of a slow shock, $U_1 = c_a$, and Eq. (8.35d), for $X_0 > 1$, implies $B_{1yz} \neq 0$, but $B_{2z} = 0$. The crossing of the shock turns off the tangential component of $\boldsymbol{B}$: we have a so-called *switch-off shock*. In the case of a fast shock the condition $U_1^2 = X_0 c_a^2$ is equivalent to the statement that the upstream speed (or shock speed in the laboratory frame) must be higher than the Alfvén speed ($X_0 > 1$). In tis case, Eq. (8.35d) tells us that $B_{1z} = 0$ (parallel shock), but $B_{2z} \neq 0$ so that crossing the shock a tangential component of $\boldsymbol{B}$ appears, and we recover the *switch-on shock*, already encountered in the discussion of parallel shocks (Fig. 8.2).

Eliminating $P_2$ between Eqs. (8.30) and (8.32), it is possible to deduce the following relationship::

$$\frac{B_{2z}^2}{B_{2x}^2} = \tan^2\theta_2 = (X_0 - 1)\left[(\gamma + 1) - (\gamma - 1)X_0 - 2\frac{c_s^2}{c_a^2}\right],$$

where $\theta_2$ is the angle between $\boldsymbol{B}_2$ and the shock normal. Since the lhs is a positive definite quantity, we obtain a limitation of the values of $X_0$ that may possibly give rise to a switch-on shock:

$$1 < X_0 \leqslant \frac{\gamma + 1 - 2c_s^2/c_a^2}{\gamma - 1}.$$

We thus see that to have such type of a shock $c_a > c_s$ must hold. The maximum value of the compression ratio is obtained for $c_a \gg c_s$, in which case we have

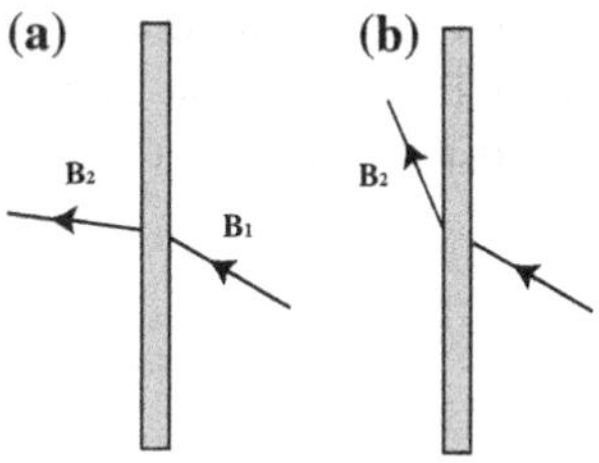


**Fig. 8.2** The refraction of the lines of $\boldsymbol{B}$ for the oblique MHD shocks: **a** slow shock, **b** fast shock

$X_0 = (\gamma + 1)/(\gamma - 1)$, i.e. the same result found for a hydrodynamic shock. The value of $\theta_2$ varies from zero for $X_0 = 1$ to a maximum of $4(1 - c_s^2/c_a^2)/(\gamma - 1)^2$ for $X_0 = (\gamma - c_s^2/c_a^2)/(\gamma - 1) < (\gamma + 1)/(\gamma - 1)$.

## 8.3 Shock Thickness

In the preceding sections shocks have been considered as discontinuities, i.e. as geometrical surfaces of zero thickness. In reality shocks are thin transition layers between two regions of different flow characteristics, with the thickness of a shock resulting from a delicate balance between nonlinear effects, that tend to steepen the wave front, and dissipative effects that oppose that tendency. That balance allows an estimation of the thickness of the shock. The relevant dissipative effects to be taken into account are due to viscosity and thermal conduction for neutral gases, plus Joule heating in the case of collisional plasmas. The equations of momentum and energy must therefore be modified to include viscous, thermal and resistive effects. The use of a simple one-fluid description to determine the structure of a shock, however, has to be taken with caution. In fact, especially in the case of a strong shock, the computed thickness could easily become comparable with the ion mean free path or the ion Larmor radius, a situation well beyond the limits of a fluid model. But even in a weak shock it may happen that the electrons are heated well before the ions and thermal equilibrium between species is reached only after a considerable time. A correct treatment should therefore allow for different electron and ion temperatures.

As we have seen, the jump conditions determine the flow parameters behind the shock in terms of the flow characteristics ahead of the shock. Thus, the total rate of dissipation of energy, which occurs within the shock thickness $\delta$, is in principle known. Given a particular dissipation mechanism it is then possible to estimate the shock thickness.

Consider first a hydrodynamic shock and assume that viscosity is the dominant dissipative effect. Starting from the energy equation for a viscous fluid, it is possible to show that (see e.g. Landau and Lihschitz [7]) the rate of dissipation of kinetic energy, $E_{kin}$ due to viscosity can be expressed as

$$\frac{\mathrm{d}E_{kin}}{\mathrm{d}t} = -\frac{\rho\,\nu}{2}\int\left(\frac{\partial U_i}{\partial x_j} + \frac{\partial U_j}{\partial x_i}\right)^2 \mathrm{d}V, \qquad (8.37)$$

where $V$ is the volume and $\nu$ and $\rho$ are the kinematic viscosity and the density, respectively. Introducing the kinetic energy per unit volume, $\bar{E} = E/V$, and performing a dimensional analysis of the preceding expression we obtain

$$\frac{\Delta\bar{E}}{\Delta t} \simeq \frac{\rho\,\nu}{2}\left(\frac{\Delta U}{\delta}\right)^2,$$

and, since $\Delta t \simeq \delta/U_1$,

$$\Delta\bar{E} \simeq \rho\,\nu \frac{(\Delta U)^2}{2\,U_1\,\delta},$$

from which

$$\delta \simeq \frac{\rho\,\nu(\Delta U)^2}{2\,U_1\Delta\bar{E}}. \tag{8.38}$$

In a strong hydrodynamic shock we have shown that the fraction of the initial kinetic energy converted into heat by viscosity is[2]

$$\Delta\bar{E} \simeq \bar{E} \simeq \rho_1 U_1^2.$$

On the other hand

$$(\Delta U)^2 = (U_1 - U_2)^2 = U_1^2(1 - U_2/U_1)^2 \simeq U_1^2,$$

where use has been made of Eq. (8.21). Inserting the preceding two expressions in Eq. (8.38) we finally get

$$\delta \simeq \frac{\nu}{U_1},$$

which shows that in the shock the Reynolds number is of the order of unity. Recalling the result of Problem 4.3, we write $\nu \simeq (P/\rho)\tau_c$. As a measure of $P$ within the shock we use $P = \frac{1}{2}(P_1 + P_2) \approx P_2/2$ in a strong shock, in which case we also have

$$\frac{P_2}{\rho_2} \approx U_1^2 \quad\longrightarrow\quad U_1 \approx \sqrt{P_2/\rho_2}.$$

Inserting this value in the estimate for $\delta$ we find

$$\delta \simeq \tau_c\sqrt{P_2/\rho_2} \simeq v_{th}^{(i)}\tau_c = \lambda.$$

We thus see that the shock thickness is comparable to the ion mean free path, as anticipated.

Consider now a strong MHD perpendicular shock and estimate the shock thickness assuming the dominant dissipative effect is Joule heating. Then the energy dissipation

---

[2] *In this formula and in the following ones in this Section we drop all factors containing $\gamma$, typically of order one.*

rate will be proportional to $j^2/\sigma$, with $\boldsymbol{j} = (c/4\pi)(\nabla \times \boldsymbol{B})$. The energy dissipation rate per unit volume will be

$$\frac{\Delta \bar{E}}{\Delta t} \simeq \frac{U_1 \Delta \bar{E}}{\delta} \simeq \frac{1}{\sigma}\left(\frac{c}{4\pi}\right)^2 \left(\frac{\boldsymbol{B}_2 - \boldsymbol{B}_1}{\delta}\right)^2,$$

which gives

$$\delta \simeq \frac{1}{\sigma}\left(\frac{c}{4\pi}\right)^2 \frac{(\boldsymbol{B}_2 - \boldsymbol{B}_1)^2}{U_1 \Delta \bar{E}} = \frac{\eta}{4\pi}\frac{(\boldsymbol{B}_2 - \boldsymbol{B}_1)^2}{U_1 \Delta \bar{E}},$$

with $\eta$ the magnetic diffusivity. If we use $\Delta \bar{E} \approx \frac{1}{2}\rho_1 U_1^2$ and recall Eq. (8.20), $\left[B/\rho\right] = 0$, we obtain for a strong shock

$$(\boldsymbol{B}_2 - \boldsymbol{B}_1)^2 = B_1^2(1 - \rho_2/\rho_1)^2 \simeq B_1^2$$

and

$$\delta \simeq \frac{2\eta}{U_1}\frac{B_1^2/8\pi}{\frac{1}{2}\rho_1 U_1^2}.$$

The preceding relation is actually a condition on the magnetic Reynolds number inside the shock as can be seen by writing it in the form:

$$R_M = \frac{U_1 \delta}{\eta} \approx 2\frac{B_1^2/8\pi}{\frac{1}{2}\rho_1 U_1^2}.$$

Thus, contrary to the hydrodynamical case, the magnetic Reynolds number within the shock is not always of the order unity, but is proportional to the upstream ratio of magnetic and kinetic energies.

## 8.4 Collisionless Shocks

As mentioned at the beginning of the chapter, most shocks generated in space and astrophysical plasmas are collisionless, in the sense that the processes governing shock thickness, structure and dynamics are not the classical dissipation processes due to collisional viscosities and resistivities, but involve wave-particle interactions which mediate energy dissipation. The notion of collisionless shocks became accepted by the early sixties in space, though a similar problem also confronted researchers involved in laboratory experiments. In the solar system, the mean free path of particles can be up to a few Astronomical Units (AU) yet shocks are observed in front of the magnetospheres of most planets as well as traveling in the solar wind away from the sun.

The first solutions proposed for the existence of collisionless shocks was associated with processes of anomalous dissipation, in particular anomalous resistivity, generated by the intense currents present in within the steepened gradients of the shock and associated with the relative motions of the current carrying particles (mostly electrons and protons). The plasma supporting such intense currents is generally unstable and these instabilities generate waves, which can mediate the transport of momentum and energies between particle species and within each species between particle populations with different speeds. For example, current carrying electrons can generate waves that may then be absorbed by ions, on scales much smaller than the mean free path. The current instabilities were first though to involve the generation of ion-sound waves (discussed in the previous chapter section on Landau damping) but more generally the wave-types depend in detail on the geometry of the shock and the magnetic field configuration.

Experimental studies of shocks in space and laboratory plasmas have shown that the characteristics of observed shocks can be strongly variable and depend crucially on the angle between the magnetic field and the shock normal (i.e. whether the shock is *quasi-parallel* or *quasi-perpendicular*), the ratio of plasma to magnetic field pressures upstream of the shock, and the magnetosonic Mach number, i.e. the ratio of the upstream flow speed to the speed of the (fast) magnetoacoustic wave propagating in the direction of the shock normal. The main characteristic feature of the first group of shocks is that ions reflected by the shock front can freely propagate back up into the upstream region. Such shocks correspond to magnetic field-shock normal angles below 45°. In the second case, reflected ions remain close to the shock and can gain energy thanks to the inductive electric field tangential to the shock surface but perpendicular to the magnetic field.

Low Mach number shocks then dissipate the required energy through anomalous resistivity mechanisms within the current carrying layer. Right hand fast magnetoacoustic/whistler waves have phase and group velocities that increase with decreasing wavelength beyond the fluid regime. Therefore, steepened fast mode shocks are expected to radiate short wavelength waves, and hence energy, into the un-shocked oncoming flow. The shortest wavelength capable of standing in the flow then forms a precursor wave-train that has been observed at these "sub-critical" shocks (but the process occurs for all Mach numbers). Above a critical Mach number, anomalous resistivity within the layer carrying the limited shock current is unable to convert the required amount of energy from directed bulk flow into thermal energy. At such "super-critical" quasi-perpendicular shocks, a fraction of the incident ions are reflected by the steep shock ramp as described above. They gyrate around the magnetic field and gain energy due to acceleration by the transverse motional electric field. Returning to the shock layer they have sufficient energy to pass through into the downstream shocked region. The separation of ions into two groups, crossing the front directly and after reflection, results in the dispersal of particles in velocity space. Reflected particles are separated from the bulk ion population due to an increase in peculiar velocity relative to the bulk motion. This process corresponds to the kinetic heating required by the shock jump conditions and it ensures the major part of energy dissipation necessary for directed energy transfer to thermal energy of plasma ion population. The process of ion thermalization takes place on rather large scale downstream of the shock front. The spatial length of the transition to ion thermal equilibrium can be treated in a similar fashion to that of the shock front thickness in collisional shocks. Reflection occurs on sufficiently smaller scales than thermalization due to a combination of magnetic forces and the electrostatic cross-shock potential, which results directly from the leading electron pressure gradient term in the generalized Ohms Law. In more detailed two-fluid descriptions, the quasi-perpendicular shock has fine structure that depends upon the characteristics of the nonlinear shock profile.

At high Mach number ion reflection appears to play a crucial role for dissipation processes also within the quasi-parallel shock. However, in the quasi-parallel shock, there is no single, monotonic shock ramp that develops as a combined result of reflection and gyration. At the quasi-parallel shock the characteristic length scales depend on the coupling between waves and energetic particles, which in turn depends on the mechanisms for extracting energetic particles from the thermal distribution. Spacecraft observations of the quasi-parallel shock shows they they have intrinsically unsteady behaviors with strong magnetic pulsations, and the pulsation shock layer is embedded in turbulence between the upstream foreshock and the downstream flow region. The structure of such shocks is studied mostly via computer simulation and is still the study of considerable research. In the Earth's magnetosphere, beacause of the changing direction of the Earth's magnetic field towards the direction of the incoming solar wind, there are both regions of quasi-perpendicular and quasi-parallel shocks, and this is most likely true of general bow shocks in astrophysics.

We refer the readers to two reviews on quasi-parallel and quasi perpendicular shocks for a summary of contemporary understanding and research on this topic (Burgess and Scholer [18]; Krasnoselskikh et al. [19]).

# Chapter 9
# Magnetic Reconnection

**Abstract** The phenomenon of magnetic reconnection and its importance and role in many plasma dynamic phenomena are introduced and described in detail. The stationary reconnection models of Sweet-Parker and Petschek are described before analyzing the resistive instabilities of current sheets and, in particular, the tearing mode. The chapter concludes with reconnection instabilities of thin sheets, or plasmoid instabilities, and the consequent general nonstationary behavior of reconnection processes.

The previous chapters have been devoted to the dynamics of ideal plasmas, neglecting all effects connected with their finite resistivity; in this chapter we shall be concerned precisely with those effects. Recalling the discussion of Chap. 5, rewrite the induction equation describing the temporal variation of $\boldsymbol{B}$, [see Eq. (5.25)]:

$$\frac{\partial \boldsymbol{B}}{\partial t} = \nabla \times (\boldsymbol{U} \times \boldsymbol{B}) + \eta \nabla^2 \boldsymbol{B}. \tag{9.1}$$

The two terms on the rhs vary over different timescales which can be defined as $\tau_f = \mathcal{L}/\mathcal{U}$ (the *fluid* or *convective* scale) and $\tau_d = \mathcal{L}^2/\eta$ (the *diffusive* or *resistive* scale), where $\mathcal{U}$ is a typical value of the fluid velocity, $\mathcal{L}$ is the spatial scale of variation of the magnetic field and $\eta = (c^2/4\pi\sigma)$ is the plasma magnetic diffusivity. The relative importance of the two terms is measured by the value of the magnetic Reynolds number, $\mathcal{R}_m = \tau_d/\tau_f = (\mathcal{U}\mathcal{L})/\eta$.

The choice of the timescales is, at least to a certain extent, arbitrary. If, as often the case, one is interested in dynamical situations dominated by the magnetic field, a logical choice would be $\mathcal{U} = c_a$. As far as the scale $\mathcal{L}$ is concerned, it seems natural at first sight to use the "macroscopic" scale typical of the fluid variables, $\mathcal{L} = L$. With these choices $\tau_f$ becomes identical to $\tau_a$, the time needed to cover the distance



C. Chiuderi and M. Velli, *Basics of Plasma Astrophysics*,
UNITEXT for Physics, DOI 10.1007/978-88-470-5280-2_9

$L$ at the Alfvén speed; the magnetic Reynolds number is then called the *Lundquist number*, $S$:

$$S = \frac{Lc_a}{\eta}. \tag{9.2}$$

Table 5.1 in Chap. 5 displays the characteristic values of the spatial and temporal scales of some plasmas of interest for the laboratory and for astrophysics, together with the corresponding Lundquist numbers. It is important to distinguish the diffusive times shown in the Table with the effective lifetimes of the corresponding magnetic fields. In fact, if these two scales were equal, we should conclude that a sunspot survives for a few million years, while its effective lifetime is never longer than a few months. The Earth's magnetic field, on the other hand, should have been extinguished a long time ago. We shall return to this point in Chap. 11, when dealing with the problem of the build-up and maintenance of magnetic fields.

The simple considerations above force us to conclude that there must be processes present in nature, different from simple diffusion, that modify the decay of magnetic fields. The existence of such mechanisms is extremely important both for laboratory and astrophysical plasmas in all circumstances where magnetic energy is converted into different forms of energy, in particular into thermal energy, as, for instance, in connection with the problem of the heating of the solar corona, to which we shall return in the following chapter.

So given the huge values of the Lundquist number, that seem to indicate that the ideal description of a plasma should be perfectly adequate, what processes might actually cause magnetic field dissipation and/or regeneration on such vastly faster timescales? First, observe that the ideal limit of the induction equation can be obtained not only by letting the magnetic diffusivity tend to zero, but also by allowing the spatial scale $\mathcal{L}$ to become arbitraryly large, because of the different dependence on the scale of the two terms of that equation. Secondly the induction equation has a vector character, i.e. spatial anisotropy, which is lost in the simple dimensional analyses on which the definition of the Lundquist number is based. So, for example, the *local* importance of the resistive term is increased in regions where the convective term vanishes, as, for instance, when the velocity $\boldsymbol{U}$ is parallel to $\boldsymbol{B}$. In such regions, the use of the macroscopic scale $L$ is not justified and a *local* scale, typically of the same order of the dimensions of the regions of vanishing convective term (potentially much smaller than $L$), seems better suited. The value of $S$ then decreases locally, the Alfvén theorem ceases to be valid and the magnetic field topology may vary, giving access to configurations of lower energy unattainable in an ideal MHD regime. In such circumstances, we expect that magnetic fields dynamics develop on timescales intermediate between $\tau_d$ and $\tau_f$, as we shall verify in the following.

Before considering processes violating Alfvén's theorem, consider once more the purely diffusive case, $S \ll 1$. In this situation the Eq. (9.1) reduces to

$$\frac{\partial \boldsymbol{B}}{\partial t} = \eta \nabla^2 \boldsymbol{B}.$$

Let's assume a constant $\eta$ and a unidirectional magnetic field given by $\boldsymbol{B} = B(x,t)\boldsymbol{e}_y$, with $B(-x,t) = -B(x,t)$. The current density will then have the form $\boldsymbol{J} = J(x,t)\boldsymbol{e}_z$. We shall assume that at the time $t = 0$ the current is localized in a narrow region around $x = 0$ and write

$$J(x,0) = J_0 \mathrm{e}^{-x^2/a^2},$$

The equation $\nabla \times \boldsymbol{B} = (4\pi/c)\boldsymbol{J}$ then reduces to

$$\frac{\partial B}{\partial x} = \frac{4\pi}{c} J, \tag{9.3}$$

so that

$$B(x,0) = \frac{4\pi J_0}{c} \int_0^x \mathrm{e}^{-x^2/a^2} \mathrm{d}x = \left(\frac{2\pi^{3/2} J_0 a}{c}\right) \mathrm{erf(x/a)} = \mathrm{B_0\, erf(x/a)},$$

where

$$\mathrm{erf(z)} = \frac{2}{\sqrt{\pi}} \int_0^{\mathrm{z}} \mathrm{e}^{-\mathrm{x}^2} \mathrm{dx}$$

is the error function and $B_0$ is the asymptotic value of $B$ when $x \to \infty$.

The temporal evolution of $J(x,t)$ can be found by deriving Eq. (9.3):

$$\frac{\partial J}{\partial t} = \eta \frac{\partial^2 J}{\partial x^2},$$

which is identical to the equation for $B(x,t)$.

In Chap. 6 we have seen that the solution of the preceding equation can be found via Fourier transforms. Recalling Eq. (5.26), we may write:

$$J(x,t) = \int \mathrm{d}k\, \mathrm{e}^{-\eta k^2 t} J(k)\, \mathrm{e}^{\mathrm{i}(k\,x)}, \tag{9.4}$$

where $J(k)$ is the Fourier transform of the current distribution at time $t = 0$, namely

$$J(k) = \frac{J_0\, a}{2\sqrt{\pi}} \exp\left(-\frac{a^2 k^2}{4}\right).$$

Using this expression and performing the integral in Eq. (9.4) we get

$$J(x,t) = J_0\, a\, (4\eta\, t + a^2)^{-1/2} \exp\left(-\frac{x^2}{4\eta\, t + a^2}\right).$$

In terms of the non-dimensional variables $\xi = x/a$ e $\tau = 4\eta t/a^2$ the current density reads:

$$J(\xi, \tau) = J_0 \frac{\exp[-\xi^2/(1+\tau)]}{(1+\tau)^{1/2}},$$

and the magnetic field,

$$B(\xi, \tau) = B_0 \, \mathrm{erf}[\xi/(1+\tau)^{1/2}].$$

The profiles of $J$ e $B$ as functions of $\xi$ for different values of $\tau$ are shown in Figs. 9.1 and 9.2.

From the figures it is seen that the current distribution widens as time progresses, while the value of $B$ and that of its gradient decrease. The magnetic energy per unit

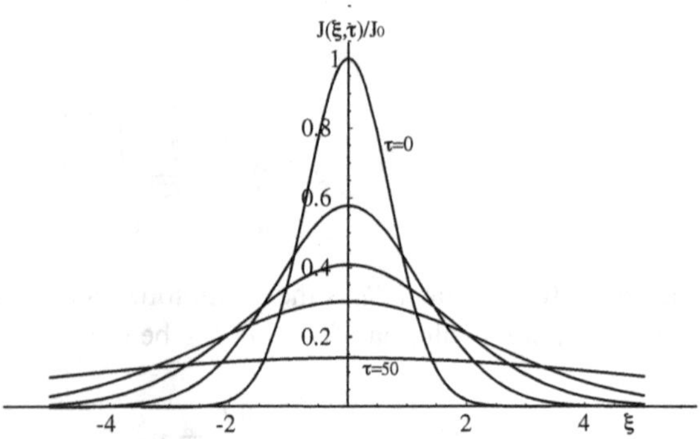


**Fig. 9.1** $J(\xi, \tau)$ as a function of $\xi$, for $\tau = 0, 2, 10, 50$

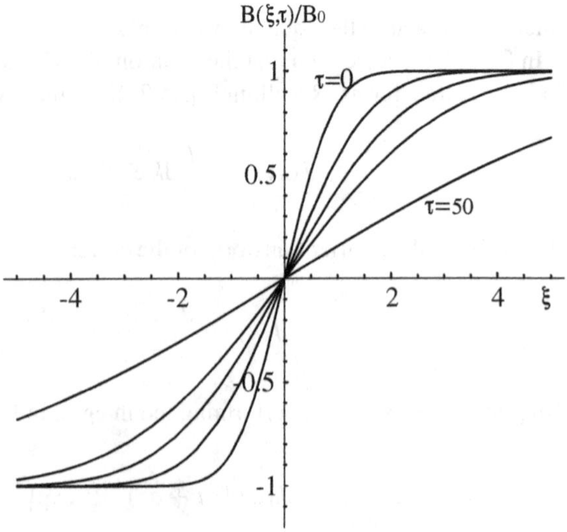


**Fig. 9.2** $B(\xi, \tau)$ as a function of $\xi$, for $\tau = 0, 2, 10, 50$

length in the $y$-direction varies with time as

$$\frac{\partial}{\partial t}\int_{-\infty}^{\infty}\frac{B^2}{8\pi}\mathrm{d}x = \frac{1}{4\pi}\int_{-\infty}^{\infty}B\frac{\partial B}{\partial t}\mathrm{d}x.$$

Inserting the expression of $\partial B/\partial t$ given by the induction equation in the preceding integral and performing an integration by parts, it is easy to obtain:

$$\frac{\partial}{\partial t}\int_{-\infty}^{\infty}\frac{B^2}{8\pi}\mathrm{d}x = \frac{\eta}{4\pi}\int_{-\infty}^{\infty}B\frac{\partial^2 B}{\partial x^2}\mathrm{d}x = -\frac{\eta}{4\pi}\int_{-\infty}^{\infty}\left(\frac{\partial B}{\partial x}\right)^2\mathrm{d}x = -\frac{1}{\sigma}\int_{-\infty}^{\infty}J^2\mathrm{d}x.$$

As expected, the magnetic energy is dissipated as a result of the finite value of the electrical conductivity and converted into heat. Even when the Lundquist number is large, and in the presence of flows, the configuration discussed above, consisting in a so-called *neutral sheet* or *current sheet*, and schematically represented in Fig. 9.3, may be considered a fundamental starting point to study resistive effects.

The magnetic field has only one component $B_y \neq 0$, but its magnitude depends on the $x$-coordinate. Let $B_y(x)$ be an odd function of $x$, ($B_y(0) = 0$) tending to $\pm B_0$ when $x \to \pm\infty$. Such a field is that generated by a homogeneous current sheet directed along $z$, of infinite extension in the $y$-direction and piling up around $x = 0$. The Lorentz force $\boldsymbol{J} \times \boldsymbol{B}$ pushes the plasma towards the neutral sheet from both sides. In the absence of resistivity the action of the Lorentz force is opposed by pressure gradients, which counteract the natural tendency of the sheet to collapses on itself. But the presence of a finite magnetic diffusivity, no matter how small, breaks the frozen-in condition and allows the decoupling of field lines from the motion of the plasma, eventually giving rise to a configuration similar to that shown in Fig. 9.4.

This transition can be "pictorially" described by saying that two lines of force are first "cut" and then "reconnected" to field lines of opposite polarity, a descrip-

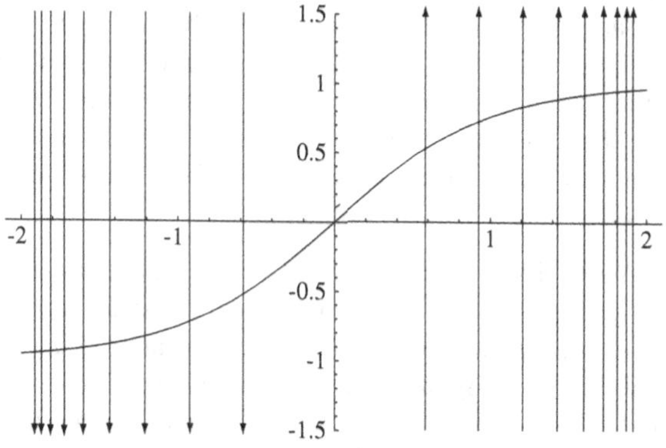


**Fig. 9.3** Neutral sheet: lines of $\boldsymbol{B}$. The *solid line* represents $B_y(x)$. The scale on the *vertical axis* refers to both the values of the $x$-coordinate and of the magnetic field

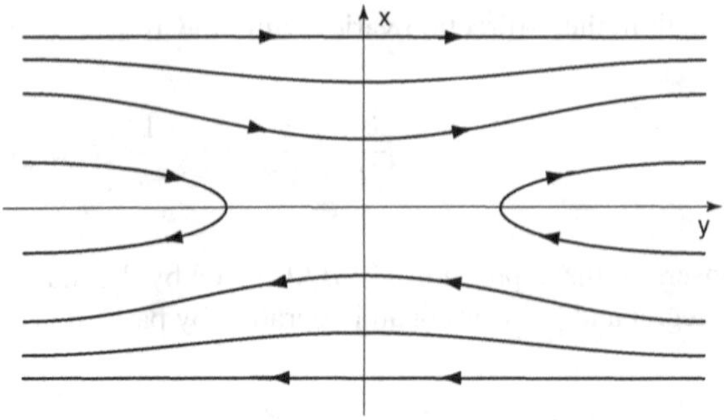


Fig. 9.4 Magnetic field lines after reconnection

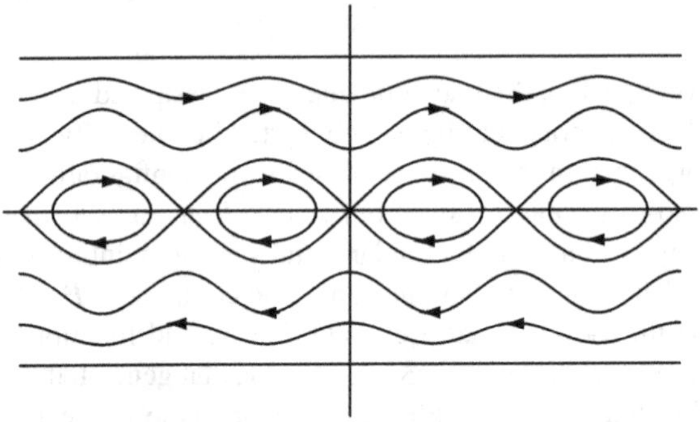

Fig. 9.5 The formation of magnetic islands

tion at the origin of the name *magnetic reconnection* usually given to this class of processes. If the magnetic reconnection of field lines occurs in several points, the final configuration resembles the one shown in Fig. 9.5.

The field shows two distinct topologies: closed field lines, so-called *magnetic islands*, and open field lines, separated by two limiting curves, known as *separatrices*. The initial *neutral line*, i.e. the line corresponding to $B_y = 0$, breaks into a discrete succession of null points, said to be of the *O-type* when they are located within closed field lines, and of *X-type* when they lie at the intersection of separatrices. To get an intuitive understanding of why this final configuration is reached, think of the initial current sheet as formed by a continuous collection of parallel line currents. Since parallel currents attract each other, such a system turns out to be unstable and the sheet evolves toward a configuration of distinct parallel currents, encircled by magnetic field lines. It is possible to show that the energy of the final configuration is lower than that of the initial one.

When discussing resistive processes it is necessary to distinguish two broad categories of models referring to different physical situations. In the first, named *driven reconnection* reconnection is "driven" by appropriate plasma flows, while in the second, known as *spontaneous reconnection* reconnection takes place in the absence of driving plasma motions, rather, the motions are initiated self-consistently by the reconnection process itself.

## 9.1 Driven Reconnection

Consider the situation where non-magnetic forces, for instance pressure gradients, induce systematic converging flows bringing together regions with different magnetic structure. Since resistivity has practically no effect at large spatial scales, the plasma can be considered as ideal and the field lines are dragged by the motion of the plasma. If the magnetic field carried by the colliding fluxes has opposite directions, a current sheet will be formed, caused by the polarity inversion of $\boldsymbol{B}$. Inside this very narrow sheet $\boldsymbol{B} \approx 0$, and resistivity plays a fundamental role allowing the breaking and subsequent reconnection of magnetic field lines. The relaxation of the magnetic tension in the newly bent field lines pushes the plasma out of the resistive region, thus reducing the local total pressure, and this in turn attracts new plasma from the outside. A stationary state will be established if the magnetic flux entering the resistive region per unit time will be exactly balanced by the annihilation and diffusion rates of the field. However, notice that while it is possible to destroy the magnetic field locally, the mass of the plasma itself is conserved, so mass has to flow out of the resistive region. The problem is well illustrated by the following model due to Parker.

Consider the two-dimensional incompressible stationary flow of a plasma described by the system of equations:

$$\rho(\boldsymbol{v} \cdot \nabla)\boldsymbol{v} = -\nabla P + \frac{1}{4\pi}\boldsymbol{J} \times \boldsymbol{B} = -\nabla\left(P + \frac{B^2}{8\pi}\right) + (\boldsymbol{B} \cdot \nabla)\boldsymbol{B} \tag{9.5}$$

$$\boldsymbol{E} + \frac{1}{c}\boldsymbol{v} \times \boldsymbol{B} = \frac{1}{\sigma}\boldsymbol{J} = \frac{\eta}{c}\nabla \times \boldsymbol{B}, \tag{9.6}$$

where

$$\nabla \cdot \boldsymbol{B} = 0; \quad \nabla \cdot \boldsymbol{v} = 0.$$

Notice that in this chapter the fluid velocity will be indicated by $\boldsymbol{v}$, instead of $\boldsymbol{U}$. One way to satisfy the condition of vanishing divergence for $\boldsymbol{B}$ and $\boldsymbol{v}$ is by taking the following structure for the vector fields in $-L \leq x, y \leq L$:

$$\boldsymbol{B} \equiv [0, B(x), 0], \quad \boldsymbol{v} \equiv [-v_0\frac{x}{a}, v_0\frac{y}{a}, 0],$$

with constants $v_0$ and $a$. This flow has a *stagnation point* ($\boldsymbol{v} = 0$) in $[x = y = 0]$. Equation (9.6) shows that the electric field is directed along $z$ and, since in a stationary situation $\nabla \times \boldsymbol{E} = 0$, $\boldsymbol{E} = E\boldsymbol{e}_z$ with $E = const$.

With this field structure, Eq. (9.6) reduces to

$$cE - v_0\frac{x}{a}B = \eta\frac{\mathrm{d}B}{\mathrm{d}x}, \tag{9.7}$$

a first order linear equation for B.

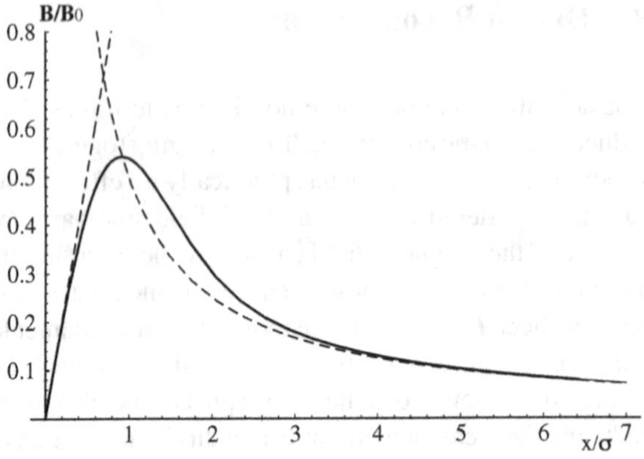


**Fig. 9.6** Profile of $B/B_0$, the *dashed lines* indicate the approximate solutions for small and large values of $x$

The behavior of the solution may be qualitatively inferred by noticing that for large values of $x$ we expect that $\mathrm{d}B/\mathrm{d}x \to 0$ and thus $B \approx E(c/v_0)(a/x)$, while for $x \to 0$, the second term on the lhs will be negligible and therefore $B \approx E(cx/\eta)$. The two approximations give the same value of $B$ for $x = x_0 = (a\eta/v_0)^{1/2}$. The fact that the magnetic field produced by a current flowing in a wire at a point *external* to the current at a distance $x$ from the wire's axis is proportional to $1/x$, suggests considering $x_0$ as an estimate of the thickness of the current sheet.

These conclusions are confirmed by the exact analytical solution of Eq. (9.7). Using standard solution methods for first order linear differential equations one obtains:

$$B = B_0 \mathrm{e}^{-(x/\sigma)^2} \int_0^{x/\sigma} \mathrm{e}^{u^2} \mathrm{d}u,$$

where

$$B_0 = 2E\frac{ca}{v_0\sigma}; \quad \sigma = (2\eta a/v_0)^{1/2} = \sqrt{2}\,x_0.$$

A plot of the exact solution, together with the approximations for small ad large values of $x$ is shown in Fig. 9.6.

## 9.1.1 The Sweet-Parker Model

The Sweet-Parker model illustrates in a schematic, extremely simple, way the basic features of stationary reconnection, without specifying the precise analytical form of $\boldsymbol{B}$ and $\boldsymbol{v}$. It is a two-dimensional, incompressible model similar to the one we have just described: $\boldsymbol{B}$ and $\boldsymbol{v}$ have non vanishing components only in the $(x, y)$ plane, while $\boldsymbol{E}$ is a constant vector, directed along $z$. The equation governing the dynamics is still Eq. (9.6). It is assumed that reconnection takes place in a narrow current sheet, whose length and thickness are $2L$ and $2\ell$, respectively. The sheet separates two regions of oppositely directed magnetic field, as shown in Fig. 9.7.

The magnetic field vanishes in the center of the resistive layer. Outside of it, the plasma can be considered as ideal and therefore the magnetic field, $B_i$ is frozen into the plasma and is convected towards the diffusive region with the speed of the flowing plasma, $v_i$. The assumption of stationarity is a very stringent one: it implies a perfect agreement between the entrance speed of the plasma and the velocity of diffusion and annihilation of the magnetic field brought into the region by the plasma from opposite directions. This implies that the incoming speed $v_i$ cannot be fixed at will. On the other hand the diffusion velocity is controlled by the gradient of $\boldsymbol{B} \approx [B_i - B(0)]/\ell = B_i/\ell$ and time-independence then imposes limitations also on the thickness of the diffusive layer, $\ell$.

The model is defined by the entrance values of $v_i$ and $B_i$ and by the corresponding outflowing values, $v_o$ and $B_o$, in addition to the characteristic dimensions of the resistive layer, $L$ and $\ell$ and the value of magnetic diffusivity $\eta$. If the latter quantity is assumed to be known, the quantities to be determined are six, while, as we shall see, the equations connecting them are only four. We must therefore fix the values of two of the unknown quantities: fixing $B_i$ and $L$ is a useful choice.

We can now estimate the values of the current density, as given by Eq. (9.6) on the border of the resistive region (where $\boldsymbol{J} = 0$) and in its center, where $\boldsymbol{B} \simeq 0$), obtaining, respectively

$$E = \frac{v_i}{c} B_i,$$

and

$$E = \frac{J_c}{\sigma},$$

where the index $c$ indicates values referring to the center of the resistive layer. Since, on the other hand, $\boldsymbol{J} = (c/4\pi)(\nabla \times \boldsymbol{B}) \to J_c \simeq (c/4\pi)B_i/\ell$, we may eliminate $E$ and $J_c$ from the preceding equations and obtain

$$v_i = \frac{\eta}{\ell}. \tag{9.8}$$

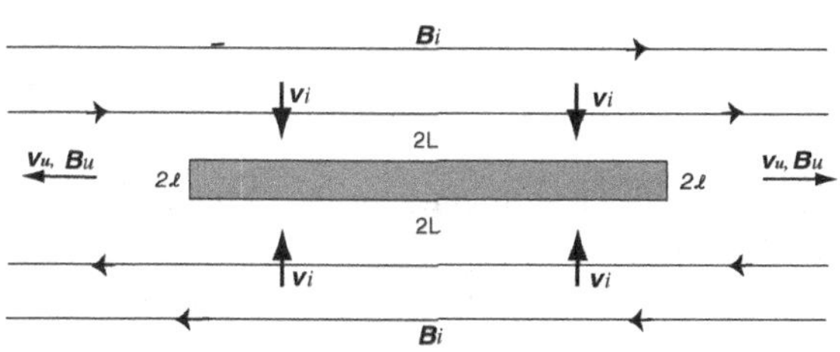


**Fig. 9.7** Sketch of the Sweet-Parker model

Mass conservation, requiring equality of the ingoing mass flux, $4L\rho v_i$, with that of the outgoing one, $4\ell\rho v_o$, may be written as:

$$L v_i = \ell v_o \quad \rightarrow \quad \ell = L\frac{v_i}{v_o}, \tag{9.9}$$

where $v_o$ is the velocity of the plasma leaving the diffusive layer. On the other hand, the magnetic flux conservation law tells us that

$$v_i B_i = v_o B_o \quad \rightarrow \quad B_o = B_i\frac{v_i}{v_o} = B_i\frac{\ell}{L}, \tag{9.10}$$

where Eq. (9.9) has been used to obtain the last equality, which can be also deduced from a dimensional analysis of $\nabla \cdot \boldsymbol{B} = 0$. Eqs. (9.8), (9.9) and (9.10) allow us to express all quantities in terms of the outflow velocity $v_o$. To determine the latter and close the system, consider the momentum equation and estimate the order of magnitude of the various terms entering it. For the moment, we shall assume that the pressure is the same everywhere and thus neglect the term $\nabla P$ in the momentum equation. We have already seen that the current density is approximately equal to $J_c \simeq (c/4\pi)B_i/\ell$, which implies that the magnetic force along the resistive layer is $J_c B_o/c \simeq (1/4\pi)(B_i B_o/\ell)$. This force accelerates the plasma within the resistive layer, bringing its velocity from zero, at the stagnation point, to the final value $v_o$ when it leaves the diffusive region. The momentum Eq. (9.5) therefore tells us that

$$\rho\frac{v_o^2}{L} \simeq \frac{1}{4\pi}\frac{B_i B_o}{\ell}.$$

Making use of the last equality in Eq. (9.10), we finally get

$$v_o = \frac{B_i}{\sqrt{4\pi\rho}} \equiv c_{ai},$$

dove $c_{ai}$ is the Alfvén speed of the incoming flux.

In conclusion, the plasma enters the resistive layer with a velocity $v_i$ and is expelled from it at a speed $c_{ai}$. The ratio of these two speeds, denoted by $R_i$, may be considered as a non-dimensional measure of the *reconnection rate*. The index $i$ indicates that $R_i$ depends only on the parameters of the ingoing flux. $R_i$ can be conveniently expressed in terms of the Lundquist number, making use of Eqs. (9.8) and (9.9):

$$R_i \equiv \frac{v_i}{c_{ai}} = (\frac{\eta}{L c_{ai}})^{1/2} = S^{-1/2}. \tag{9.11}$$

Notice that the length entering the definition of $S$ is the length of the sheet, $L$. Moreover

$$\ell = L\,S^{-1/2} \quad \text{and} \quad B_o = B_i\,S^{-1/2}.$$

Since $S \gg 1$, $v_i \ll c_{ai}$, $\ell \ll L$ and $B_o \ll B_i$.

Notice also that the length $\ell$, giving the spatial scale in the direction of the thickness of the resistive layer, has been defined without any reference to the scale of the magnetic field in the same direction. However, it seems natural to assume that the two scales are correlated and this suggests identifying the scale $\ell$ with the thickness of the current sheet generating the magnetic field. In fact, since the dissipation is proportional to $\eta J^2$, resistivity acts wherever a current flows. These considerations will turn out to be useful in the following.

In the version of the Sweet-Parker model so far presented, it has been assumed, without justification, that the plasma pressure was the same everywhere, mostly to underline the fact that the outgoing plasma acceleration can be produced under the action of the sole magnetic tension. The addition of a pressure gradient alters the outflow speed and consequently the reconnection rate. To estimate the effects connected with the presence of pressure gradients, we first observe that in our long and thin resistive layer, the acceleration and the magnetic tension can be neglected in the $y$-component of the momentum equation, that therefore reduces to

$$\frac{\partial}{\partial y}\left(\frac{B^2}{8\pi}+P\right)=0.$$

Integrating the preceding equation between 0 and $\ell$ we get:

$$P_c = P_i + \frac{B_i^2}{8\pi}.$$

The pressure at the center of the resistive layer is therefore larger than that at the entrance, contrary to what assumed so far. Let's now look at the momentum equation in the $x$-direction:

$$\rho v_x \frac{\partial v_x}{\partial x} = \frac{1}{2}\frac{\partial v_x^2}{\partial x} = \frac{1}{c} J\, B_y - \frac{\partial P}{\partial x}.$$

Recalling that $J=(c/4\pi)(\partial B_x/\partial y)\approx(c/4\pi)(2B_i/2\ell)$ and integrating the preceding equation between 0 and $L$ we obtain:

$$\rho\frac{v_o^2}{2} = \frac{1}{4\pi}\frac{B_i}{\ell}\int_0^L B_y \mathrm{d}x - (P_o - P_c).$$

If $B_y$ is assumed to vary linearly between zero in $x=0$ and $B_o$ in $x=L$, i.e. $B_y(x)=(B_o/L)x$. an estimate of the integral is easily obtained, allowing us to write:

$$v_u^2 = \frac{B_o B_i}{4\pi\rho}\frac{L}{\ell} - \frac{2(P_o-P_c)}{\rho} = c_{ai}^2\frac{B_o}{B_i}\frac{L}{\ell} - \frac{2(P_o-P_c)}{\rho} = c_{ai}^2 - \frac{2(P_o-P_c)}{\rho},$$

where Eq. (9.10) has been used. If the central pressure, $P_c$, is expressed in terms of $P_i$ we arrive at:

$$v_u^2 = 2c_{ai}^2 + \frac{2(P_i - P_o)}{\rho}. \tag{9.12}$$

Defining once again the reconnection rate as $v_i/v_o$ and using Eq. (9.12) we obtain:

$$R_i \equiv \frac{v_i}{v_o} = \frac{c_{ai}}{v_o} S^{-1/2}.$$

If the outflow velocity is less than $c_{ai}$, namely if the exit pressure is sufficiently larger than the entrance one, and precisely if $\Delta P = P_o - P_i > \rho c_{ai}^2 = B_i^2/4\pi$, the reconnection rate turns out to be larger than that of the Sweet-Parker model which, as we have seen, equals $S^{-1/2}$.

### *9.1.2 An Outline of the Petscheck Model*

The weakest point of the Sweet-Parker model (and of its generalizations as well) is the modest value of the reconnection rate. This seems to preclude the possibility of identifying magnetic reconnection as the mechanism capable of transforming the magnetic energy in other forms of energy, such as thermal energy or kinetic energy of accelerated particles. On the other hand, in the laboratory and especially in astrophysics, phenomena are observed that necessarily imply such transformations.

A typical example is given by solar flares that exhibit a sudden increase of the emissivity over the whole electromagnetic spectrum, localized in regions where the magnetic field is particularly intense, the so-called active regions. This phenomenon is associated with the presence of accelerated particles and, occasionally, the ejection of large amounts of matter. The total energy released by a large flare is estimated to be of the order of $10^{32}$ erg, on time scales of the order of hours. A careful analysis of the magnetic configuration in the flare's site before and after the flaring brings to the conclusion that the only conceivable source of the energy released during the flare in various forms is the magnetic energy stored in the magnetic field prior to the flare. Observations further indicate that most of the energy is emitted during the initial phase of the flare, on timescales of the order of minutes or less, implying a high conversion efficiency of the magnetic energy. It is therefore natural to inquire if the limited theoretical efficiency of the process could not be connected to the particular geometrical configuration chosen and if other configurations exist allowing more substantial reconnection rates.

To try to answer this question, let's first notice that the analysis of Sweet ans Parker is a local analysis, in the sense that it focuses only on the magnetic configuration in the immediate surroundings of the diffusive layer, a region whose dimensions are generally small as compared to the global size of the system. On the other hand, it seems likely that the input values of the velocity and of the magnetic field, $v_i$ and $B_i$, are actually determined by the prevailing conditions in a region whose spatial scale is larger than the one so far analyzed. This suggests to extend our analysis to a wider region, where the spatial scale and the characteristic values of $\boldsymbol{v}$ and $\boldsymbol{B}$ are, respectively, $L_e$, $v_e$ and $b_e$, the index $e$ denoting the fact that those values are, in fact, referring to a region *external* to the Sweet-Parker sheet. As a consequence, we shall have a new value of the Lundquist number $S_e = L_e c_{ae}/\eta$, and, according to the previously introduced definition, a new reconnection rate, $R_e = v_e/c_{ae}$. Then, the problem to solve is the determination of $v_i$ in terms of $v_e$ and the evaluation of $R_e$ in terms of $S_e$.

A first relationship is immediately obtained from the magnetic flux conservation, $v_e B_e = v_i B_i$, that can be cast in the form:

$$\frac{R_i}{R_e} = \frac{c_{ae}}{c_{ai}}\frac{v_i}{v_e} = \frac{B_e^2}{B_i^2}. \tag{9.13}$$

The ratio of the spatial scales is easily found by using Eqs. (9.13), (9.11) and (9.9):

$$\frac{L_i}{L_e} = \frac{S_i}{c_{ai}}\frac{c_{ae}}{S_e} = \frac{S_i}{S_e}\frac{B_e}{B_i} = \frac{S_i}{S_e}\left(\frac{R_i}{R_e}\right)^{1/2} = \frac{1}{S_e}R_i^{-3/2}R_e^{-1/2} \tag{9.14}$$

and

$$\frac{\ell}{L_e} = \frac{\ell}{L_i}\frac{L_i}{L_e} = R_i\frac{1}{S_e}R_i^{-3/2}R_e^{-1/2} = \frac{1}{S_e}R_i^{-1/2}R_e^{-1/2}. \tag{9.15}$$

If, starting from a particular configuration, it is possible to determine the ratio $B_i/B_e$, Eqs. (9.13), (9.14) and (9.15) give us the possibility to deduce the dimensions of the diffusive region as functions of $R_e$ and $S_e$. Moreover, if the model provides further relations among the quantities defining the model, for instance between $B_e$, $B_i$ and $L_e/L_i$, we may arrive at an estimate of $R_e$ in terms of $S_e$ alone, as in the case of the Sweet-Parker model. The prototype of this kind of models is that proposed by Petschek in 1964. The basic starting point of this model is the deep analogy between the interaction of a supersonic gas flow with an obstacle and that of a plasma impinging on the diffusive region with a speed larger than a characteristic velocity of the system. In both cases shocks are formed and this allows the dissipation to take place not only in the resistive region, but also on the shock fronts, thus increasing the efficiency of the energy conversion.

The detailed treatment of the Petschek model falls beyond the scope of this book: we therefore will restrict ourselves to the simple description of its main results. The characteristic velocity mentioned above is that of a "slow" shock, $B_n/\sqrt{4\pi\rho}$, where $B_n$ is the magnetic field component normal to the shock plane. For a geometry of the Sweet-Parker type, it can be shown that a system of four steady shocks forms, emanating from the vertices of the resistive region. The dissipation is mostly concentrated on the shock fronts. For a specific heats ratio $\gamma = 5/3$, two fifths of the incoming magnetic energy are converted into heat and three fifths into kinetic energy of accelerated particles. The maximal reconnection rate turns out to be:

$$R_e(max) \propto (\ln\, S_e)^{-1},$$

much larger than the Sweet-Parker one for most plasmas. Moreover, since the dependence of $R_e$ on $S_e$ is only logarithmic, the reconnection rate is only weakly sensitive even to large variations of $S_e$.

In spite of these positive features, the Petschek model does not seem to be able to correctly describe the physics of the resistive processes, except when an anomalous resistivity manifests itself in small localized regions, as demonstrated by numerical simulations. Notice that the resistivity is only one of the parameters capable of invalidating the frozen-flux theorem: in the generalized Ohm equation, besides the collisional resistive term, at least the Hall term and the gradient of the electronic pressure term are present. Numerical simulations have proved that the Hall effect can indeed help establishing a reconnection regime faster than Sweet-Parker. Without going into details, a way to understand this speed-up of the reconnection rate is to realize that the latter depends on the fact that the evacuation of the reconnected plasma from the Sweet-Parker layer is caused by waves generated by the relaxation of the tension of the reconnected magnetic field lines, that proceeds at the Alfvén speed. On the other hand, if the resistivity is sufficiently small, the current sheet can shrink to dimensions comparable to the gyration radius of the ions, whose dynamics will become decoupled from that of the electrons, still frozen in the magnetic field. At those scales, as shown in Chap. 7, the Hall term is no longer negligible. As we have seen, the Hall effect introduces, among the possible modes, a whistler wave, obeying a quadratic dispersion relation in the wave vector $k$ that can propagate at a speed larger than the Alfvén velocity. The emission of whistler waves from current sheet in Hall-MHD simulations shows that these waves may induce reconnection rates faster than those predicted by the Sweet-Parker model.

It is natural to enquire if other ways to speed-up the reconnection process exist, even within the framework of a classical collisional resistivity, also by considering configurations different from the isolated and steady current sheet so far considered. The last few years have witnessed a growing evidence of the existence of a more complex dynamics in the observed or numerically simulated current sheets. They tend to break and to form multiple magnetic islands, or plasmoids, which in turn may coalesce or move away from each other, forming increasingly narrow current sheets. This calls for the study of the possible instabilities of a current layer, the subject of a subsequent section.

## 9.2 Spontaneous Reconnection

The reconnection processes so far considered assumed the presence of appropriate material flows dragging the field towards one or more dissipative regions. The origin of those flows was left unspecified and the possibility that the system would eventually reach a steady state was implicitly assumed. At this point it seems natural to inquire whether a resistive process could develop *spontaneously* without the need of imposing a velocity field from the outside. In this case we speak of *resistive instabilities*: they can have a great influence both in laboratory and astrophysical plasmas, provided that their reconnection rates are considerably higher than the rate of pure diffusion. The latter, as we know, develop on timescales of the order of $\tau_d = L\, c_a/\eta = S\, \tau_a$, where $\tau_a = L/c_a$, is the time needed to cover the distance $L$ at the Alfvén speed. Since in practice all plasmas have $S \gg 1$ (see Table 5.1), the diffusive times are generally too long to be of interest. As we shall see, resistive instabilities develop on timescales proportional to $\tau_d(\tau_a/\tau_d)^p = \tau_d\, S^{-p}$, with $0 < p < 1$, which might allow for growth rates $\gamma = \tau^{-1}$ sufficiently fast to be of physical interest.

The origin of these instabilities is connected with the tendency of homogeneous current sheets to evolve into a series of current filaments, which in turn diffuse and release magnetic energy. This evolution may originate from several causes. Density variations at small spatial scales may cause the formation of local structures, but have little effect on the global field structure. These instabilities are known as the *gravitational modes* and *rippling modes* and may act as local sources of turbulence. The most important instability is known as *the tearing instability*. Like all instabilities it is characterized by a wavelength $\lambda$ (or by a wave number $k = 2\pi/\lambda$) and by a growth time $\tau$ (or by a growth rate $\gamma = 1/\tau$). It is a *long wavelength* instability, capable of producing important global changes in the field structure, with a growth rate given by

$$\tau \simeq \tau_d \left(\frac{\tau_a}{\tau_d}\right)^{2/5} (kL)^{-2/5}.$$

The tearing instability develops even in the absence of a neutral line, since the dynamics associated to it is substantially unaltered by the presence of a field component parallel to the initial current sheet. As we shall see, in this case the instability is localized close to the points where

$$\boldsymbol{k} \cdot \boldsymbol{B}_0 = 0.$$

The tearing instability was first analyzed in a fundamental paper by Furth, Killeen and Rosenbluth in 1963 [20]. We shall now illustrate the most relevant points of that paper.

### 9.2.1 Tearing Mode Instability

The model to be discussed has the following characteristic features. In the unperturbed state the velocity vanishes and the magnetic field is given by:

$$\boldsymbol{B}_0 = B_{0x}(y)\, \boldsymbol{e}_x + B_{0z}\, \boldsymbol{e}_z,$$

with $B_{0x}(0) = 0$, $B_{0x}(\pm\infty) = \pm\bar{B}$. The Alfvén speed is defined by

$$c_a = \frac{\bar{B}^2}{4\pi\rho_0}.$$

Even if not strictly necessary, we shall assume for simplicity that $B_{0z}$ is constant and that the plasma is incompressible, $\rho = \rho_0 = const$. At zeroth order the induction equation reduces to

$$\nabla \times (\eta \nabla \times B_0) = 0.$$

If we further assume that the resistivity $\eta$ is also constant, we see that the preceding equation implies that

$$\frac{\mathrm{d}^2 B_{0x}}{\mathrm{d}y^2} = 0,$$

a condition that would impose too strong a limitation on the form of the equilibrium state. In reality, the unperturbed state will not be an equilibrium state in a strict sense, but will evolve on time scales of the order of $\tau_d = L^2/\eta$, exceedingly long with respect to the dynamical evolution time of the instability, which, as already remarked, develops on time scales intermediate between $\tau_a$ and $\tau_d$. In other words, the regime under consideration is defined by

$$\tau_a \ll \tau \ll \tau_d \qquad \text{or} \qquad \gamma\tau_a \ll 1 \ll \gamma\tau_d.$$

In our configuration the current density is given by:

$$\boldsymbol{J}_0 = \frac{c}{4\pi}(\nabla \times \boldsymbol{B}_0) = -\frac{c}{4\pi} B'_{0x}\, \boldsymbol{e}_z,$$

where the prime indicates the derivative with respect to $y$,

The starting point are the resistive MHD equations. Following the usual procedure, we shall consider velocity and field perturbations given by:

$$\boldsymbol{B}_1 = B_{1x}(x, y)\,\boldsymbol{e}_x + B_{1y}(x, y)\,\boldsymbol{e}_y; \quad \boldsymbol{v}_1 = v_{1x}(x, y)\boldsymbol{e}_x + v_{1y}(x, y)\boldsymbol{e}_y.$$

By linearizing all the equations we get:

$$\nabla \cdot \boldsymbol{v}_1 = 0 \qquad \text{continuity equation,}$$

$$\rho_0 \frac{\partial \boldsymbol{v}_1}{\partial t} = -\nabla P_1 + \frac{1}{c}\left(\boldsymbol{J}_0 \times \boldsymbol{B}_1 + \boldsymbol{J}_1 \times \boldsymbol{B}_0\right) \qquad \text{momentum equation,}$$

$$\frac{\partial \boldsymbol{B}_1}{\partial t} = \nabla \times \left(\boldsymbol{v}_1 \times \boldsymbol{B}_0\right) + \eta \nabla^2 \boldsymbol{B}_1 \qquad \text{induction equation,}$$

together with the condition $\nabla \cdot \boldsymbol{B}_1 = 0$.

The effect of the instability will be the generation of a magnetic field with a component $B_{1y} \neq 0$, that was absent in the equilibrium field and of plasma flows in the $x, y$ plane dragging the magnetic field $\boldsymbol{B}_0$ towards the dissipative region on both sides of the line $y = 0$. It then follows that the $y$-component of the plasma velocity $v_{1y}$ is an *odd* function of $y$, with $v_{1y} < 0$ for $y > 0$. This observation will turn out to be useful in the following.

By taking the curl of the momentum equation to eliminate the pressure term we obtain the system:

$$\gamma\rho_0(\nabla \times \boldsymbol{v}_1) = \frac{1}{c}\nabla \times \left(\boldsymbol{J}_0 \times \boldsymbol{B}_1 + \boldsymbol{J}_1 \times \boldsymbol{B}_0\right), \tag{9.16a}$$

$$\gamma\boldsymbol{B}_1 = \nabla \times \left(\boldsymbol{v}_1 \times \boldsymbol{B}_0\right) + \eta\nabla^2\boldsymbol{B}_1, \tag{9.16b}$$

in addition to the equations $\nabla \cdot \boldsymbol{B}_1 = 0$ and $\nabla \cdot \boldsymbol{v}_1 = 0$. At this point it turns out to be convenient to perform a Fourier expansion with respect to the ignorable coordinates $x$ and $t$. Since we aim at determining the growth rate of the instability, we shall replace $-\mathrm{i}\omega$ with $\gamma$ in the temporal dependence of the Fourier expansions. A generic first order quantity $f_1(x, y, t)$ will thus be represented by:

$$f_1(x, y, t) = f(y)\mathrm{e}^{\mathrm{i}kx+\gamma t}.$$

$\gamma > 0$ corresponds to an unstable situation. We therefore have:

$$v_{1x} = v_x(y)\mathrm{e}^{\mathrm{i}kx+\gamma t}; \quad v_{1y} = v_y(y)\mathrm{e}^{\mathrm{i}kx+\gamma t},$$

and an analogous representation for $\boldsymbol{B}_1$. The two equation for the divergence of $\boldsymbol{B}_1$ and $\boldsymbol{v}_1$ that now read:

$$\mathrm{i}kb_x(y) + b_y'(y) = 0 \quad \text{and} \quad \mathrm{i}kv_x(y) + v_y'(y) = 0,$$

allow us to eliminate $v_x$ e $b_x$ from the system.

Referring to the $z$-component of Eq. (9.16a), introduce in the rhs the definition of $\boldsymbol{J}$ in terms of $\boldsymbol{B}$. A long, although straightforward, calculation produces the following equation

$$\gamma(v'' - k^2v) = \frac{\mathrm{i}k}{4\pi\rho_0}\left[B_{0x}(b'' - k^2b) - B_{0x}''b\right] \tag{9.17}$$

where, to simplify the notation we have replaced $v_y$ e $b_y$ with $v$ and $b$, respectively. Finally, the $y$-component of Eq. (9.16b) gives us

$$\gamma b = \mathrm{i}kB_{0x}v + \eta(b'' - k^2b). \tag{9.18}$$

The fourth order differential system formed by Eqs. (9.17) and (9.18) for the unknowns functions $b$ and $v$ is the basic system for the analysis of the tearing instability.

Let's first observe that the effect of resistivity will be confined, as usual, to a thin layer, of thickness $\ell$, close to $y = 0$, where the $x$-component of the equilibrium field vanishes. Notice that the $z$-component of the field, $B_{oz}$, has disappeared from the system and therefore does not influence the development of the instability. The magnetic field, however, does not vanish in $y = 0$, where in fact $\boldsymbol{B} = B_{oz}\boldsymbol{e}_z$, but, in that plane, $\boldsymbol{k} \cdot \boldsymbol{B}_0 = 0$, as anticipated. Moreover, Eq. (9.18) tells us that within the resistive layer $b''$ ($\approx \gamma b/\eta$) has to be very large because of the smallness of $\eta$.

Outside the resistive layer the plasma behaves as an ideal plasma and this allows us to put $\eta = 0$ almost everywhere. At first sight then, it would seem possible to consider the term proportional to $\eta$ in Eq. (9.18) as a perturbation applied to the ideal system. A more detailed analysis, however, shows that actually we are dealing with a *singular* perturbation, meaning that when we move from a situation in which $\eta = 0$ to one having $\eta \neq 0$ the order of the differential system changes from second to fourth order and this gives rise to difficulties with the number of initial conditions to be imposed on the system. The approach to be followed with this kind of problems is based on *boundary layer* theory, which basically consists in the separation of the region of integration into two distinct domains: that *external* to the boundary layer, where we put $\eta = 0$ and the *internal* one where resistivity has to be taken into account. Let $a$ be the spatial scale of the external region, namely the macroscopic scale for the variation of the magnetic field in the direction *transverse* to the extension of the resistive layer, which is connected to the tickness of the current layer generating $\boldsymbol{B}_0$. The quantities $\tau_a$ and $\tau_d$ will now be defined as $\tau_a = a/c_a$ and $\tau_d = a^2/\eta$. The arguments presented in the section on the Sweet-Parker model allow us to identify $a$ with the quantity $\ell$ of that section.

We expect $a$ to be much larger than the thickness of the boundary layer, $\delta$, an unknown quantity whose value will be determined together with the actual solution of the problem. Making use of the boundary layer technique, we shall first solve the external problem and evaluate the unknown functions and their derivatives at the border of the boundary layer. Next, we shall solve the internal problem and connect smoothly the internal and external solutions at the border of the boundary layer. As we shall see, the conditions that allow such a smooth connection determine both the thickness of the resistive layer $\delta$ and the value of the growth rate $\gamma$. This technique turns out to be very efficient since in the external region only the solution of a second order equation is required, while in the internal one the smallness of the spatial scale usually allows important simplifications to be introduced. Furthermore, this procedure is essential whenever the problem has to be solved numerically. In fact, it would be impossible to solve the resistive problem everywhere, because of the huge value of the ratio $a/\delta$ which would force the use of an extremely fine integration grid with a consequent prohibitive number of integration points. On the contrary, the boundary layer technique greatly reduces the number of integration points, both in the external region, where the spatial scale is large, and in the internal one, of modest extension.

The actual solution of the system of Eqs. (9.17) and (9.18) requires the knowledge of a specific profile for $B_{0x}(y)$ and presents substantial technical difficulties even for relatively simple profiles, such as the so-called Harris profile: $B_{0x} = \bar{B}\tanh(y/a)$. However, it is possible to find the correct expressions for $\delta$ and $\gamma$ by using simple dimensional considerations and heuristic arguments, as we shall now do,

The diffusive layer extends in $y$ from $-\delta/2$ to $\delta/2$ and *in its interior* the resistive term in Eq. (9.18) will dominate over the convective one, while the latter will be the dominant one outside. The point $y = \delta/2$ can therefore be taken as the point where the two terms have the same order of magnitude. Since, as already anticipated, the regime of interest is one of "long"wavelengths, $\lambda \gg a \gg \delta$, in the internal region the term $b'' \simeq b/\delta^2$ will be much larger than $k^2 b \simeq b/\lambda^2$, which, in turn, will be much larger than $B''_{0x} b$ since the equilibrium field varies slowly on the resistive scale. In a similar way, the term $v''_{1y}$ will be dominant with respect to $k^2 v$. Equation (9.17) thus reduces to:

$$\gamma v'' = \frac{ikB_{0x}}{4\pi\rho_0} b'' \quad \text{for} - \delta/2 < y < \delta/2 \tag{9.19}$$

and Eq. (9.18) becomes simply:

$$\gamma b = ikB_{0x} v + \eta b'' \tag{9.20}$$

Equations (9.19) and (9.20) are the equations for the internal region.

To determine the external solution, let $\eta \to 0$ in Eq. (9.18) and eliminate $v$ between this equation and Eq. (9.17). The resulting equation reads

$$b'' - k^2 b - \frac{B''_{0x}}{B_{0x}} b - (\gamma\tau_a)^2 (ka)^{-2} \left(\frac{\bar{B}}{B_{0x}}\right)^2 \left(\frac{B'_{0x}}{B_{0x}}\right) \left(b' - \frac{B'_{0x}}{B_{0x}} b\right) = 0.$$

Putting

$$\gamma\tau_a = \frac{\gamma\tau_d}{S},$$

and keeping in mind that the growth rate of pour instability $\gamma = 1/\tau$ must have a finite value also for $S \gg 1$, we see that the terms proportional to $\gamma\tau_a$ can be neglected and therefore the equation for the magnetic field in the ideal case is:

$$b'' - k^2 b - \frac{B''_{0x}}{B_{0x}} b = 0. \tag{9.21}$$

If the resistivity could be neglected *everywhere*, or equivalently, if Eq. (9.21) were valid for *all values* of $y$, we would get a solution continuous in $y = 0$, but with a discontinuous derivative in that point. To illustrate this general property of the solutions of Eq. (9.21), let's consider the case $B''_{0x} = 0$, which, as already noted, would be a rigorous equilibrium solution for a constant $\eta$. The ideal equation then reduces to

$$b'' - k^2 b = 0,$$

whose solution, with the appropriate convergence properties at $\pm\infty$, is:

$$b(y) = b(0)\mathrm{e}^{-|ky|},$$

exhibits a discontinuous derivative in $y = 0$.

The jump of the solution going from the external to the internal region is normally expressed in terms of the non dimensional parameter

$$\Delta' = a\,\frac{b'(\delta/2) - b'(-\delta/2)}{b(\delta/2)}.$$

The assumed continuity of both $b(y)$ and its derivative in $y = \delta/2$, implies that the value of $\Delta$ can be computed using only the *external* solution. In general $\Delta$ must be computed numerically, but can be given an analytical expression for particular choices of the profile of $B_{0x}$. An important case is that of the already mentioned Harris profile, $B_{0x} = \bar{B}\tanh(y/a)$ for which it is possible to show that

$$\Delta' = 2\frac{1-(ka)^2}{ka}.$$

To procedure to obtain solutions continuous together with their first derivative is the following. We first put $\eta = 0$ only in the region $|y| > \delta/2$. In the internal region, where $\eta \neq 0$, the second derivative of $b$ must be sufficiently large to eliminate the discontinuity of the first derivative. In that region we may write:

$$b''(y) = \lim_{\delta\to 0}\frac{b'(\delta/2) - b'(-\delta/2)}{\delta} \approx \frac{b'(\delta/2) - b'(-\delta/2)}{b(\delta/2)}\,\frac{b(\delta/2)}{\delta} = \frac{\Delta'\,b(\delta/2)}{a\,\delta}, \tag{9.22}$$

An approximate estimate of Eq. (9.19) at $y = \delta/2$ is obtained making use of Eq. (9.22), namely

$$v''(\delta/2) \simeq -\frac{v(\delta/2)}{\delta^2}.$$

The presence of the minus sign in the preceding expression is due to the fact that $v$ is an odd function of $y$ while $v''(\delta/2) > 0$, as can be realized considering, for instance, a function of the type $\sin(y)$. We finally get:

$$-\gamma\left(\frac{v}{\delta^2}\right) = \frac{\mathrm{i}k B_{0x}}{4\pi\rho_0}\frac{b}{a\delta}\Delta',$$

where all the functions are evaluated in $y = \delta/2$.

On the other hand, Eq. (9.20), always for the external solution, gives

$$\gamma b \simeq \mathrm{i}k B_{0x}\, v,$$

and, combining together the last two equations,

$$\gamma^2 = \frac{(k B_{0x})^2}{4\pi \rho_0} \frac{\delta}{a} \Delta',$$

Since $B_{0x}(\delta/2) \simeq B_{0x}(0) + (\delta/2) B'_{0x}(0) = (\delta/2) B'_{0x}(0)$ we finally arrive at:

$$\gamma^2 \simeq \frac{1}{4} \frac{(k B'_{0x}(0))^2}{4\pi \rho_0} \frac{\delta^3}{a} \Delta' = A \left( \frac{(ka)}{\tau_a} \right)^2 \left( \frac{\delta}{a} \right)^3 \Delta', \tag{9.23}$$

with

$$A = \left( \frac{a B'_{0x}(0)}{2\bar{B}} \right)^2$$

This equation shows that in order to have a real $\gamma$ (and thus an instability) we must have $\Delta' > 0$.

Let's now turn to the internal equation, which, using Eq. (9.22), can be written as

$$\gamma b(\delta/2) \simeq \eta b'' \simeq \eta \frac{\Delta'}{a\delta} b(\delta/2),$$

or

$$\gamma \simeq \eta \frac{\Delta'}{a\delta} \quad \text{implying} \quad \frac{\delta}{a} \simeq \frac{\Delta'}{\gamma \tau_d}.$$

By eliminating $\delta/a$ between the above equation and Eq. (9.23) we obtain an estimate of the growth rate:

$$\gamma\, \tau_d \simeq A^{1/5} (ka)^{2/5}\, S^{2/5} (\Delta')^{4/5}. \tag{9.24}$$

Alternatively, if we eliminate $\gamma$ between the same equations, we obtain an estimate of the thickness of the resistive layer:

$$\frac{\delta}{a} \simeq A^{-1/5} (ka)^{-2/5} S^{-2/5} (\Delta')^{1/5}. \tag{9.25}$$

If the values of $A$ and $\Delta'$ just found are used for the Harris current sheet, $B_{0x} = \tanh(y/a)$, and the diffusive time, instead of the Alfvén time, is used to normalize the times, we arrive at the dispersion relation

$$\gamma\,\tau_a \simeq 2^{2/5}(ka)^{-2/5}\,S^{-3/5}(1-k^2a^2)^{4/5}. \tag{9.26}$$

A more detailed analysis shows that the approximations used limit the validity of our result to wavelengths satisfying the condition $kaS^{1/4} \gg 1$. The maximum growth rate, obtained for $kaS^{1/4} \simeq 1$, scales as $\gamma\,\tau_a \simeq S^{-1/2}$. Notice that such a growth rate, associated with a timescale which is orders of magnitude smaller than the diffusive timescale for large Lundquist numbers, can still be extremely slow. Consider the solar flare phenomenon, which has timescales of an hour or less, with a coronal Lundquist number of $S \geq 10^{13}$: for a typical Alfvén crossing time say of 10 s, the timescale associated with the fastest growing tearing mode becomes over a year! A possible way out of this timescale problem may come from considering current sheets with a thickness which is much smaller than the macroscopic scale, as, for example, was found in the Sweet Parker current sheet.

### *9.2.2 The Plasmoid Instability of Thin Current Sheets*

The analysis carried out for the tearing mode illustrated that current layers are unstable under quite general conditions, over timescales scaling as $\tau/\tau_a \sim S^{1/2}$. This timescale seems to be similar to that associated with the reconnection time for the Sweet-Parker current sheet discussed previously (Eq. (9.11)), and this might lead to believe there is a connection between the two results. Careful consideration however shows this coincidence to be accidental: the internal, dissipation layer of the tearing mode was given by Eq. (9.25), which together with the relation $ka \simeq S^{-1/4}$, leads to $\delta/a \sim S^{-3/10}$, or identifying the current shear scale $a$ with a macroscopic scale $L$, leads to a singular current layer which is thicker than the Sweet-Parker sheet $a/L \sim S^{-1/2}$. This suggests that Sweet-Parker's stationary reconnection solution might also be unstable to tearing modes, which might have even faster growth rates than that of the classical tearing mode.

To see if this might happen, one must start by considering the same normalizations for both the tearing mode and the Sweet-Parker model, specifically as concerns the Lundquist number, which was based on the current sheet length $L$ for the Sweet-Parker case, but normalized to a current sheet thickness $a$ for the tearing mode. As a consequence, the definitions of $\tau_a$ and $\tau_d$ as well as $S = \tau_d/\tau_a$ were different. Using an asterisk to denote the values as defined in the section on the tearing mode, and leaving the Sweet-Parker values as they are, we find that:

$$\tau^* = \tau_a^*(S^*)^{1/2} = \frac{a}{c_a}\left(\frac{ac_a}{\eta}\right)^{1/2} = \frac{L}{c_a}\left(\frac{Lc_a}{\eta}\right)^{1/2}\left(\frac{a}{L}\right)^{3/2}.$$

Now remembering that the Sweet-Parker aspect ratio is $a/L \sim S^{-1/2}$, the following hypothetical result is found for a tearing mode on the Sweet-Parker current sheet:

$$\tau^* \sim \tau_a S^{1/2} S^{-3/4} = \tau S^{-1/4}. \tag{9.27}$$

This means that if the Sweet-Parker current sheet were found to be unstable, the instability would grow on a timescale which would tend to vanish as the Lundquist number $S \to \infty$. Before discussing this result we analyze the tearing instability for the Sweet-Parker current sheet following the reasoning of Loureiro et al. (2007) [22].

Again, we must linearize the equation of motion and the induction equation, but this time including the background flow in the equilibrium of the Sweet-Parker model and therefore generalizing the Eqs. (9.17) and (9.18). In this case, the linearized equations take the form

$$\rho_0(\nabla \times [\frac{\partial \boldsymbol{v}_1}{\partial t} + \boldsymbol{v}_1 \cdot \nabla \boldsymbol{v}_0 + \boldsymbol{v}_0 \cdot \nabla \boldsymbol{v}_1] = \frac{1}{c} \nabla \times \left(\boldsymbol{J}_0 \times \boldsymbol{B}_1 + \boldsymbol{J}_1 \times \boldsymbol{B}_0\right), \quad (9.28a)$$

$$\frac{\partial \boldsymbol{B}_1}{\partial t} = \nabla \times \left(\boldsymbol{v}_1 \times \boldsymbol{B}_0\right) + \nabla \times \left(\boldsymbol{v}_0 \times \boldsymbol{B}_1\right) + \eta \nabla^2 \boldsymbol{B}_1, \quad (9.28b)$$

where $\boldsymbol{v}_0$ is the background flow, defined by $\boldsymbol{v}_0 \equiv [v_0 x/l, -v_0 y/l, 0]$. Proceeding as in the static case, taking the $z$ component of Eqs. (9.28a) and (9.28b), care must be taken to include the spatial gradients of $\boldsymbol{v}_0$. Differentiating the equation of motion along the $x$ direction, and using incompressibility, one finds

$$\rho_0\left(\frac{\partial}{\partial t} + \boldsymbol{v}_0 \cdot \nabla\right) \nabla_\perp^2 v_{1y} + \rho_0 \frac{v_0}{l} \nabla_\perp^2 v_{1y} = \frac{B_{0x}}{4\pi} \frac{\partial}{\partial x} \nabla_\perp^2 b_{1y} - \frac{1}{4\pi} \frac{\partial b_{1y}}{\partial x} \frac{\partial^2 B_{0x}}{\partial y^2} \quad (9.29a)$$

$$\frac{\partial b_{1y}}{\partial t} + \boldsymbol{v}_0 \cdot \nabla b_{1y} = B_{0x} \frac{\partial b_{1y}}{\partial y} - \frac{v_0}{l} b_{1y} + \eta \nabla^2 b_{1y}, \quad (9.29b)$$

where $\nabla_\perp^2 = \partial^2/\partial x^2 + \partial^2/\partial y^2$ and nothing has yet been stated about the time-dependence of perturbations. The equations above depend explicitly on both the coordinates $y$, as before, but also $x$, because of the background flow. It is still possible, however, to eliminate this dependence completely by still taking the ansatz $v_{1y} = v(y, t)\mathrm{e}^{ikx}$ (and the same for $B_{1y}$) but allowing for a time dependence in the wavevector $k = k(t)$. This is because the background acceleration along the current sheet has the effect of stretching the wavelength in the $x$ direction, so that one expects $k(t) = k_0 \exp(-v_0 t/a)$. In this way the equations for $v$, $b$ reduce, as previously, to having only an explicit $y$ dependence. Now if we search for eigenmodes with growth rates $\gamma$, i.e. solutions of the form $v(y, t) = v(y)\mathrm{e}^{\gamma t}$, with $\gamma \gg v_0/a$, terms proportional to $v_0/a$ in the linearized equations may be neglected, and Eqs. (9.17) and (9.18) are recovered.

In this way, the instability of the Sweet-Parker current sheet has been shown more rigorously, since the scale of the equilibrium is precisely the thickness of the Sweet-Parker current sheet $x_0 = (a\eta/v_0)^{1/2}$, so, as long as the $\Delta'$ is positive for this equilbrium, the sheet is unstable with the fast timescale given by (9.27). We will not calculate $\Delta'$ explicitly for the Sweet-Parker equilibrium, but the result indeed shows that the Sweet-Parker current sheet is unstable to this *super-tearing* mode or

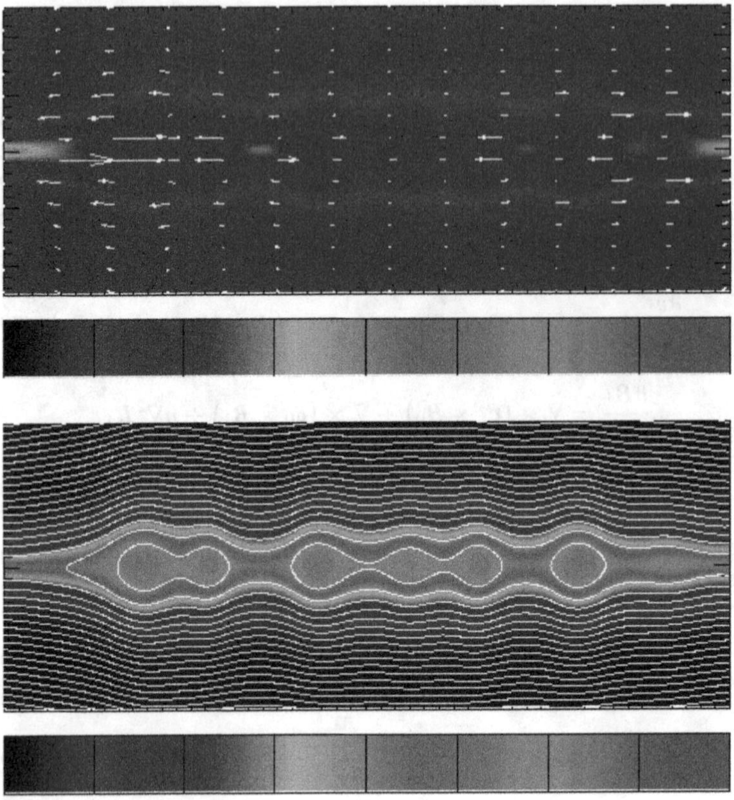

**Fig. 9.8** Island formation in a Sweet-Parker current sheet with Lundquist $S \simeq 10^5$ after about 2 Alfvén times showing multiple island formation. *Top*, current intensity (peaking at x-points and separatrices). *Bottom*, density and magnetic field lines. Units arbitrary, color contours increasing from *blue* to *red*

*plasmoid* instability, so called for its fast evolution leading to the formation of many islands, as shown in Fig. 9.8.

From the linear point of view, the existence of instabilities with a growth rate which increases with increasing Lundquist number is problematical. In ideal MHD, magnetic reconnection is prohibited, and the finding of an "infinitely unstable" mode in the limit of infinite Lundquist number points to a strong singularity in the behaviour of the MHD equations. Diffusive terms change the order of the equations, even in the linearized equations, and though it would not be too surprising to find a different physical behaviour between systems described by ideal MHD and those described by the resistive equations with $S \to \infty$, the presence in the latter case of infinitely fast growing instabilities has only one possible interpretation: the un-realizability of the corresponding equilibrium configurations (one could also argue that an infinite growth rate is non-causal and unphysical). Pucci and Velli (2013) [23] have shown that in the asymptotic $S \to \infty$ regime an effectively "ideal" tearing mode survives, in the sense that its growth rate does not depend on the Lundquist number.

The instability also defines a maximum critical aspect ratio (or ratio between current sheet width and length) $L/a \sim S^{1/3}$, above which any current configuration must be intrinsically unstable and turbulent, and for which a laminar configuration is impossible to attain. In other words, if the MHD configuration leads to thinning sheets, an instability with an ideal growth rate arises before the current sheet collapses to the Sweet-Parker aspect ratio. This type of instability naturally leas to the formation of many islands and resulting nonlinear dynamics, which are most appropriately described within a theory of magnetohydrodynamic turbulence, which we will discuss in the following chapter.

## Questions and Problems

**9.1.** Compute the values of density, temperature, and magnetic field for the plasmas of Table 5.1. Then compute the values of the resistivity and verify that the corresponding Lundquist numbers are those given in the Table.

**9.2.** Show that for the Harris current sheet $B_{0x} = \bar{B} \tanh(y/a)$

$$\Delta' = 2\frac{1-(ka)^2}{ka}.$$

*Hint*

Show that Eq. (9.21) can be written as

$$b'' - k^2\, b + \frac{2}{a^2} sech^2(y/a)\, b = 0.$$

Look for solutions of the above equation of the form:

$$b = b_0\, e^{\mp ky}\left(1 \pm \frac{\tanh(y/a)}{ka}\right),$$

for $y > 0$ and $y < 0$, respectively.

Show that the solutions present a discontinuity in $y = 0$

$$\Delta' = \frac{b'(0^+) - b'(0^-)}{b(0)}$$

that has the required value.

# Chapter 10
# MHD Turbulence

**Abstract** Turbulent plasmas are extremely common in the astrophysical context. The phenomenological theory of isotropic and homogeneous hydrodynamical turbulence is presented and the fundamental results of Kolmogorov deduced. Their generalization to plasmas in the MHD regime is then discussed. The relevance of turbulence theory for the problem of solar coronal heating is the subject of the last section.

As illustrated in the preceding Chapters the dissipative coefficients, both resistive and viscous, are as a rule extremely small in astrophysical plasmas. Equivalently, the Reynolds and Lundquist numbers are extremely large. We know from hydrodynamics that in these circumstances fluids enter a regime where vortices of different sizes and lasting over different times, smaller vortices typically lasting less than larger ones, appear in the flow, and generally chaotic, unpredictable motions result. When the chaotic motions, associated with the vortices, are present over a wide range of spatial scales we speak of a regime of *developed turbulence*. The distinctive feature of this regime is that energy appears to be *transported* from the largest spatial scales to those sufficiently small to allow dissipation to set in. Indeed, the energy dissipation rate as measured experimentally in many different fluids of varying Reynolds numbers appears to be independent from the viscosity of the fluid, i.e., it reaches a finite value in the limit of zero viscosity.

The appearance of stronger and stronger gradients in the flow, already remarked in Chap. 8, is clearly borne out from inspection of the equation of motion of a viscous fluid, described by the Navier-Stokes equation,

$$\frac{\partial \boldsymbol{U}}{\partial t} + (\boldsymbol{U} \cdot \nabla)\boldsymbol{U} = -\frac{1}{\rho}\nabla P + \nu \nabla^2 \boldsymbol{U}, \tag{10.1}$$

where incompressibility has been assumed, $\nabla \cdot \boldsymbol{U} = 0$, and $\nu$ is the *kinematic viscosity*. The importance of the nonlinear convective term, $(\boldsymbol{U} \cdot \nabla)\boldsymbol{U}$, increases when the spatial scale $l$ decreases, (but not faster than $1/l$), and the nonlinearity of the term leads to the formation of harmonics in Fourier space and sharper gradients. For compressible motions, the convective derivative acting on sound waves is the cause of the formation of increasingly steep wave fronts. In the incompressible case,



C. Chiuderi and M. Velli, *Basics of Plasma Astrophysics*,
UNITEXT for Physics, DOI 10.1007/978-88-470-5280-2_10

the same term remains responsible for the generation of harmonics and reduced spatial scales. However, in this case vortex structures are formed, rather than steep wave fronts. On the other hand, the importance of the viscous term, $\nu\nabla^2\boldsymbol{U}$, increases as $1/l^2$, and therefore, at sufficiently small scales, the latter always dominates. Let $L$ be the spatial scale where energy is injected into the system, and $l_d$ the spatial scale where the viscous term (or, in general, the dissipative terms) is no longer negligible. If the scales $L$ and $l_d$ are separated by many orders of magnitude, $L \gg l_d$, as usually happens in astrophysical plasmas, we speak of *completely developed turbulence*.

In natural plasmas energy sources for turbulence are present either in connection with global motions or as a consequence of instabilities arising in non-potential magnetic configurations. If the spatial and temporal scales involved are such that the plasma can be described by the MHD equations, we speak of MHD turbulence. MHD turbulence plays a fundamental role in the dynamics of astrophysical plasmas in very different environments and over a very wide range of scales, from kilometers to kiloparsecs and beyond. The understanding of turbulence generated by nonlinear dynamics of plasmas and magnetic fields is essential to attack problems such as the physics of the outer layers of convective stars, like the Sun, the heating of the solar corona and the acceleration of the solar wind, or the dynamics of the molecular clouds and of the star-forming regions, where turbulence generates the viscosity needed to remove angular momentum. The interaction of turbulence with higher energy particles is a determining factor in understanding the properties of cosmic rays.

In this chapter we describe some of the main results of the modern theory of MHD turbulence, taking a phenomenological approach, based on invariance principles and on a dimensional analysis of the equations. For simplicity, we shall start from the fundamental results established by Kolmogorov for homogeneous and isotropic hydrodynamical turbulence. These results remain the basis for the further generalizations required to include the effects arising from the presence of magnetic fields.

Even if hydrodynamics and magnetohydrodynamics share several common features, the development of turbulence in a magnetized plasma is considerably different from the purely hydrodynamical case, because of the intrinsic anisotropy created by the magnetic field, which allows the existence of Alfvén waves propagating along field lines. We shall restrict our study to the case of an incompressible plasma, even though strictly incompressible plasmas don't really exist, though they remain a good approximation as long as the fluctuations are subsonic or subalfvénic, i.e. are sufficiently smaller than the characteristic speeds of propagation in the plasma. Our analysis must therefore be understood as a model for the evolution of the fluctuations at small Mach numbers, as remarked in Chap. 7. We shall then discuss the specific case of the solar corona and of the solar wind. In the latter environment, MHD turbulence has been observed and studied in great detail over the last fifty years, thanks to the many spacecraft launched into interplanetary space.

## 10.1 Homogeneous and Isotropic Hydrodynamical Turbulence

The observations of the spectra of the fluctuations of velocity, magnetic field and density measured indirectly in distant astrophysical environments, or directly *in situ* in the solar wind, have produced a general consensus on the nature of the processes operating in those contexts. The emerging picture is one where energy is injected at macroscopic scales and transferred to structures of increasingly reduced scales, until these are so small that the dissipative terms become comparable to the other terms, no matter how small the viscous and resistive coefficients might be.

The source of free energy at the injection scale may be connected, for instance, to the kinetic energy associated with the velocity difference between two streams, as in the Kelvin-Helmholtz instability, or from the difference of thermal energy between two fluid layers, as in convection, or from the free energy stored in a non-potential magnetic field, as in the case of the tearing instability. As a consequence of the non-linear terms of the equations, the energy is re-distributed over different scales, in a way that does not correspond to an equipartition, but rather to a power-law energy distribution over spatial scales. The observed spectral index of this distribution is close to 1.67. According to the theory originally proposed by Kolmogorov in 1941, there are two properties that characterize a completely developed turbulence. First: the dissipation within the fluid is independent of the exact value of the viscosity and **does not** tend to zero when the viscosity vanishes. Second: the energy transfer from the injection scales to the dissipative ones does not take place directly, but through a nonlinear cascade which transfers the energy from larger to smaller eddies. This transfer is due to the interactions between those modes that, at each step, have comparable scales. Already at scales slightly smaller than the injection ones, the nonlinear interactions that cause the cascade lose memory of the macroscopic anisotropies at the so-called injection scales, so that the turbulence acquires a homogeneous and isotropic character, at least in a statistical sense.

To put these considerations in a more quantitative form, we start again from the Navier-Stokes Eq. (10.1). The nonlinear term appears to be due to the convective derivative, however the pressure here is an unknown, and its role is to ensure incompressibility of the flow. One way to remove it, assuming a constant density, is to take the curl of that equation. Defining the *vorticity* $\boldsymbol{\omega} = \nabla \times \boldsymbol{U}$, we get

$$\frac{\partial \boldsymbol{\omega}}{\partial t} = \nabla \times (\boldsymbol{U} \times \boldsymbol{\omega}) + \nu \nabla^2 \boldsymbol{\omega}. \tag{10.2}$$

Notice that Eq. (10.2) is apparently identical to Eq. (5.25), with the vorticity $\boldsymbol{\omega}$ replacing $\boldsymbol{B}$ and the viscosity replacing the magnetic diffusivity $\eta$. However, the analogy is not complete, since Eq. (10.2) depends only on one vector field, i.e. $\boldsymbol{U}$, $\boldsymbol{\omega}$ being defined in terms of $\boldsymbol{U}$, while two independent vector fields, $\boldsymbol{U}$ and $\boldsymbol{B}$, enter Eq. (5.25).

The Reynolds number, that measures the ratio between the convective and the viscous terms on the rhs of Eq. (10.2) at the macroscopic scales or injection scales, is defined by $R = \mathcal{U}L/\nu$, where $\mathcal{U}$ e $L$ are a typical value of the velocity and its scale of variation, respectively. It is therefore clear that if the viscous term dominates

$R \ll 1$ and the fluid motions will be *laminar*, while in the opposite limit, $R \gg 1$, the motions will be *turbulent* with extremely irregular variation of the velocity at every point. We may thus identify $\mathcal{U}$ with the average, over very long times, of the effective velocity at each point of the fluid. In other words, the velocity of each fluid particle can be written as $\boldsymbol{U} = \boldsymbol{U}_0 + \boldsymbol{w}$, where $\boldsymbol{w}$ refers to the fluctuating part of the velocity, while $|\boldsymbol{U}_0| \simeq \mathcal{U}$. In an intuitive way, the flow can be described as the superposition of *turbulent eddies* of different dimension, or *scale*, the latter being defined as the order of magnitude of the distance that separates two points with substantially different speeds, $\Delta U \simeq |\boldsymbol{w}|$. If $\ell$ is the scale just defined and $w_\ell$ the corresponding velocity variation, we may associate a Reynolds number $R_\ell = w_\ell \ell/\nu$ with each scale.

In spite of the complexity and chaotic nature of the turbulent motions, a large amount of information concerning their characteristics can be deduced in a simple way making use of dimensional considerations as shown by Kolmogorov and Obukhov. When a turbulent flow sets in, vortices of large dimensions first appear, then giving rise to vortices of increasingly smaller size. In the large vortices, $\ell$ can be identified with $L$ and, on this scale, $\Delta U \simeq \mathcal{U}$ and consequently $R_L \simeq R$. For these vortices viscosity is not important and no dissipation of energy takes place. The subsequent decrease of the size of the vortices, reduces $\ell$ down to values where dissipation sets in. This happens at the scale $\ell_d$, the *dissipative scale*, corresponding to a local Reynolds number $R_d \simeq 1$. The energy, injected in kinetic form at the scale $L$, is transferred unaltered to smaller scales until the dissipative scale is reached, where it is converted into heat.

The fact that the energy transfer takes place in a conservative way follows directly from the properties of the nonlinear terms of the Navier-Stokes equation, namely the convective derivative and the pressure gradient. As mentioned above, the role of the pressure term is to ensure incompressibility of the fluid, $\nabla \cdot \boldsymbol{U} = 0$. It can be calculated by taking the divergence of the Navier-Stokes equation and imposing incompressibility, which leads to the Poisson-type equation for the pressure:

$$\nabla^2\left(\frac{P}{\rho}\right) = -\nabla \cdot (\boldsymbol{U} \cdot \nabla \boldsymbol{U}). \tag{10.3}$$

Therefore, the pressure field necessary to maintain the fluid's incompressibility is also a *nonlinear* function of the convective derivative. Introducing the Fourier transform of the velocity field in three spatial dimensions as:

$$\boldsymbol{U}(\boldsymbol{x}, t) = \int d^3\boldsymbol{k}\, \boldsymbol{U}_{\mathbf{k}}(t) e^{i\boldsymbol{k}\cdot\boldsymbol{x}}, \tag{10.4}$$

and applying a Fourier transform to both the Navier-Stokes equation and the pressure Eq. (10.3), the pressure may then be eliminated using the latter, to obtain

$$\frac{\partial U_{i\mathbf{k}}}{\partial t} = \int_{\mathbf{p}+\mathbf{q}=\mathbf{k}} A_{ilm}\, U_{l\mathbf{p}} U_{m\mathbf{q}} d\boldsymbol{q} - \nu \mathrm{k}^2 \mathrm{U}_{\mathrm{i}\mathbf{k}}, \tag{10.5}$$

where the tensor $A_{ilm}$ is given by

$$A_{ilm} = -ik_m\left(\delta_{il} - \frac{k_i k_l}{k^2}\right).$$

The fluid energy is conserved by nonlinear interactions among modes of different wavenumbers in the following sense. For every set of vectors $\mathbf{k}$, $\mathbf{p}$, $\mathbf{q}$ satisfying $\mathbf{p}+\mathbf{q}=\mathbf{k}$, the nonlinear term associated with the interactions between them is such that the time derivatives of the energies in each mode, defined as $E_{\mathbf{k}} = |\boldsymbol{U}_{\mathbf{k}}|^2/2$, when calculated from the interaction with the other two modes, obeys the relationship:

$$\partial E_{\mathbf{k}}/\partial t + \partial E_{\mathbf{p}}/\partial t + \partial E_{\mathbf{q}}/\partial t = 0.$$

In other word, the nonlinear interactions conserve energy between each triad of interacting modes. Because the full nonlinear term is nothing but a sum over triads, the nonlinear terms conserve energy, but redistribute it over different wavenumbers. According to Kolmogorov, therefore, energy flows in *wavenumber space* until it reaches the smallest scales where it can be dissipated. Since viscous dissipation only takes place at the scale $\ell_d$, all quantities characterizing the fluid at scales $\ell > \ell_d$ don't depend on viscosity. This is also true of the energy per unit mass transferred from one scale to the next per unit time that we shall indicate by $\varepsilon$. The dimensions of $\varepsilon$ are evidently $[l^2 t^{-3}]$. In a steady situation we can estimate $\varepsilon$ both at the injection scale and at the dissipation scale. At injection $\varepsilon$ can depend only on the density $\rho$, on the macroscopic velocity $U_0 \simeq \mathcal{U}$ and on the scale $L$. A simple dimensional estimate shows that $\varepsilon$ can be written as h

$$\varepsilon \simeq \frac{\mathcal{U}^3}{L}. \tag{10.6}$$

Similarly, at the dissipation scale we may write:

$$\varepsilon \simeq \frac{w_d^3}{\ell_d},$$

from which follows:

$$\left(\frac{\mathcal{U}}{w_d}\right)^3 \simeq \frac{L}{\ell_d}. \tag{10.7}$$

Equation (10.6) allows us to define the concept of *turbulent viscosity*. Recalling the expression for energy dissipation in a viscous fluid, Eq. (8.37), we see that the energy dissipation rate per unit mass, our $\varepsilon$, can be written as

$$\varepsilon = \frac{1}{M}\frac{\mathrm{d}E}{\mathrm{d}t} = -\frac{1}{2}\frac{\tilde{\nu}}{V}\int\left(\frac{\partial U_i}{\partial x_j} + \frac{\partial U_j}{\partial x_i}\right)^2 \mathrm{d}V,$$

where $\tilde{\nu}$ is the viscosity relevant to the case under study. A dimensional estimate of the above expression shows that $\varepsilon \simeq \tilde{\nu}\,(\mathcal{U}/L)^2$. A comparison with Eq. (10.6) then shows that

$$\tilde{\nu} \simeq \mathcal{U}\,L = \nu\,R. \tag{10.8}$$

The ratio of the turbulent viscosity to the kinematic viscosity is thus equal to $R \gg 1$: the turbulent viscosity exceeds the kinematical one by a very large factor. The condition $R_d \simeq 1$ can now be used to obtain an estimate of the ratio between the dissipative scale $\ell_d$ and the injection scale $L$. By combining this condition with Eq. (10.7) we get:

$$\ell_d = \frac{\nu}{w_d} = \frac{\nu}{\mathcal{U}}\left(\frac{L}{\ell_d}\right)^{1/3} \quad \Longrightarrow \quad \ell_d^{4/3} = \frac{\nu}{\mathcal{U}} L^{1/3} = \frac{L^{4/3}}{R},$$

and finally

$$\frac{\ell_d}{L} = R^{-3/4},$$

which shows that the scale $\ell_d$ is much smaller than the injection scale $L$. The intermediate scales, $L \gg \ell \gg \ell_d$, are associated with the so-called *inertial range* where a straightforward transfer of energy from one scale to the next takes place. In this range it is easy to evaluate the order of magnitude of the turbulent velocity associated with a particular scale $\ell$. In fact, if for each scale $\ell$, we write $\varepsilon \propto \tilde{\nu}(w_\ell/\ell)^2$ and use Eq. (10.8) we obtain

$$\varepsilon \propto w_\ell^3/\ell \quad \Longrightarrow \quad w_\ell \propto (\varepsilon\ell)^{1/3},$$

known as the *Kolmogorov-Obukhov law*. The relationships just found can be used to define the timescale for the transfer of energy from one scale to the next, $\tau^*$, that in this case coincides with the so-called *turnover time*: $\tau^* = \tau_{nl} \equiv \ell/w_\ell$.

The preceding results can be cast in a different form by introducing the spectral density (per unit mass), $E(k)$, where $k$ is the wavenumber associated with the scale $\ell$, $k \simeq 1/\ell$. Let's write now the total energy per unit mass as

$$E/m \propto \int E(k)\mathrm{d}k,$$

from which we deduce that the dimensions of $E(k)$ are $[l^3\,t^{-2}]$. On the other hand, we know that in the inertial range $E(k)$ can depend only on $\varepsilon$ and $k$. Therefore

$$E(k) \propto \varepsilon^{2/3}\,k^{-5/3}, \tag{10.9}$$

which is the celebrated *Kolmogorov spectrum*.

Actually, Kolmogorov proved a more rigorous result, the so-called *four-fifths law*. To describe it, it is first necessary define the velocity increments calculated between two points of the fluid separated by a distance given by the vector $\boldsymbol{r}$:

$$\delta \boldsymbol{U}(\boldsymbol{r}) = \boldsymbol{U}(\boldsymbol{x}+\boldsymbol{r}) - \boldsymbol{U}(\boldsymbol{x}). \tag{10.10}$$

The concepts of isotropy and homogeneity are summarized by the statistical properties of powers of velocity increments $\delta \boldsymbol{U}$ (called the structure functions of the relevant order $n$)

$$S_i^n(r) = \left\langle \delta U_i^n \right\rangle \tag{10.11}$$

The structure functions depend, for an isotropic and homogeneous turbulence, only on the distance between the two points, $r$, and not on the origin chosen for the increments $\boldsymbol{x}$ (homogeneity), nor on the direction of the vector $\boldsymbol{r}$ (isotropy). Statistical averages performed over the components of $\delta \boldsymbol{U}$ can consist either in time averages (in a steady state case) or spatial averages (in a region where the turbulence can be considered homogeneous, in which case frequency structure functions must not depend on the origin in time). Structure functions are the quantities that are generally used in the experimental study of the properties of turbulence, both in terrestrial labs and when *in situ* measurements are performed in space.

Kolmogorov's law is connected to the dependence of the third-order structure function on scale and establishes that, under the assumptions of homogeneity, isotropy and vanishing viscosity, the longitudinal components (i.e. along $\boldsymbol{r}$) of the third order structure function is proportional to both the energy dissipated in the fluid and the scale $l$

$$S^3_{long}(l) = -\frac{4}{5}\epsilon\, l. \tag{10.12}$$

A more detailed discussion of hydrodynamic turbulence can be found in the textbooks by Landau and Lifschitz [7], Frisch [9].

## 10.2 Magnetohydrodynamic Turbulence

Consider now a plasma described by the MHD equations, immersed in a uniform magnetic field $\boldsymbol{B}_0$, and suppose that velocity and magnetic field fluctuations are present in the form of Alfvén waves of arbitrary amplitude. The first thing to remember, as we have seen in the chapter on waves, is that Alfvén waves propagating in one direction along the field are an exact, nonlinear solution of the MHD equations. In terms of the Elsasser variables, $z^{\pm} = \boldsymbol{u} \pm \boldsymbol{b}/\sqrt{4\pi\rho}$ the incompressible MHD equations may be written:

$$\frac{\partial \mathbf{z}^{\pm}}{\partial t} \mp \mathbf{c}_a \cdot \nabla \mathbf{z}^{\pm} = -\frac{1}{\rho}\nabla p^T - \left(\mathbf{z}^{\mp} \cdot \nabla \mathbf{z}^{\pm}\right). \tag{10.13}$$

where $\boldsymbol{c}_a = \boldsymbol{B}_0/\sqrt{\rho}$ is the Alfvén velocity and $p^T = p + |\boldsymbol{B}_0 + \boldsymbol{b}|^2/8\pi$ is the total pressure, this time satisfying the Poisson-type equation

$$\nabla^2 \frac{p^T}{\rho} = -\nabla \cdot (\mathbf{z}^{\mp} \cdot \nabla \mathbf{z}^{\pm}) \tag{10.14}$$

which guarantees the incompressibility of $\mathbf{z}^{\pm}$, again found by taking the divergence of Eq. (10.13). In the above we have always assumed a constant density.

Taking the Fourier transform in space, Eq. (10.5) is generalized to

$$\left( \frac{\partial \mathbf{z}_{\mathbf{k}}^{\pm}}{\partial t} \mp i\mathbf{k}\cdot\mathbf{c}_a \mathbf{z}_{\mathbf{k}}^{\pm} \right) = -i\mathbf{P}(\mathbf{k}) \int_{\mathbf{p}+\mathbf{q}=\mathbf{k}} \mathbf{z}_{\mathbf{p}}^{\mp}\, \mathbf{z}_{\mathbf{q}}^{\pm}\, \mathrm{d}^3\mathbf{q} \tag{10.15}$$

$$\mathbf{P}(\mathbf{k})_{ilm} = k_m \left( \delta_{il} - \frac{k_i k_l}{k^2} \right). \tag{10.16}$$

The difference between hydrodynamics and magnetohydrodynamics now appears clearly, in the sense that in MHD the nonlinear interactions couple two different fields $\mathbf{z}^{\pm}$, which are nothing but Alfvén waves propagating in opposite directions along the mean magnetic field $\boldsymbol{B}_0$. If we imagine ensembles of localized wave-packets traveling along the field in opposite directions, it is clear that the nonlinear interactions among them will be limited, in time, to the effective duration over which they cross paths, and this, in general, will slow down the nonlinear cascade. A different way of understanding this slowing down effect consists in moving to an "interaction" representation of the fields, by incorporating the Alfvén wave-like temporal oscillations into the definition of the fields themselves:

$$\mathbf{z}_{\mathbf{k}}^{\pm}(t) \equiv \mathbf{z}_{\mathbf{k}}^{\pm}(t) \exp(\mp \mathrm{i}\mathbf{k}\cdot\mathbf{c}_{\mathrm{a}} t),$$

so that Eq. (10.15) becomes

$$\frac{\partial \mathbf{z}_{\mathbf{k}}^{\pm}}{\partial t} = -i\mathbf{P}(\mathbf{k}) \int_{\mathbf{p}+\mathbf{q}=\mathbf{k}} \mathbf{z}_{\mathbf{p}}^{\mp}\, \mathbf{z}_{\mathbf{q}}^{\pm} \exp(\mp 2i\, \mathbf{p}\cdot\mathbf{c}_a\, t)\, \mathrm{d}^3\mathbf{q} \tag{10.17}$$

In the hydrodynamic case both $\mathbf{z}$ fields simplify into the simple velocity fluctuations $\mathbf{u}$, and the kernel in the convolution integral does not oscillate in time, (in other words, the hydrodynamic interactions of vortices in homogeneous turbulence is always *resonant*). In MHD however the kernel oscillates in time, decreasing the intensity of nonlinear interactions, and there are no resonant interactions between modes unless the vectors $\mathbf{p}$ and $\mathbf{c}_a$ are orthogonal. This introduces a directional anisotropy in the nonlinear cascade due to the presence of the mean field which has important implications when generalizing Kolmogorov's law to MHD.

This MHD effect which slows down the cascade was introduced, even in the absence of a mean field $B_0$, by Iroshnikov and Kraichnan independently in the early sixties. The dimensional argument may be made as follows: at any given scale in the inertial range, the Elsasser field on scales larger than the one in consideration may be seen as a mean stochastic field $c_a^s$, upon which wave packets at the smaller scale under consideration propagate while interacting nonlinearly to give rise to packets on smaller scales still, even though there still is a global isotropy in a statistical sense. This means that the cascade is still delayed by this Alfvén effect. A new time-scale $\tau_a \simeq l/c_a^s$ therefore appears and the nonlinear transfer time $\tau^*$ will no longer be equal to the *turnover time*

$$\tau_{nl} \simeq l/u(l) \simeq l\sqrt{4\pi\rho}/b(l) \simeq l/z^+(l) \simeq l/z^-(l),$$

where we have generalized from hydrodynamics to MHD by introducing the rms values of magnetic field fluctuatins, $b(l)$, and supposed there is equipartition between velocity fluctuations, magnetic field fluctuations in velocity units and Elsasser field fluctuations. The collisions between Alfvén wave-packets on a scale $l$ is limited by the crossing time $\tau_a$, and supposing that the collisions between individual wave-packets are independent stochastic events, the value of the effective time $\tau^*$ increases by a factor

$$\tau^* \simeq \tau_{nl}\left(\frac{\tau_{nl}}{\tau_a}\right).$$

Substituting this value in the equation for the energy flux Equation, we find

$$\Pi(l) \sim \frac{1}{2}\frac{u(l)^2}{\tau^*} \simeq \frac{1}{2}\frac{u(l)^4}{lc_a^s} \simeq \epsilon \tag{10.18}$$

from which the energy on scale $l$ may be found as $E(l) \simeq (\epsilon c_a^s l)^{1/2}$, or in terms of energy epctra $E_k \simeq (\epsilon c_a^s)^{1/2} k^{-3/2}$ known as the Iroshnikov-Kraichnan spectrum of MHD turbulence.

In the above we have assumed that the amplitudes of the fields $z^\pm$ were about the same. However, from the MHD Eq. (10.13) it is easily seen that the *turnover* time for the fields $z^\pm$ depend on the amplitude of the opposite field $z^\mp$:

$$\tau_{nl}^\pm = l/z^\mp. \tag{10.19}$$

In the case of magnetohydrodynamics it may be shown that the global integral invariants are different from the ones in hydrodynamics: in MHD nonlinear interactions separately conserve the energies per unit mass of the individual Elsasser fields $E^\pm = 1/2|\mathbf{z}^\pm|^2$ (corresponding physically to the conservation of total energy in the fluctuations as well as the mixed or cross-helicity $H_m = <\boldsymbol{u} \cdot \boldsymbol{b}>$). This implies that separate cascades with independent energy fluxes $\Pi^+$ and $\Pi^-$ are possible. However, repeating the phenomenological arguments from Iroshnikov and Kraichnan separately on the fields $z^+$ and $z^-$ we find the extremely interesting result that the two fluxes $\Pi^+$ and $\Pi^-$ are identical:

$$\Pi_k^+ = \Pi_k^- = k^3 E_k^+ E_k^- / c_a^s. \tag{10.20}$$

Note that for equal amplitudes in the fields we are once again led to the Iroshnikov-Kraichnan spectrum. If however the large scale amplitudes of the two fields are initially different, Eq. (10.20) leads to a new conclusion: because the energy fluxes, and therefore the dissipation rates, of the two fields are identical, for long times the field with initially lower energies will tend to be depleted with respect to the dominant one. This will eventually quench the turbulence, slowing down nonlinear interactions and leaving a nonlinear Alfvén wave, with a well-defined power spectrum, as an exact solution of the MHD equations, which in the presence of an even small mean field will end up propagating along the field in one direction. This evolution towards a configuration in which only one Elsasser field dominates is called dynamic alignment, since it involves an increase in the correlation between velocity and magnetic field fluctuations with time (corresponding to the decrease towards zero of one of the Elsasser fields). Incompressible MHD simulations have indeed shown that dynamic alignment occurs, while observations of the evolution of Alfvénic turbulence in space, in the solar wind, show that for the case of turbulence propagating away from the sun the opposite happens: there is a tendency for the development of the Elsasser mode corresponding to Alfvén waves propagating towards the sun, with respect to those propagating away, as the distance from the Sun increases, so that at large distances the two fields $\mathbf{z}^{\pm}$ become equivalent.

We have said nothing at present about the possible directional anisotropy of the cascade in the presence of a mean field, yet we have seen that the nonlinear fluxes depend on the angle between the wave-vectors and the mean field via the corresponding Alfv'en times. If we imagine taking wave-packets in 3D that are elongated along the mean field, so that the dominant wave-vectors are quasi-perpendicular to the field, i.e. $\mathbf{k}\cdot\mathbf{c}_a \simeq 0$ then propagation effects among wave-packets become negligible and we should return to a hydrodynamic or Kolmogorov type cascade (provided there is equipartition between the fields $\mathbf{z}^{\pm}$). In cases where the mean field is strong $B_0 \gg |\mathbf{z}^{\pm}|$, and remaining within the approximation where characteristic scales along the field are much greater than those transverse to the mean field, the effective reduction of nonlinear interactions along the field can be such as to quench parallel nonlinear interactions entirely, and a further simplification of the MHD equations, to so called reduced MHD or RMHD is possible. These equations describe two-dimensional fluctuating fields, polarized in planes orthogonal to the mean magnetic field and whose nonlinear interactions are confined in this orthogonal plane. Communication across planes occurs only via the propagation along the mean field $B_0$. Reduced MHD corresponds therefore to a series of 2D incompressible MHD planes, coupled together by the linear propagation of the fluctuations as Alfvén waves along $B_0$ from one plane to another. In this approximation the spectral anisotropy that may develop can be calculated precisely. In the 2D planes where nonlinear interactions occur via the 2D incompressible MHD equations, we expect a cascade completely analogous to a Kolmogorov cascade with hydrodynamic type spectra. However, the evolution at a scale $l_\perp$ in two different planes separated by a distance $L$ along the field will be completely independent only if the propagation time between the two planes, $\tau_a = l_\parallel / c_a$, is longer than the nonlinear time $\tau^*(l_\perp)$. There will therefore be an anisotropy set by the region where $\tau_a = \tau_{nl}(l_\perp)$. Considering the Kolmogorov value for the nonlinear time $\tau^* = \tau_{nl}$, this region will be defined by:

$$\frac{l_\parallel}{c_a} \simeq \frac{l_\perp}{u(l_\perp)} \sim \frac{l_\perp}{\epsilon l_\perp^{1/3}} \tag{10.21}$$

which in terms of parallel and perpendicular wavevectors $k_{||}, k_\perp$ may also be written as

$$k_{||}^c = \frac{u}{c_a} k_\perp{}^{2/3} k_{\perp 0}^{1/3}.$$

This condition is known as the critical balance condition for $k_\perp$ and $k_{||}$. A spectrum will develop in the parallel direction as a consequence of uncorrelated perpendicular planes only for wavevectors satisfying $k_\parallel \leq k_{||}^c$.

Magnetohydrodynamic turbulence in more general regimes is still an extremely active area of research, with vast astrophysical applications: we mention here the problems of stellar convection (where the role of the magnetic field is secondary, but the problem of maintaining and generating the field is fundamental, (see the chapter on dynamo theory), the heating and dynamics of the interstellar medium, the stability of molecular clouds, viscosity in accretion disks, and the heating of stellar coronae and acceleration of stellar winds. Let us briefly discuss this latter problem within the framework of MHD turbulence, with specific application to the case of our star, the Sun.

## 10.3 Turbulence and Coronal Heating

The solar corona is a hot, very low density plasma with a temperature of about $10^6$ K. All stars, with the possible exceptions of A and B type, have external atmospheric regions where the temperature is much higher than that of the respective photosphere. The photosphere-corona interface is an extremely complex and dynamic region, developing over a few thousand kilometers in the chromosphere, where the temperature rises slowly, and then through the very thin transition region where the temperature jumps from about $10^4$ K to above $10^6$ K over just a few hundred kilometers. The chromosphere, transition region and corona are characterized by a substantial energy loss: from the chromosphere and the transition region this occurs mostly via radiation in visible and EUV, while losses from the corona occur through radiation, conduction back towards the photosphere, and in the open magnetic field energy flux transported into the solar wind. A mechanical energy source is required to keep this temperature imbalance active, since by the second law of thermodynamics heat can not flow from the colder photosphere upwards. So there must be a viable mechanism allowing energy to be transported between different regions in the solar atmosphere, and then releasing and dissipating it in the higher temperature corona. The source is found in the abundant mechanical energy of the turbulent photospheric velocity fluctuations, which must be transported upwards and then dissipated within one or two solar radii and most probably within a few tens of thousand of kilometers above.

White light images of the solar corona show that is highly structured, with fine loops and plumes jetting outwards. EUV and x-ray images show how the strongest energy is released above photospheric regions where the magnetic field is strong, sunspots and active regions, so that the association of coronal heating with the photospheric magnetic field is a strong one. Even dark coronal regions, known as coronal holes, and usually confined to the polar regions of the sun, show remarkably high temperatures. Their relative darkness is due to the presence of open magnetic fields, leading to smaller densities as these regions give rise to the expanding solar wind. An estimate of the energy flux required to sustain the coronal energy losses is about $\epsilon \simeq 10^7$ erg/cm$^2$/sec for active regions, $\epsilon \simeq 8\ 10^5 - 10^6$ erg/cm$^2$/sec for the quiet sun and $\epsilon = 5\ 10^5 - 8\ 10^5$ erg/cm$^2$/sec, for coronal holes (Fig. 10.1).

It is natural to imagine that the solar magnetic field might play an essential role both in the transfer of energy upwards from the photosphere into the chromosphere and corona and in the subsequent dissipation of this energy. Photospheric motions buffet and displace the footpoints of magnetic field lines extending into the corona, therefore launching fast slow and Alfvén waves upwards with a wide range of periods, ranging from a few hours (lifetime of supergranulation cells) to a few minutes (granulation cells, solar 5 min oscillations) and below. The Lundquist number calculated for typical densities and magnetic fields in the solar corona and scales of order $10^4$ km turns out to be huge, around $10^{12}$, so the only way that dissipation can occur in the corona is that dynamical evolution involve the formation of small scales. Applying MHD turbulence theory to the solar corona and solar wind however requires taking into account the characteristics of the system, which involve the boundary conditions (how photospheric motions couple to the atmosphere above), the global magnetic filed topology (whether the coronal field is open or closed) and finally the gravitational stratification and inhomogeneous nature of the corona, which is finely structured by the magnetic field itself.

The simplest way to consider the photosphere-corona interface is to completely neglect any details of chromospheric dynamics and consider it a sharp boundary between two separate ideal plasmas with different densities and temperatures (but essentially constant pressure across the boundary). The density of the photosphere is so much greater than that of the corona that photospheric inertia is strong enough to neglect the back reaction of the field on the plasma. This means that one can assume that the photospheric velocity field is a given assigned flow, determined by convective motions below. In the corona however, it is the magnetic field that dominates, i.e. the plasma from photosphere to corona goes from $\beta \gg 1$ to $\beta \ll 1$. Using $\boldsymbol{n}$ to denote the normal to the photosphere, which points radially outwards from the sun, the energy flux associated with photospheric motions is dominated by the Poynting vector

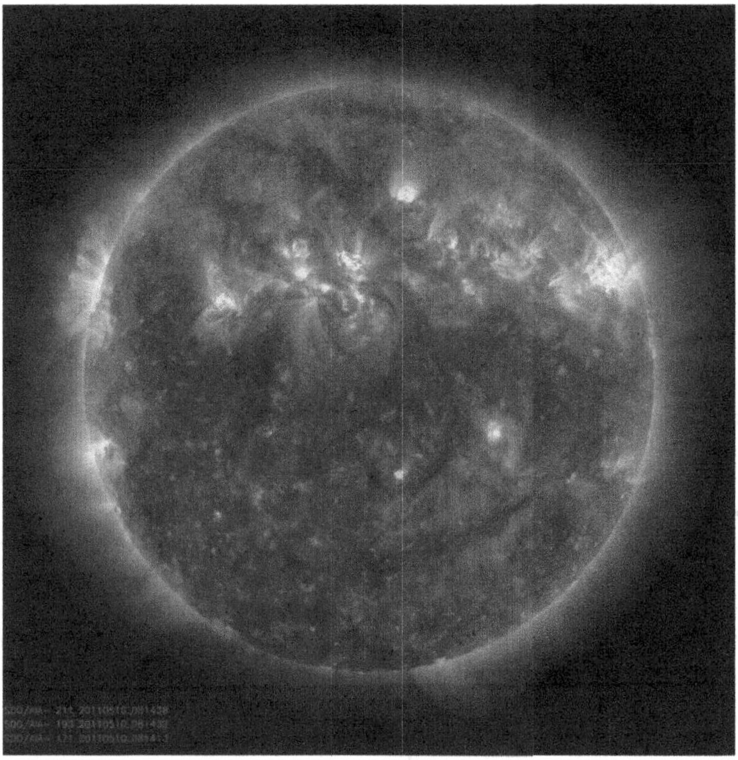

**Fig. 10.1** An image of the solar corona superposing three different images taken in different wavelengths. Different colors correspond to different temperatures: *blue* $\simeq 6 \times 10^5\ K$, *green* $\simeq 1,3 \times 10^6\ K$, and *red* $\simeq 2 \times 10^6\ K$ (Courtesy NASA-SDO)

$$\boldsymbol{S} = \frac{c}{4\pi}(\boldsymbol{E} \times \boldsymbol{B}), \tag{10.22}$$

where $\boldsymbol{E}$ is the Electric field. Eliminating it using the ideal Ohm's law we find

$$\boldsymbol{S} \cdot \boldsymbol{n} = (b_\perp^2(\boldsymbol{u}_{ph} \cdot \boldsymbol{n}) - B_0(\boldsymbol{b}_{\perp ph} \cdot \boldsymbol{u}_{ph}))/4\pi, \tag{10.23}$$

where the photospheric velocity field is given by $\boldsymbol{U} = \boldsymbol{u}_{ph}$ and the magnetic field at the photosphere - corona boundary is given by $\boldsymbol{B} = B_0\boldsymbol{n} + \boldsymbol{b}_{\perp ph}$ ($\perp_{ph}$ indicates vector components parallel to the photospheric surface). In this approximation, we have neglected the energy flux associated with density fluctuations, present in magnetoacoustic waves, or the possibility that plasma jets (observed to exist, i.e. spicules) may contribute significantly to coronal energy balance. The reason we neglect the compressible component of plasma motions is that in the limit of very small $\beta$ slow modes essentially become sound waves ducted along the magnetic field, which steepen to form shocks already in the chromosphere, before arriving into the corona. Fast waves

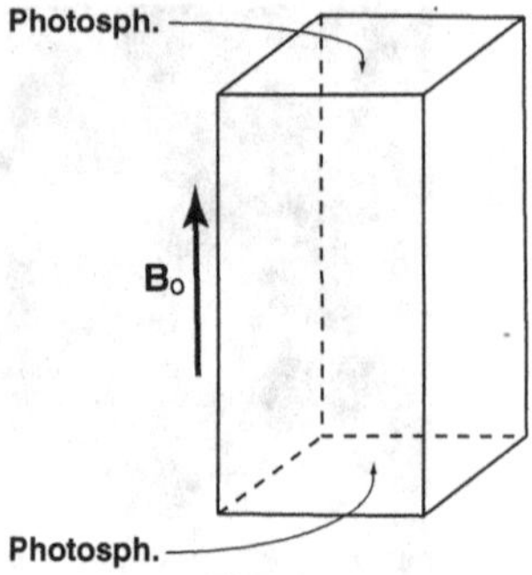


**Fig. 10.2** Cartesian approximation of a coronal loop

feel the strong density gradients with height and suffer total reflection, except for quasi-parallel propagation where they become identical to Alfvén waves.

Equation (10.23) shows that energy flux into the corona is made up of two terms. The first term is the energy flux associated with emerging or sub-merging magnetic field, and either carried into the corona or away from it by the radial component of the velocity field. The second term, proportional to the scalar product of the velocity and magnetic field components parallel to the photosphere times the radial component of the magnetic field, describes the energy flux associated with waves along the field crossing into the corona, or simply the dragging of that field by photospheric motions. It is hard to separate the contributions from the two terms observationally, though they typically are of the same order of magnitude. In what follows, we will limit our consideration to the problem of an assigned velocity field parallel to the photospheric surface, neglecting for simplicity the dynamics associated with emerging flux.

The coronal response to photospheric motions depends strongly on the frequency of the motions themselves as well as the coronal magnetic field configuration, for example if it opens into the heliosphere or closes back down into the photosphere. Consider the case of the closed solar corona, and consider a configuration of a coronal loop of length $L$, transverse cross section diameter $l$: observationally one has $l/L \ll 1$. This allows us to approximate the coronal loop in cartesian geometry, neglecting the curvature of the field and assuming that it is simply a box delimitated on top and bottom by photospheric planes, as shown in Fig. 10.2. The transverse, incompressible oscillations driven by photospheric motions are then made up of Alfvén waves which, in the linear approximation, obey the equations

$$\rho \frac{\partial \boldsymbol{u}_\perp}{\partial t} = \frac{B_0}{4\pi} \frac{\partial \boldsymbol{b}_\perp}{\partial z}, \tag{10.24}$$

$$\frac{\partial \boldsymbol{b}_\perp}{\partial t} = B_0 \frac{\partial \boldsymbol{u}_\perp}{\partial z}, \tag{10.25}$$

where $B_0$ is the strong, uniform axial magnetic field and $\boldsymbol{u}_\perp, \boldsymbol{b}_\perp$ are the transverse oscillation of velocity and magnetic field. We have neglected gravitational stratification along the loop in the $z$ direction, because the scale height at a temperature of $10^6$ K is of the same order of magnitude of the height $L$ of the loop itself. These equations are solved assigning the velocity field $\boldsymbol{u}_\perp = \boldsymbol{u}_{ph}$ at the boundaries in $z = 0, L$, and supposing that at $t = 0$ the transverse field vanishes everywhere, i.e. the field lines are all initially straight and perpendicular to the photosphere. To simplify the problem further, consider photospheric motions to be assigned only on one of the two boundaries, while the other will be kept anchored with so that $\boldsymbol{u}_\perp = 0$, in $z = 0$, while $\boldsymbol{u}_\perp = \boldsymbol{u}^0_{ph}(x, y)\cos(\omega_{ph}t)$ in $z = L$. $\omega_{ph} = 2\pi/T_{ph}$ is a typical frequency associated with photospheric motions, with $T \geq 3\mathrm{m}$, much longer than the propagation time of an alfvén wave along a loop which is of the order of tens of seconds or less.

A solution of Eqs. (10.24) and (10.25) with the assigned initial and boundary conditions is therefore given by

$$\boldsymbol{u}_\perp = \boldsymbol{u}^0_{ph}(x, y)\cos(\omega_{ph}t)\frac{\sin(\omega_{ph}z/c_{a0})}{\sin(\omega_{ph}L/c_{a0})}, \tag{10.26}$$

$$\frac{\boldsymbol{b}_\perp}{\sqrt{4\pi\rho}} = \boldsymbol{u}^0_{ph}(x, y)\sin(\omega_{ph}t)\frac{\cos(\omega_{ph}z/c_{a0})}{\sin(\omega_{ph}L/c_{a0})}, \tag{10.27}$$

where $c_{a0} = B_0/\sqrt{\rho_c}$ is the coronal Alfvén speed.

This is a valid solution only if the frequencies associated with photospheric motions are such that $\sin(\omega_{ph}L/c_{a0}) \neq j\pi$, with $j$ an integer (positive, negative, or 0), because of the denominators in Eqs. (10.26) and (10.27). The zeros of the denominators identify frequencies where the photospheric driver is in resonance with Alfvén wave oscillations in the loop, in which case linear solutions give rise to a secular growth of the amplitude of oscillations with time. In other words, at resonant frequencies a coronal loop may store significant amounts of energies, that will be saturated or limited only by dissipative processes: so the idea comes naturally that the photospheric driving of closed coronal loops will lead to the spontaneous development of MHD turbulence in terms of Alfvénic-type fluctuations within the loop, which may provide an explanation for the heating of the loop itself. This because we have identified a process which leads to a secular increase of energy in the loop, with amplitudes of the fields $\boldsymbol{u}_\perp$, $\boldsymbol{b}_\perp$ growing without limit; if nonlinear interactions and a cascade did not develop, the saturation of growth due simply to the *linear* resistive and viscous damping with coronal values of the Lundquist and Reynolds numbers would lead to enormous oscillations, with field amplitudes much greater than those observed, albeit indirectly, in the corona itself. The heating of the loop and saturation of amplitudes will therefore occur via a nonlinear cascade.

As mentioned above, the characteristic propagation time of Alfvén waves along a loop, a few seconds or tens of seconds, is typically much smaller than that associated with the variations of photospheric motions. It is therefore important to distinguish the resonances with $j > 0$ from the resonance at null frequency $j = 0$, corresponding to stationary photospheric flows (i.e. stationary as measured using the coronal crossing time clock). This case, for which $\omega_{ph}t \to 0$, can be found taking the corresponding limit of Eqs. (10.26) and (10.27): we find that $\sin(\omega_{ph}L/c_{a0}) \simeq \omega_{ph}L/c_{a0}$, $\sin(\omega_{ph}z/c_{a0}) \simeq \omega_{ph}z/c_{a0}$, from which we obtain velocity field fluctuations

$$u_\perp = u^0_{ph}(x, y)z/L, \tag{10.28}$$

while for the magnetic field,

$$\frac{b_\perp}{\sqrt{4\pi\rho}} = u^0_{ph}(x, y)\frac{c_{a0}}{L}t. \tag{10.29}$$

Therefore, with stationary photospheric flows, the coronal velocity field remains limited in amplitude with growing time, while the transverse magnetic field grows secularly. Another way of interpreting this result is to consider the original coronal field lines as strings, which are tied on the plane where the velocity vanishes, but are dragged about by the stationary flow on the other photospheric plane. It is this dragging which leads to a secular growth of the transverse field. Parker, in 1972, [24], proposed that this kind of *coronal dynamo* might be the main cause of coronal heating. It is important to point out that the slow, stationary photospheric motions are not homogeneous but depend on position, therefore the growing perturbed coronal magnetic field will not be in equilibrium, so a coronal dynamics will ensue where the stresses in the transverse field attempt to relax towards an equilibrium through the action of the Lorentz force on the plasma. Because the plasma $\beta$ of the corona is very small, the corresponding coronal equilibrium configurations would be expected to be force-free states. Parker conjectures however that for generic motions, the force-free equilibrium states attainable could not be continuous, but would contain singularities in the topology of the field, or current sheets, whose stability we have discussed in the previous chapter. In the language of turbulence, Parker conjecture that nonlinear interaction would naturally lead to extremely thin structures, whose destruction via reconnection would lead to coronal heating.

The energy flux into the corona resulting from the field-line dragging by stationary flows results grows in time without bound (within the linear approximation considered here) as can be understood by calculating the Poynting flux using expressions Eqs. (10.28) and (10.29): $S = c_{a0}\rho|u^0_{ph}|^2 c_{a0}t/L$. Parker estimated the time it would take for such flux to reach a value that could balance coronal energy losses, $S = \epsilon$, and from this estimate calculated the corresponding amplitude of fields and types of current sheets that would have to develop. For a typical active region, Parker found a time required to reach a stationary flux of $10^7$ ergs/cm$^2$/sec, starting from typical supergranular estimates of the photospheric velocity field, $u_{ph} \simeq 0.5$ km/s, an axial magnetic field $B_0 \simeq 100$ Gauss, and a loop of length $L \simeq 10^5$ km, of about $t = 5.0\ 10^4$ sec, compatible with the correlation time of supergranulation itself.

Numerical simulations of this problem (within the approximations we have made) indeed prove that the coronal volume develops a regime of MHD turbulence where the energy contained in magnetic field oscillations dominate the kinetic energy, and dissipation does occur through the spontaneous development of turbulent current sheets, resulting in a dissipation rate that does not depend on the Lundquist number. The energy spectra however differ from those estimated using the models of Iroshnikov–Kraichnan or Kolmogorov. As may be expected, detailed studies of the problem in geometries closer to those of the real Sun and with a more appropriate treatment of the photosphere-corona boundaries are the subject of intense and active research.

## Questions and Problems

**10.1.** Derive Eq. (10.5), by multiplying the Navier-Stokes equation (10.1) and the Poisson equation for the pressure (10.3) by $e^{-i\boldsymbol{k}\cdot\boldsymbol{x}}$, and using the definition of the Fourier transformed velocity field (10.4). Recall that

$$\int d^3\boldsymbol{x} e^{-i(\boldsymbol{k}-\boldsymbol{q})\cdot\boldsymbol{x}} = (2\pi)^3\delta^3(\boldsymbol{k}-\boldsymbol{p})$$

where $\delta$ is the Dirac distribution.

**10.2.** Show that Eqs. (10.26) and (10.27) are solutions of Eqs. (10.24) and (10.25).

**10.3.** Show that a solution of the system (10.24) and (10.25) in the case where $\sin(\omega_{ph}L/B_0) = j\pi$ with $j$ a relative integer of any value, is given, for the velocity field, by

$$\boldsymbol{u}_\perp = (-1)^j \boldsymbol{u}^0_{ph}(x,y)\left(\frac{z}{L}\cos\left(\omega_{ph}t\right) - \frac{c_{a0}t}{L}\sin\left(\omega_{ph}t\right)\sin\left(\frac{\omega_{ph}z}{c_{a0}}\right)\right). \quad (10.30)$$

Find the corresponding solution for the magnetic field fluctuation which satisfies the initial condition $\boldsymbol{b}_\perp = 0$ everywhere as well as the boundary conditions. This problem shows that the resonances with non-zero frequencies lead to secular growth of both the magnetic and velocity fields.

## Solutions

**10.3.** To eliminate the magnetic field from Eqs. (10.24), (10.25), differentiate Eq. (10.24) with respect to time and Eq. (10.25) with respect to $z$. The magnetic field may then be found by solving Eq. (10.25) with the given solution for $\boldsymbol{u}_\perp$:

$$\boldsymbol{b}_\perp = (-1)^j\,\boldsymbol{u}^0_{ph}\,(x,y)\left[\frac{B_0}{\omega_{ph}L}\sin\left(\omega_{ph}t\right)\left(1-\cos\left(\frac{\omega_{ph}z}{c_{a0}}\right)\right) + \frac{B_0 t}{L}\cos\left(\frac{\omega_{ph}z}{c_{a0}}\right)\cos(\omega_{ph}t)\right]. \quad (10.31)$$

# Chapter 11
# Build-Up of Magnetic Fields

**Abstract** The problem of the build-up and maintenance of cosmical magnetic fields is presented. The basic phenomenological theory of astrophysical dynamos is outlined. The more formal approach of mean-field electrodynamic is also exposed. Simple solutions of the dynamo equations are discussed. Finally, the problem of the creation of magnetic fields in an unmagnetized plasma is briefly considered.

## 11.1 The Dynamo Problem

Magnetic fields appear to be ubiquitous in the Universe: almost all celestial bodies, from planets and stars to galaxies and galaxy clusters host magnetic fields of different intensities. The magnetic field on Earth is about one gauss and most of the planets, with the exception of Venus and possibly Mars, also exhibit magnetic fields of varying strength. The average magnetic field of the Sun is of the order of a few gauss, but locally it can be much stronger, reaching values up to a few thousand gauss in sunspots. Magnetic fields of normal stars also reach values around 1,000 G, but those of compact stars may be extremely intense by comparison: the field of a magnetized white dwarf is typically of the order of a few million gauss and the magnetic fields of pulsars fall in the range of $10^{12}$ G. The galactic magnetic field, on the contrary, is extremely weak, of the order of a few microgauss. To a first approximation, the external fields of planets and stars are well approximated by magnetic dipoles, but, as we shall see, the internal fields are likely to have a very different structure. At first, one might think that magnetic fields in different contexts throughout the Universe are all evolved remnants of primordial fields, present since the very beginning of the cosmic history. A little thought, however, shows that this assumption must be false. Cosmic plasmas are not perfect conductors and are therefore subject to ohmic decay, even if their resistivity is very small, as shown by the induction equation describing the time evolution of magnetic fields (see Eq. (5.25)):

$$\frac{\partial \boldsymbol{B}}{\partial t} = \nabla \times (\boldsymbol{U} \times \boldsymbol{B}) + \eta \nabla^2 \boldsymbol{B}. \tag{11.1}$$



C. Chiuderi and M. Velli, *Basics of Plasma Astrophysics*,
UNITEXT for Physics, DOI 10.1007/978-88-470-5280-2_11

The two terms on the right are associated with very different timescales given, respectively, by $\tau_f = \mathcal{L}/\mathcal{U}$, the *fluid* or *convective* scale and $\tau_d = \mathcal{L}^2/\eta$, the *diffusive* or *dissipative* scale. Here $\mathcal{U}$ is a typical value of the fluid speed, $\mathcal{L}$ is the scale of the spatial variation of the magnetic field and $\eta = (c^2/4\pi\sigma)$ is the magnetic diffusivity of the plasma which we will take here to be uniform.

Using the typical dimensions of the system under consideration for $\mathcal{L}$, $\tau_d$ turns out to be extremely long for stars ($\tau_d \simeq 10^{10}$ years) and galaxies ($\tau_d \geq 10^{26}$ years), but much smaller, about $10^5$ years for the Earth. The problem of how to maintain the magnetic field, however, is not limited to ours and other planets, since the field of the Sun is known to reverse its sign over a period of about eleven years (yielding an average 22 year Solar cycle) and such a rapid oscillation (with respect to the computed $\tau_d$) requires a dynamic explanation. Another aspect of the problem is that the existence of a primordial magnetic field at the beginning of the Universe is itself an unanswered question. In the absence of a primordial field, when, how and on which timescales has the magnetic field been generated?

The solution to the problem of the build-up of magnetic fields can only come from the convective term of the induction equation that couples fluid motion and magnetic field and that, at least in principle, might not only contrast but overcome the decay due to the diffusive term. The effect of velocity on magnetic fields, which defines *dynamo action*, may result in a reduction of the spatial scales entering the definition of the dissipative scale, thus allowing changes to occur over timescales considerably shorter than $\tau_d$.

It is often useful to describe the magnetic field in terms of the *vector potential* $\boldsymbol{A}$, related to $\boldsymbol{B}$ by

$$\boldsymbol{B} = \nabla \times \boldsymbol{A},$$

a representation made possible by the fact that $\nabla \cdot \boldsymbol{B} = 0$. A further *gauge condition*, i.e. $\nabla \cdot \boldsymbol{A} = 0$, is necessary to completely specify $\boldsymbol{A}$. As discussed in Chap. 5, $\boldsymbol{A}$ is defined modulus the gradient of an arbitrary function $\phi$, and thanks to this gauge invariance $\boldsymbol{B}$ remains the same.

Expressing $\boldsymbol{B}$ in terms of $\boldsymbol{A}$ in Eq. (11.1) we obtain

$$\nabla \times \left[\frac{\partial \boldsymbol{A}}{\partial t} - \boldsymbol{U} \times (\nabla \times \boldsymbol{A}) - \eta\nabla^2\boldsymbol{A}\right] = 0,$$

which implies that

$$\frac{\partial \boldsymbol{A}}{\partial t} = \boldsymbol{U} \times (\nabla \times \boldsymbol{A}) + \eta\nabla^2\boldsymbol{A} + \nabla\phi. \tag{11.2}$$

Taking the divergence of the above equation and using the gauge condition $\nabla \cdot \boldsymbol{A} = 0$ we get:

$$\nabla^2\phi = \nabla \cdot (\boldsymbol{U} \times \boldsymbol{B}).$$

The gauge condition therefore eliminates the arbitrariness of $\phi$.
Equations (11.1) and (11.2) constitute the basis for the discussion of the dynamo problem.

The complete solution of the problem would require the induction equation to be solved at the same time and in parallel with the momentum equation, an extremely difficult non linear problem. At the heart of the dynamo problem lies understanding which velocity fields, sustained by the available forces, are capable of supporting a growing, oscillating or steady magnetic field against ohmic decay. To limit the difficulties, the problem is often restricted to the *kinematic dynamo problem*, where the induction equation is actually decoupled from the equation of motion by *assuming* a particular form for the velocity field. Even so, the determination of a flow with the appropriate characteristics is not an easy task. We begin by examining the simplest possible flow configurations. Unfortunately, as we shall see, none of these basic flows are able to generate or sustain a magnetic field.

## 11.2 The Anti-Dynamo Theorems

A certain number of two- and three-dimensional theorems can be proven that deny the possibility of producing a magnetic field lasting in time. Common assumptions to all theorems are the incompressibility of the flow, $\nabla \cdot \boldsymbol{U} = 0$, and the vanishing of $\boldsymbol{U}$, $\boldsymbol{B}$ and $\boldsymbol{A}$ at infinity. We shall now illustrate the most relevant ones.

**Theorem 1** *In cartesian coordinates* $[x, y, z]$ *it is impossible to generate by dynamo action a* $z$*-independent field that vanishes at infinity.*

Assume $z$-independence both for $\boldsymbol{U}$ and $\boldsymbol{B}$. Then write

$$\boldsymbol{B} = \boldsymbol{B}_z \boldsymbol{e}_z + \boldsymbol{B}_\perp = \boldsymbol{B}_z \boldsymbol{e}_z + \nabla \times (A\boldsymbol{e}_z),$$

and

$$\boldsymbol{U} = U_z \boldsymbol{e}_z + \boldsymbol{U}_\perp,$$

with $A = A(x, y)$. Equations (11.1) and (11.2) then transform into

$$\frac{\partial B_z}{\partial t} + (\boldsymbol{U}_\perp \cdot \nabla) B_z = \eta \nabla^2 B_z + (\boldsymbol{B}_\perp \cdot \nabla) U_z,$$

$$\frac{\partial A}{\partial t} + (\boldsymbol{U}_\perp \cdot \nabla) A = \eta \nabla^2 A.$$

Now multiply the last equation by $A$ and integrate over the whole space

$$\frac{1}{2}\frac{\partial}{\partial t}\int A^2 \mathrm{d}V = -\int A(\boldsymbol{U}_\perp \cdot \nabla) A \mathrm{d}V - \eta \int (\nabla A)^2 \mathrm{d}V.$$

But

$$A(\boldsymbol{U}_\perp \cdot \nabla)A = \frac{1}{2}(\boldsymbol{U}_\perp \cdot \nabla)(A^2) = \frac{1}{2}\nabla \cdot (A^2\boldsymbol{U}_\perp) - \frac{1}{2}A^2(\nabla \cdot \boldsymbol{U}_\perp).$$

The last term of the preceding equation vanishes (incompressibility!) and therefore the first integral on the rhs can be transformed into a surface integral at infinity where all the fields vanish. The integral of $A^2$ therefore decays to zero (unless $A = const.$ in which case $\boldsymbol{B}$ would be identically zero). Since $\boldsymbol{A}$ decays in time, $\boldsymbol{B}_\perp$ must also tend to zero. But then the equation for $B_z$ no longer has a source term, so that $B_z$ will vanish as well. This theorem states that dynamos are intrinsically three-dimensional, and no 2D field can be maintained by dynamo action.

**Theorem 2** *In cartesian coordinates* $[x, y, z]$ *a planar flow cannot produce a dynamo effect.*

The velocity field now has the form $\boldsymbol{U} = [U_x(x, y, z, t), U_y(x, y, z, t), 0]$. Notice that the field components are allowed to depend on $z$. The $z$-component of Eq. (11.1) is

$$\frac{\partial B_z}{\partial t} + (\boldsymbol{U} \cdot \nabla)B_z = \eta\nabla^2 B_z,$$

since $U_z = 0$. Multiplying this equation by $\boldsymbol{B}_z$, integrating over the volume and repeating the reasoning of the previous theorem, we see that the integral of the convective term $(\boldsymbol{U} \cdot \nabla)B_z$ vanishes. Thus,

$$\frac{1}{2}\frac{\partial}{\partial t}\int \boldsymbol{B}_z^2 \mathrm{d}V = -\eta\int (\nabla \boldsymbol{B}_z)^2 \mathrm{d}V,$$

showing that $\boldsymbol{B}_z$ must tend to zero in time. If $\boldsymbol{B}_z = 0$, it is easy to see that the vector potential becomes simply $\boldsymbol{A} = A(x, y)\boldsymbol{e}_z$, and the remaining field components are given by $B_x = (\partial A/\partial y)$ and $B_y = -(\partial A/\partial x)$. Moreover, defining $\boldsymbol{Q} = \boldsymbol{U} \times \boldsymbol{B}$, we have $Q = (\boldsymbol{U} \cdot \nabla)\boldsymbol{A}$. Taking the curl of Eq. (11.1) and evaluating the $z$-component we get:

$$\frac{\partial}{\partial t}(\nabla \times \boldsymbol{B})_z = [\nabla \times (\nabla \times \boldsymbol{Q})]_z + \eta\nabla^2(\nabla \times \boldsymbol{B})_z.$$

But,

$$(\nabla \times \boldsymbol{B})_z = \frac{\partial B_y}{\partial x} - \frac{\partial B_x}{\partial y} = \frac{\partial^2 A}{\partial x^2} + \frac{\partial^2 A}{\partial y^2} \equiv \nabla_\perp^2 A,$$

and

$$[\nabla \times (\nabla \times \boldsymbol{Q})]_z = -\nabla^2 Q_z = -\nabla_\perp^2(\boldsymbol{U} \cdot \nabla)A,$$

Introducing these expressions in the induction equation , we finally obtain

$$\nabla_\perp^2\Big[\frac{\partial A}{\partial t} + (\boldsymbol{U}\cdot\nabla)A - \eta\nabla^2 A\Big] = 0.$$

Since this equation is linear in $A$, we can Fourier transform it. The operator $\nabla_\perp^2$ becomes $-k_\perp^2 = -(k_x^2 + k_y^2)$ and the Fourier transform of the expression in square brackets must therefore vanish. This implies

$$\frac{\partial A}{\partial t} + (\boldsymbol{U}\cdot\nabla)A = \eta\nabla^2 A,$$

namely the same equation already encountered in Theorem 1. Following the same procedure we can prove that $A \to 0$: the magnetic field therefore decays. No dynamo action is possible for planar flows.

Moving now to a more realistic axisymmetric configuration, we can prove that

**Theorem 3** *An axisymmetric magnetic field vanishing at infinity cannot be maintained by dynamo action*

This is Cowling's celebrated theorem (1934) [25]. The proof, which is different from the original one, is analogous to that of Theorem 1. Any axisymmetric vector field can be represented in a cylindrical coordinate system $(r, \theta, z)$ as the sum of a *toroidal* (i.e. azimuthal) field and a *poloidal* field lying in the meridional planes. The latter, in turn, is the sum of the radial and vertical components of the field. Adopting such a representation for the magnetic field $\boldsymbol{B} = B_\theta \boldsymbol{e}_\theta + \boldsymbol{B}_p$, it is easy to see that the poloidal magnetic field can be expressed as $\boldsymbol{B}_p = \nabla\times(A\boldsymbol{e}_\theta)$, with $A = A(r, z, t)$. Because of the assumed axial symmetry $(\partial/\partial\theta) = 0$, $\nabla\cdot\boldsymbol{A} = 0$ and the $\theta-$component of Eq. (11.2) becomes

$$\frac{\partial A}{\partial t} + \frac{1}{r}\boldsymbol{U}_p\cdot\nabla(rA) = \eta\Big(\nabla^2 A - \frac{A}{r^2}\Big). \tag{11.3}$$

We now multiply the above equation by $r^2 A$ and integrate over the volume. Again, the integral of the convective term can be transformed into

$$\frac{1}{2}\int \nabla\cdot(r^2A^2\boldsymbol{U})\mathrm{d}V,$$

which vanishes if $A$ has "good" convergence properties at infinity. We therefore arrive at

$$\frac{1}{2}\frac{\partial}{\partial t}\int (r^2A^2)\mathrm{d}V = -\eta\int |\nabla(r\,A)|^2\mathrm{d}V,$$

which shows that $(rA)$ must decay in time, unless $rA = const$. But in this case $A$ would diverge on the axis and therefore also $\boldsymbol{B}_p$. After $A$ has gone to zero, $\boldsymbol{B}_p = 0$. On the other hand, using the toroidal-poloidal representation of $\boldsymbol{B}$, we find that the $\theta$-component of Eq. (11.1) is

$$\frac{\partial B_\theta}{\partial t} = (\boldsymbol{B}_p \cdot \nabla) U_\theta - (\boldsymbol{U}_p \cdot \nabla) B_\theta + U_r \frac{B_\theta}{r} - U_\theta \frac{B_r}{r} + \eta \left( \nabla^2 B_\theta - \frac{B_\theta}{r^2} \right),$$

or equivalently,

$$\frac{\partial B_\theta}{\partial t} + r(\boldsymbol{U} \cdot \nabla)\left(\frac{B_\theta}{r}\right) = r(\boldsymbol{B}_p \cdot \nabla)\left(\frac{U_\theta}{r}\right) + \eta \left( \nabla^2 B_\theta - \frac{B_\theta}{r^2} \right). \tag{11.4}$$

If $\boldsymbol{B}_p = 0$ the above equation reduces to

$$\frac{\partial B_\theta}{\partial t} + r(\boldsymbol{U} \cdot \nabla)\left(\frac{B_\theta}{r}\right) = \eta \left( \nabla^2 B_\theta - \frac{B_\theta}{r^2} \right).$$

Multiplying by $B_\theta / r^2$ and integrating over the volume we obtain:

$$\frac{1}{2}\frac{\partial}{\partial t} \int (B_\theta / r)^2 \mathrm{d}V + \frac{1}{2} \int \nabla \cdot [(B_\theta / r)^2 \boldsymbol{U}] \mathrm{d}V = -\eta \int |\nabla (B_\theta / r)|^2 \mathrm{d}v,$$

and, discarding the second integral, we conclude that also $B_\theta$ has to decay, unless $B_\theta \propto r$ which is unacceptable since $B_\theta$ is assumed to vanish at infinity. Steady axisymmetric dynamos do not exist.

Finally, it is also possible to prove that purely toroidal flows cannot generate steady magnetic fields.

The above theorems, and in particular Cowling's theorem, show that even relatively simple field topologies, such as the approximate dipole structure of the Earth's magnetic field or of a sunspot, are not directly amenable to simple symmetric fluid motions, which must have a more complicated structure. These rigorous results prevented substantial progresses on dynamo theory for a long time.

## 11.3 Phenomenological Theory of Astrophysical Dynamos

It is clear that understanding the structure of cosmic magnetic fields and of the systems of currents producing them requires clever interpretations of the available data. Just to summarize the most relevant features deduced from observation, we first remark that all magnetized bodies rotate. If the magnetic fields configurations may be approximately described by dipoles at lowest order, their axes turn out to be more or less aligned with the rotation axes, although there are exceptions to this. Terrestrial planets are composed of a non-conducting solid mantle, enclosing highly conducting liquid iron cores, where convective motions carry heat from the center, most likely composed of solid iron, to the mantle and outwards. Jupiter, a gaseous planet, is mainly composed of an alloy of metallic hydrogen and helium. All the magnetic decay timescales of these planets are very short compared with their lifetimes, spanning from $10^4$ to $10^8$ years. The Sun is known to rotate and to possess a convection zone extending from about $0.7\, R_\odot$ almost to the photosphere.

Unfortunately, in most cases we cannot observe the motions that generate the external magnetic fields we see in celestial bodies. This is certainly true for the Earth and the planets, where observations can at most suggest the pattern of the internal motions. The situation with the Sun has greatly improved with the advent of the *helioseismology*, the detailed study of the spectrum of the oscillations seen on the surface, interpreted as global oscillation modes. A notable exception to the shortage of direct observations is provided by the galactic disc, where we are able to actually see the the dynamo mechanisms at work. An important conclusion, based on all the available evidence, both direct and indirect, is that, independently of the form of the *external* magnetic field, the *internal* field is essentially *azimuthal*, as we are now going to illustrate, limiting for simplicity our discussion to the representative cases of the Earth, the Sun and the Galaxy.

## 11.3.1 The Generation of Toroidal Fields from Poloidal Ones

The dipole field of the Earth, measured on the surface, has a strength of about 0.3 G at the equator and 0.6 G at the poles. If one extrapolates these values to the surface of the core, $r = R_c \simeq R_E/2$, where $R_E$ is the radius of the Earth, one arrives at $B_p(R_c) \simeq 5\,G$. On the other hand, a careful analysis of the observed anomalies of the terrestrial magnetic field suggests that the surface of the molten metal core does not rotate uniformly. Its angular velocity, $\Omega$, is higher at the poles than at the equator, which implies that $\Omega$ is a decreasing function of the distance from the rotation axis. It is clear that any non-uniform rotation of the core would stretch the magnetic field lines and drag them out of the meridional planes, thus generating a toroidal component of the magnetic field that was originally absent. It is also clear that the generated toroidal field would have opposite directions in the northern and southern hemispheres.

To be more quantitative, consider again the azimuthal component of the induction Eq. (11.4):

$$\begin{aligned} r\left[\frac{\partial}{\partial t}\left(\frac{B_\theta}{r}\right) + (\boldsymbol{U}\cdot\nabla)\left(\frac{B_\theta}{r}\right)\right] &= r(\boldsymbol{B}\cdot\nabla)\left(\frac{U_\theta}{r}\right) + \eta\left(\nabla^2 B_\theta - \frac{B_\theta}{r^2}\right) \\ &= r(\boldsymbol{B}\cdot\nabla\Omega) + \eta\left(\nabla^2 B_\theta - \frac{B_\theta}{r^2}\right), \end{aligned} \tag{11.5}$$

where the angular velocity $\Omega = U_\theta/r$ has been explicitly introduced. The left hand side of the equation is simply the lagrangian derivative of $B_\theta/r$, multiplied by the factor $r$. The first term on the right could, in principle, produce an increase of the azimuthal field, while the last one is as usual the diffusive term responsible for the field's decay. Consider a case in which $B_\theta = 0$ at $t = 0$, so that $\boldsymbol{B} = \boldsymbol{B}_p$. Equation (11.5) then becomes

$$\frac{\partial B_\theta}{\partial t} = r(\boldsymbol{B}_p \cdot \nabla\Omega),$$

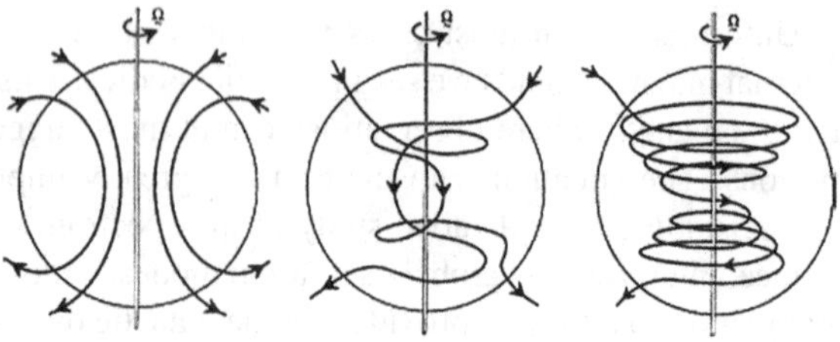

**Fig. 11.1** Sketch illustrating the generation of a toroidal field from a purely poloidal one

showing how a toroidal field can be generated from a purely poloidal one, see Fig. 11.1.

The existence of differential rotation, $\nabla\Omega \neq 0$, thus appears to be a *necessary* condition for the growth of toroidal fields. Given an appropriate $\Omega$ profile, $B_\theta$ would increase with time until either the decay term sets in or the field grows sufficiently that a back reaction on the flow via the Lorentz force begins to inhibit differential rotation, preventing any further growth. In the case of the Earth, since the value of the core's lag with respect to the mantle is known, it is possible to estimate the core-mantle speed difference at the core radius, which turns out to be very small, only $3 \times 10^{-2}$ cm/s, corresponding to a lag of a complete turn every $2 \times 10^3$ years. On the other hand, the decay time of the terrestrial magnetic field is of the order of $10^5$ years, during which period the differential rotation progresses five times or more around the core, thus allowing a substantial growth of the azimuthal field.

According to Bullard [26] we can envisage that the increase of $B_\theta$ stops when the Lorentz force, $(\boldsymbol{j}\times\boldsymbol{B})/c = [(\nabla\times\boldsymbol{B})\times\boldsymbol{B}]/4\pi \simeq (B_p/4\pi\, R_c)B_\theta$, balances the effect of the Coriolis force that acts on the convective motions, $2\,\rho\,\boldsymbol{\Omega}_E \times \boldsymbol{v}_c \simeq 2\,\rho\,\Omega_E\, v_c$, where $\Omega_E$ is the angular velocity of the Earth, $\rho$ the density and $\boldsymbol{v}_c$ the typical speed of convective motions. Therefore

$$(B_p/4\pi\, R_c)B_\theta \simeq 2\,\rho\;\Omega_E\, v_c \quad \text{or} \quad \frac{B_\theta}{B_p} \simeq 8\pi\,\rho\;\Omega_E\, v_c\, R_c/\mathcal{B}_p^2.$$

Using $\rho \simeq 10\,\mathrm{g\,cm^{-3}}$, $v_c \simeq 10^{-2}\,\mathrm{cm\,s^{-1}}$ and $\Omega_E \simeq 7\times 10^{-5}\,\mathrm{s^{-1}}$ we find $B_\theta/B_p \simeq 200$, which proves that indeed the toroidal component is the dominant part of the internal terrestrial field. In spite of this, that field is inevitably hidden, because the poloidal currents associated with it cannot flow upwards through the non-conducting mantle.

As mentioned previously, a lot more is known about the Sun, as a result of helioseismologic observations. From the detailed study of the spectrum of the global oscillations and, in particular, from the rotational splitting of the low order modes, we are now able to infer the behavior of the angular velocity of the Sun from the surface to the bottom of the convection zone and, with a somewhat larger uncertainty all the way down to $\approx 0.2R_\odot$. Figure 11.2 shows the profile of solar rotation rate as a function of the radius from the sun's center, as deduced form the analysis of the solar oscillations.

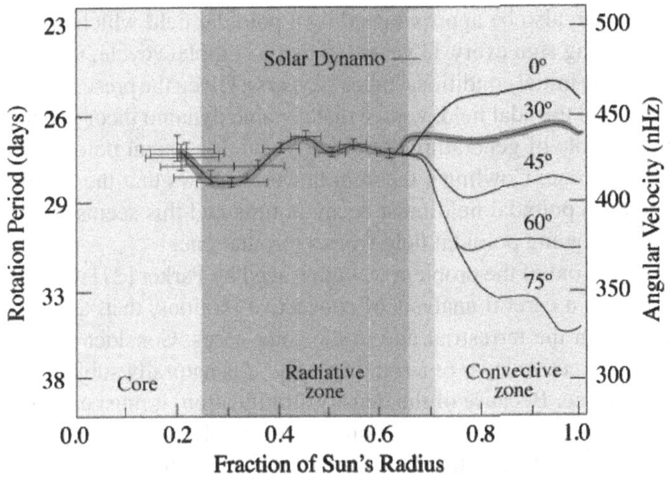


**Fig. 11.2** Internal rotation of the Sun from helioseismology

The visible solar surface does not rotate rigidly, with the photospheric angular velocity, unlike the terrestrial case, being higher at the equator than the poles. Moving inward, the angular velocities at all latitudes are seen to converge towards a common value, so that the rotation appears to be uniform below $\approx 0.6 R_{\odot}$.

The general scenario, therefore, is similar to that of the Earth's interior and we expect to find a strong toroidal field in the convection zone. The magnetic pressure inside the flux tubes increases, making them buoyant and forcing them to emerge through the photosphere in the form of bipolar magnetic regions, whose strongest manifestations are sunspots. Given the mechanism that generates the toroidal field, we expect the leading magnetic spots to have different polarities in the northern and southern hemispheres. This is precisely what observations of bipolar sunspots show, thus confirming the correctness of our deductions. Finally, in the gaseous galactic disc the main magnetic field is observed to be azimuthal. The poloidal component, which is certainly present, is sufficiently small to be hidden by the fluctuations of the much stronger toroidal field.

## 11.3.2 The Generation of Poloidal Fields from Toroidal Ones

So far, we have seen that differential rotation is capable of generating a toroidal field from a poloidal one. But this not only does not solve the dynamo problem, but rather complicates it. The fact is that the fields that we see are, in most cases, poloidal rather than toroidal: the general mean dipolar field of the Earth, apart from local and secular variations, is substantially steady, maintaining both its strength and direction for periods of $10^5$ years or longer; the mean solar field, to a first rough approximation, can also be approximated by a poloidal field which however varies with time, reversing sign every 11 years (so that a complete cycle, i.e. the complete regeneration of the initial conditions, takes 22 years). Given the presence in both cases of a strong internal toroidal field, we see that a viable dynamo theory should provide mechanisms capable of generating both steady and dynamical poloidal fields from an initial toroidal one: Cowling's theorem however shows that the vector potential $\boldsymbol{A}$ that generates a poloidal field must decay in time and this seems to preclude the possibility of a growing poloidal field from a toroidal one.

A brilliant way out of the problem was envisaged by Parker [27] in 1955. Parker's starting point was a careful analysis of convective motions, that, as we have seen, are present both in the terrestrial and in the solar cases. Consider a rising blob of plasma in the convective layer of a rotating body: it is naturally subject to the action of the Coriolis force. Because of the density stratification, it must expand as it rises so that its velocity developes non vanishing tangential components. In a spherical coordinate system, with $\Omega$ defining the polar axis, $\boldsymbol{\Omega} = \Omega(\cos\theta\, \boldsymbol{e}_r - \sin\theta\, \boldsymbol{e}_\theta)$, the tangential components of the Coriolis force, $-2\,\rho(\boldsymbol{\Omega} \times \boldsymbol{U})$ are

$$2\rho\,\Omega[U_\phi \cos\theta\, \boldsymbol{e}_\theta - (U_\theta \cos\theta + U_r \sin\theta)\boldsymbol{e}_\phi]$$

To simplify the discussion let's neglect the radial motion completely by putting $U_r = 0$. The motion of a fluid particle in the plane tangent to the sphere at a given point is also subject to a Coriolis force which in this case is simply

$$2\rho\,\Omega \cos\theta(U_\phi \boldsymbol{e}_\theta - U_\theta \boldsymbol{e}_\phi).$$

The effect of such a force on a diverging flow (corresponding to a rising blob of plasma) can be easily found from the preceding formula considering, in turn, all possible signs for the components $U_\theta$ and $U_\phi$. In the northern hemisphere, $\theta < \pi/2$, the fluid rotates in a clockwise direction with respect to the local vertical, while an anti-clockwise vortex forms in the opposite hemisphere. Of course, for a descending blob, giving rise to a convergent flow, the senses of rotation are the opposite. Such *cyclonic* motions are well known in meteorology, where they control the local climate and occasionally develop into hurricanes. If these cyclonic motions carry magnetic flux tubes with them, the twist converts toroidal fields into poloidal ones. Let's in fact consider a piece of a toroidal magnetic field line and assume that the freezing-in condition applies. A rising blob of plasma will carry the field with it forming a loop, approximately shaped as the greek capital letter $\Omega$, that, for a $\pi/2$ twist, will lie in the meridional plane. A local, small-scale poloidal field has thus been formed from the original toroidal field, see Fig. 11.3

As we have seen, the sense of rotation of the twisting loop changes going from one hemisphere to the other, but the direction of the toroidal field changes as well, so that the circulation sense of the poloidal fields of the newly formed loops is *the same* in both hemispheres. The same conclusion can be reached also by considering the current associated with the magnetic loops generated by the twisting motions. The current is *parallel* to the toroidal field in the northern hemisphere and antiparallel in

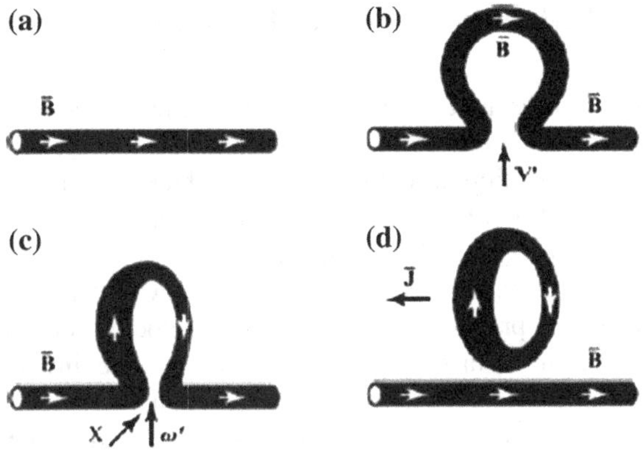


**Fig. 11.3** Sketch of the formation of a local poloidal loop from a toroidal field. **a** Segment of a toroidal field line. **b** Field distortion due to a rising flow. **c** $\pi/2$ twist due to the Coriolis force. **d** Final configuration after a magnetic reconnection

the southern one, but because of the reversal of the toroidal field, the sense of the currents is the *same* in both hemispheres.

Each convective cell thus generates a small scale poloidal field. If their number is large enough, two adjacent loops may come in close contact and their oppositely directed fields may reconnect forcing the loops to coalesce and to form larger structures. At the end of the process a poloidal field of the same scale of the toroidal field results.

So far we have considered only rising motions, but evidently sinking motions are also present, which would tend to produce a poloidal field of opposite sense. Therefore, to have a net effect, some asymmetry must exist. There can be several causes for such an asymmetry, the most obvious being the density stratification which forces rising cells to expand and falling ones to contract. Magnetic buoyancy also tends to favor rising motions. The existence of motions of overall definite handedness thus appear to be a necessary ingredient for the generation of poloidal fields from toroidal ones.

Since the rate of generation of $\boldsymbol{B}_p$ must somehow be proportional to the toroidal component $B_\theta$, the simplest possible model of poloidal field generation is obtained, as suggested by Parker, by adding a *source* term $\alpha B_\theta$ to the rhs of Eq. (11.3), where $\alpha$ has the dimensions of a velocity and, in general is a function of position and time $\alpha = \alpha(\boldsymbol{r}, t)$. Equations (11.3) now reads

$$\frac{\partial A_p}{\partial t} + \frac{1}{r}\boldsymbol{U}_p \cdot \nabla(rA_p) = \alpha B_\theta + \eta\left(\nabla^2 A_p - \frac{A_p}{r^2}\right). \tag{11.6}$$

The term $\alpha B_\theta$, responsible for the so-called $\alpha$-*effect*, is equivalent to an electric field or *mean electromotive force* $\mathcal{E}_\theta = \alpha B_\theta / c$, that counteracts the decay due to the resistive term. Depending on the relative importance of these terms, the poloidal field may grow or reach a steady state. Recalling our previous considerations, we see that in order to have the same sign of the current, $j_\theta = \sigma \mathcal{E}_\theta = (\sigma/c)\alpha B_\theta$, in both hemispheres $\alpha$ has to change sign across the equator.

## 11.4 Mean-field Electrodynamics

The pioneering approach of Parker prompted the plasma community to look for a more formal justification for the appearance of the crucial electromotive force term in the induction Eq. (11.6). This was achieved by the so-called *mean-field theory* first proposed by Steenbeck, Krause and Rädler in 1966,[1] that we shall now illustrate, closely following their treatment.

Basically, the theory relates the time evolution of the mean magnetic field to the statistical properties of the turbulent velocity field, $\boldsymbol{v}$. Turbulence is usually considered to be incompressible and isotropic, i.e. **invariant** for arbitrary *rotations* of the axes. However, as already remarked, the turbulent velocity field should posses a definite handedness, which implies that it is **not invariant** for *reflections* of the axes with respect to a plane. Reflection invariance, in fact, would mean that right-handed motions are as probable as left-handed ones, contrary to our assumption. A velocity field lacking reflection invariance is said to have a non vanishing *helicity*, $\mathfrak{h}$, defined as the average value of $\boldsymbol{v} \cdot (\nabla \times \boldsymbol{v})$:

$$\mathfrak{h} = \langle \boldsymbol{v} \cdot (\nabla \times \boldsymbol{v}) \rangle$$

Here and in the following, the average $\langle \boldsymbol{F} \rangle$ of a certain quantity $F$ has to be intended as an ensemble average, defined as the expectation value of $F$ in an ensemble of identical systems. Notice that the averaging operation commutes with both space and time derivatives.

It is a general characteristic of homogeneous and isotropic turbulence that the small-scale fluctuations at a given point, $\boldsymbol{r}$, have a correlation with another quantity at a different point, $\boldsymbol{r}'$, only if the distance between the two points is not too large: in other words, correlations vanish if $|\boldsymbol{r} - \boldsymbol{r}'| > \lambda_{corr}$, where $\lambda_{corr}$ is the *correlation length*. Similarly, the correlations at different times are zero if if they are taken between times separated by $t - t' > \tau_{corr}$, where $\tau_{corr}$ is the *correlation time*. Thus, to determine the correlation between any two fluctuating quantities, the knowledge of those quantities is required only in the neighborhood of the point considered. If the initial and boundary conditions are outside of that neighborhood, they will not influence the value of the correlation.

Vectors can be classified into two different classes according to their transformation properties for reflections with respect to a plane: *polar* vectors (or simply vectors) change their sign upon reflection, whereas *axial* vectors don't. It is clear that axial vectors must be connected with operations where handedness is involved, like cross products or curls. The velocity, electric field and current density are vectors, while the magnetic field is an axial vector.

The scalar product of a vector and an axial vector produces a quantity that changes sign under reflections: such a quantity is called a *pseudo-scalar*. An example of a

[1] An exposition in english of the original german papers can be found in the book by Krause and Rädler [13].

pseudo-scalar is the helicity. If $\mathfrak{h} \neq 0$ the turbulent velocity field is not reflection invariant. In fact, if the turbulence were reflection invariant, on the one hand $\mathfrak{h}$ should remain unaltered under reflections since all the properties of the velocity field do not change, while on the other it should change sign since it is a pseudo-scalar, Hence, $\mathfrak{h} = 0$ for such a field.

Let's now consider a small-scale turbulent velocity field, $\boldsymbol{u}$, which is statistically stationary, homogeneous and pseudo-isotropic, meaning that it is rotation invariant but not reflection invariant. Such a turbulence will both produce a fluctuating magnetic field, $\boldsymbol{b}$ on a small-scale $l$, and sustain a mean magnetic field, $B_0$, on a much larger scale $L$, so that the total field will be $\boldsymbol{B} = \boldsymbol{B}_0 + \boldsymbol{b}$. The induction equation (11.1) then becomes:

$$\frac{\partial}{\partial t}(\boldsymbol{B}_0 + \boldsymbol{b}) = \nabla \times [(\boldsymbol{U}_0 + \boldsymbol{u}) \times (\boldsymbol{B}_0 + \boldsymbol{b})] + \eta \nabla^2 (\boldsymbol{B}_0 + \boldsymbol{b}), \tag{11.7}$$

where $U_0$ is a large-scale velocity field. We now average this equation on a scale $\ell = \mathcal{O}(l)$. Taking into account that $\langle \boldsymbol{u} \rangle = \langle \boldsymbol{b} \rangle = 0$, we obtain

$$\frac{\partial \boldsymbol{B}_0}{\partial t} = \nabla \times [\boldsymbol{U}_0 \times \boldsymbol{B}_0 + \langle \boldsymbol{u} \times \boldsymbol{b} \rangle] + \eta \nabla^2 \boldsymbol{B}_0. \tag{11.8}$$

Subtracting Eq. (11.4) from Eq. (11.7) we get:

$$\frac{\partial \boldsymbol{b}}{\partial t} = \nabla \times [\boldsymbol{U}_0 \times \boldsymbol{b} + \boldsymbol{u} \times \boldsymbol{B}_0 + \boldsymbol{u} \times \boldsymbol{b} - \langle \boldsymbol{u} \times \boldsymbol{b} \rangle] + \eta \nabla^2 \boldsymbol{b}. \tag{11.9}$$

If $\boldsymbol{U}_0$, $\boldsymbol{B}_0$ and $\boldsymbol{u}$ are known, the above equation in principle determines $\boldsymbol{b}$, even if the actual solution is hard to obtain. To proceed, let's evaluate the order of magnitude of the various terms entering Eq. (11.9). The large scale velocity field can be removed by adopting a reference system moving with the local value of $\boldsymbol{U}_0$. The orders of magnitude of the other terms at the rhs are, respectively: $\mathcal{O}(uB_0/\ell)$, $\mathcal{O}(ub/\ell)$, $\mathcal{O}(\eta b/\ell^2)$. If the **small-scale** magnetic Reynolds number $(\ell u/\eta) \ll 1$, the term $\nabla \times [\boldsymbol{u} \times \boldsymbol{b} - \langle \boldsymbol{u} \times \boldsymbol{b} \rangle]$ can be neglected and Eq. (11.9) becomes a linear equation.[2] The large scale field $B_0$ is determined by Eq. (11.8), where the term $\langle \boldsymbol{u} \times \boldsymbol{b} \rangle$ appears to be the only source term capable of opposing the field decay. To allow the growth of the field this term must therefore be kept. This procedure is known as the *quasi-linear approximation* or *first-order smoothing*.

Taking into account all the above approximations, the final form of our system of equations is:

$$\frac{\partial \boldsymbol{B}_0}{\partial t} = \nabla \times \langle \boldsymbol{u} \times \boldsymbol{b} \rangle + \eta \nabla^2 \boldsymbol{B}_0, \tag{11.10}$$

---

[2] The assumption on the magnetic Reynolds number may be difficult to justify in astrophysical conditions. In spite of this it is customary to accept it.

and

$$\frac{\partial \boldsymbol{b}}{\partial t} = \nabla \times (\boldsymbol{u} \times \boldsymbol{B}_0) + \eta \nabla^2 \boldsymbol{b}. \tag{11.11}$$

Cosmic plasmas are, as a rule, excellent conductors: we shall therefore start by looking for the solutions of Eq. (11.11) with $\eta = 0$, which, in this high-conductivity limit, can be written as

$$\boldsymbol{b}(\boldsymbol{r}, t) = \int_{-\infty}^{t} \nabla \times [\boldsymbol{u}(\boldsymbol{r}, t') \times \boldsymbol{B}_0(\boldsymbol{r}, t')] \mathrm{d}t'. \tag{11.12}$$

According to our previous discussion, we identify the source term in Eq. (11.10) with an electromotive force and write

$$\mathcal{E}(\boldsymbol{r}, t) = \frac{1}{c} \langle \boldsymbol{u} \times \boldsymbol{b} \rangle = \frac{1}{c} \int_{-\infty}^{t} \langle \boldsymbol{u}(\boldsymbol{r}, t) \times \nabla \times [\boldsymbol{u}(\boldsymbol{r}, t') \times \boldsymbol{B}_0(\boldsymbol{r}, t')] \rangle \mathrm{d}t'.$$

As illustrated above, the integrand is different from zero only if $t - t' < \tau_{corr}$ and in this time interval the large scale field $\boldsymbol{B_0}$ can be assumed to be constant, $\boldsymbol{B_0}(\boldsymbol{r}, t') \simeq \boldsymbol{B_0}(\boldsymbol{r}, t)$. Changing the integration variable into $\tau = t - t'$, we finally have

$$\mathcal{E}(\boldsymbol{r}, t) = \frac{1}{c} \int_{0}^{\infty} \langle \boldsymbol{u}(\boldsymbol{r}, t) \times \nabla \times [\boldsymbol{u}(\boldsymbol{r}, t - \tau) \times \boldsymbol{B}_0(\boldsymbol{r}, t)] \rangle \mathrm{d}\tau. \tag{11.13}$$

Since

$$\nabla \times (\boldsymbol{u} \times \boldsymbol{B_0}) = (\boldsymbol{B_O} \cdot \nabla)\boldsymbol{u} - (\boldsymbol{u} \cdot \nabla)\boldsymbol{B_0} \quad (\text{for } \nabla \cdot \boldsymbol{B_0} = \nabla \cdot \boldsymbol{u} = 0),$$

the $i$-th component of Eq. (11.13), in a right-handed cartesian coordinate system $[x_1, x_2, x_3]$, can be written as

$$c\,\mathcal{E}_i(\boldsymbol{r}, t) = \epsilon_{ijk} w_{jkl} \boldsymbol{B}_{0l} - \epsilon_{ijk} w_{jl} \frac{\partial B_{0k}}{\partial x_l}, \tag{11.14}$$

where we have introduced the notations:

$$w_{jkl} = \int_{0}^{\infty} \langle u_j(\boldsymbol{r}, t) \frac{\partial u_k(\boldsymbol{r}, t - \tau)}{\partial x_l} \rangle \mathrm{d}\tau \quad \text{and} \quad w_{jl} = \int_{0}^{\infty} \langle u_j \boldsymbol{r}, t) u_l(\boldsymbol{r}, t - \tau) \rangle \mathrm{d}\tau.$$

and $\epsilon_{ijk}$ is the totally antisymmetric tensor.[3]

---

[3] $\epsilon_{ijk}$ vanishes if at least two of the indices are equal, equals $+1$ for a permutation of the indices even with respect to the fundamental permutation 123 and $-1$ for an odd permutation.

The average quantities $w_{ijk}$ and $w_{il}$ reflect the assumed properties of the turbulent velocity field: for an isotropic turbulence they must remain unchanged for arbitrary rotations of the axes. It is then easy to deduce a series of important properties of these quantities.

First of all, $w_{ijk} = 0$ if at least two indices are equal. To show this, consider a rotation of $\pi/2$ around the $x_1$ axis. If $\boldsymbol{e}_i$ are the unit vectors along the coordinate axes, as a result of the rotation $\boldsymbol{e}_2 \rightarrow \boldsymbol{e}_3$, $u_2 \rightarrow u_3$ so that $w_{112}$ transforms into $w_{113}$. But a rotation of $-\pi/2$ around the same axis would give $\boldsymbol{e}_2 \rightarrow -\boldsymbol{e}_3$ and $w_{112} \rightarrow -w_{113}$. Thus, $w_{112} = w_{113}$, but also $w_{112} = -w_{113}$ which implies $w_{112} = w_{113} = 0$. Similarly, considering rotations of $\pm\pi/2$ around the $x_3$ axis, we would find $w_{111} = w_{222} = 0$. This procedure can be applied to other values of the indices and shows that the indices of a non vanishing $w_{ijl}$ must all be different. On the contrary, by the same means, we can prove that $w_{il} = 0$ if $i \neq l$ and that $w_{11} = w_{22} = w_{33}$. If $\tilde{\eta}$ is their common value, we have

$$\tilde{\eta} = w_{11} = w_{22} = w_{33} = \frac{1}{3}(w_{11} + w_{22} + w_{33}).$$

Taking into account the above properties, we obtain for the components of the electromotive force $\boldsymbol{\mathcal{E}}$:

$$\begin{aligned} c\,\mathcal{E}_1 &= (w_{213} - w_{312})B_{01} - \tilde{\eta}(\frac{\partial B_{02}}{\partial x_3} - \frac{\partial B_{03}}{\partial x_2}) \\ &= (w_{213} - w_{312})B_{01} - \tilde{\eta}(\nabla \times \boldsymbol{B_0})_1 \\ c\,\mathcal{E}_2 &= (w_{321} - w_{123})B_{02} - \tilde{\eta}(\nabla \times \boldsymbol{B_0})_2 \\ c\,\mathcal{E}_3 &= (w_{132} - w_{231})B_{03} - \tilde{\eta}(\nabla \times \boldsymbol{B_0})_3 \end{aligned}$$

Exploiting again the invariance properties of $w_{ijk}$, it is easy to show that the coefficients of $B_{0i}$ in the preceding equations are all equal. Denoting by $\alpha$ their common value, we obtain

$$\begin{aligned} \alpha &= w_{213} - w_{312} = w_{321} - w_{123} = w_{132} - w_{231} \\ &= \frac{1}{3}(w_{213} + w_{321} + w_{132} - w_{312} - w_{123} - w_{231}). \end{aligned}$$

We have thus shown that the electromotive force $\boldsymbol{\mathcal{E}}$ can be written as

$$c\,\boldsymbol{\mathcal{E}} = \alpha \boldsymbol{B_0} - \tilde{\eta}(\nabla \times \boldsymbol{B_0}), \tag{11.15}$$

recovering the phenomenological term introduced by Parker with the addition of an extra term proportional to $\nabla \times \boldsymbol{B_0}$. The effect of the term proportional to $\tilde{\eta}$ is easily appreciated if we introduce the expression for $\langle \boldsymbol{u} \times \boldsymbol{b} \rangle$, as given by Eq. (11.15), into the equation for the mean magnetic field (11.10) that becomes

$$\frac{\partial \boldsymbol{B}_0}{\partial t} = \alpha(\nabla \times \boldsymbol{B_0}) + (\eta + \tilde{\eta})\nabla^2 \boldsymbol{B}_0, \tag{11.16}$$

showing that the presence of turbulence has increased the magnetic diffusivity.

Writing the expressions above, we have implicitly assumed that both $\alpha$ and $\tilde{\eta}$ are constants, when they should in general be functions of $\boldsymbol{r}$ and $t$. In particular we know that $\alpha$ has different signs in the two hemispheres (see the comment after Eq. (11.6)): therefore the term containing $\alpha$ in Eq. (11.16) should read $\nabla \times (\alpha \boldsymbol{B_0})$. Moreover, since in general neither $\eta$ nor $\tilde{\eta}$ are space-independent, an extra term $-\nabla(\eta + \tilde{\eta}) \times (\nabla \times \boldsymbol{B_0})$ should be added to the rhs of Eq. (11.16): our conclusions concerning the effect of turbulence thus remain unaltered.

Introducing the explicit expressions of the $w_{ijk}$ and $w_{ik}$ in the definitions of $\alpha$ and $\tilde{\eta}$, we obtain:

$$\alpha = -\frac{1}{3}\int_0^\infty \langle \boldsymbol{u}(\boldsymbol{r}, t) \cdot [\nabla \times \boldsymbol{u}(\boldsymbol{r}, t - \tau)]\rangle \mathrm{d}\tau, \tag{11.17}$$

and

$$\tilde{\eta} = \frac{1}{3}\int_0^\infty \langle \boldsymbol{u}(\boldsymbol{r}, t) \cdot \boldsymbol{u}(\boldsymbol{r}, t - \tau)\rangle \mathrm{d}\tau. \tag{11.18}$$

An order of magnitude estimate of $\alpha$ and $\tilde{\eta}$ gives:

$$\alpha \simeq -\frac{1}{3}\langle \mathfrak{h}\rangle \tau_{corr} \quad \text{and} \quad \tilde{\eta} \simeq \frac{1}{3}\langle u^2\rangle \tau_{corr}.$$

Summarizing, the mean field theory, applied to an axisymmetric configuration, results in the following form of the basic dynamo equations, Eqs. (11.1) and (11.2):

$$\frac{\partial A}{\partial t} + \frac{1}{r}(\boldsymbol{U}_p \cdot \nabla)(rA) = \alpha B_\theta + \eta_0(\nabla^2 - \frac{1}{r^2})A, \tag{11.19}$$

$$\frac{\partial B_\theta}{\partial t} + r(\boldsymbol{U}_p \cdot \nabla)\left(\frac{B_\theta}{r}\right) = r(\boldsymbol{B}_p \cdot \nabla \Omega) + \nabla \times (\alpha \boldsymbol{B}_p) + \eta_0(\nabla^2 - \frac{1}{r^2})B_\theta, \tag{11.20}$$

where for simplicity $\eta$ and $\tilde{\eta}$ are assumed to be constant and $\eta_0 = \eta + \tilde{\eta}$ is the total diffusivity.

Equation (11.20) shows that a toroidal field can be generated from a poloidal one not only by a nonuniform angular velocity ($\omega$-effect), but also by the presence of the term proportional to $\alpha$ ($\alpha$-effect). On the other hand, the generation of a poloidal field from a toroidal one, that completes the dynamo cycle, only depends on $\alpha$. Accordingly, we speak of $\alpha\omega$-dynamos when in the first phase the $\omega$-effect dominates and of $\alpha^2$-dynamos when the $\alpha$-effect dominates.

### 11.4.1 Simple Solutions of the Dynamo Equations

The dynamo equations are far too complex to allow a detailed analytical treatment, even within the kinematic dynamo theory. Models with a reasonable degree of realism can only be based on numerical simulations. However, the study of highly simplified configurations can help to understand the general behavior of the solutions of the dynamo equations. In the following we shall therefore limit ourselves to consider solutions in simple geometries, but reminiscent of the more realistic ones encountered in astrophysical contexts.

#### Slab Geometry

If we restrict ourselves to a thin spherical shell, the solution of the dynamo equations is considerably simplified in cartesian geometry by setting up a coordinate system where the $z$-axis is directed along the local vertical and the $y$-axis is directed eastwards, i.e. in the toroidal direction. In a right-handed coordinate system therefore, the $x$-axis points south. The large scale velocity field is assumed to be purely toroidal $\boldsymbol{U} = U(z)\boldsymbol{e}_y$ with a vertical shear. Moreover the problem is considered to be $y$-independent, $\partial/\partial y = 0$. The magnetic field can be represented as

$$\boldsymbol{B}_0 \equiv \left[-\frac{\partial A}{\partial z}, B_y, \frac{\partial A}{\partial x}\right],$$

with $\boldsymbol{A} = A(x, z)\boldsymbol{e}_y$

In this configuration the dynamo equations are:

$$\begin{aligned} \frac{\partial A}{\partial t} &= \alpha B_y + \eta_0 \nabla^2 A, \\ \frac{\partial B_y}{\partial t} &= \frac{\partial A}{\partial x}\frac{\mathrm{d}U}{\mathrm{d}z} - \alpha\nabla^2 A + \eta_0 \nabla^2 B_y. \end{aligned} \tag{11.21}$$

$\alpha$ is considered to be constant: the cases of $\alpha > 0$ and $\alpha < 0$ can be treated separately.

We look for solutions in the form of waves: $A \propto \exp[\mathrm{i}(k_x x + k_z z) + pt)]$, where $k_x$ is assumed to be positive and $p$ is in general complex, $p = \gamma - \mathrm{i}\omega$, as one could have anticipated from the fact that the system of dynamo equations (11.21) is not self-adjoint. For simplicity assume a linear shear for $\boldsymbol{U}$, $U(z) = U'z$, with $U' = const.$

Start from the case of an $\alpha\omega$-dynamo and therefore neglect the $\alpha$ term in the equation for $B_y$. In this framework, the dispersion relation may be written:

$$(p + \eta_0 k^2)^2 = \mathrm{i}k_x \alpha U',$$

or

$$p = (\mathrm{i}k_x\alpha U')^{1/2} - \eta_0 k^2 = -\eta_0 k^2\Big[1 - \Big(\frac{\mathrm{i}k_x\alpha U'}{\eta_0^2 k^4}\Big)^{1/2}\Big].$$

Since $\alpha U'$ can be positive or negative we have:

$$\Big(\frac{\mathrm{i}k_x\alpha U'}{\eta_0^2 k^4}\Big)^{1/2} = \frac{1 \pm \mathrm{i}}{\sqrt{2}}\Big(\frac{|k_x\alpha U'|}{\eta_0^2 k^4}\Big)^{1/2} = (1 \pm \mathrm{i})\sqrt{|N_D|},$$

where we have introduced the *dynamo number*, $N_D$:

$$N_D = \frac{\alpha U'}{2\eta_0^2 k^3},$$

and the upper sign refers to positive $N_D$ values. Notice that the properties of the solution are not affected by the value of $k_z$. Thus,

$$p = -\eta_0 k^2(1 - \sqrt{|N_D|}) \mp \mathrm{i}\,\eta_0 k^2\sqrt{|N_D|}$$

Growing solutions require that $\gamma > 0$, which implies $|N_D| > 1$. The sign of $N_D$ determines the direction of propagation of the waves. In fact, when $N_D > 0$, $\omega = -\eta_0 k^2\sqrt{|N_D|} < 0$,

$$A \propto \exp\Big[\mathrm{i}k_x(x + \frac{|\omega|}{k_x}t)\Big]$$

and the wave therefore propagates in the negative $x$−direction (i.e. northward), while for $N_D < 0$, $\omega > 0$ and the wave propagates southward.[4] These wave-like solutions of the dynamo equations have been called by Parker *migrating dynamo waves*.

We turn now to the case of $\alpha^2$−dynamos and thus neglect the effect of the velocity shear in the dynamo equations (11.21) by putting $U = 0$. The dispersion relation in this case turns out to be:

$$p = \pm k\alpha - \eta_0 k^2.$$

Since $p$ is real there are no migrating waves. We have growing or decaying solutions according to whether

$$\frac{|\alpha|}{\eta_0 k} \gtrless 1,$$

with the appropriate choice of sign. The limiting case $|\alpha| = \eta_0 k$ produces a steady state dynamo.

The interest of this slab model resides in its simplicity. Its relevance to real astrophysical situations, typically requiring a spherical geometry, is marginal: for instance,

---

[4] If $k_x < 0$, the whole reasoning can be repeated with the same final results.

in the case of the Sun, a plane geometry $\alpha\,\omega$-dynamo could be only used to describe a thin spherical shell. In spite of this, there are a few interesting consequences. As already remarked, $\alpha$ changes sign across the equator: in the case of the Sun, $\alpha > 0$ in the northern hemisphere. On the other hand, sunspots are known to form in both hemispheres at about 45° of latitude and to migrate towards the equator during the solar cycle. If we want to connect this observational fact to the properties of Parker's migrating dynamo waves, we must admit that the toroidal velocity must *increase* with depth, making $U' < 0$. Then $N_D \propto \alpha U' < 0$ in the northern hemisphere, while $N_D > 0$ in the southern one. In this situation, the waves migrate towards the equator in both hemispheres, in agreement with the observed behavior of sunspots.

Another known quantity is the duration of the solar cycle requiring 22 years for a complete restoration of the original poloidal field. If we tentatively identify the cyclic behavior of the solar magnetic field with the constant-amplitude solution of the planar model, $|N_D| = 1$, we have $|\omega| = \eta_0 k^2$. Since the dynamo action takes place within the convection zone, whose depth is $h \simeq 2 \times 10^5$ km, the *vertical* wavelength must be less than $2h$ or $k_z \geqslant \pi/h$. Let $\lambda = 2\pi/k_x$ be the *horizontal* wavelength and assume $\lambda \gg h$. Then

$$k^2 = k_x^2 + k_z^2 \geqslant \frac{4\pi^2}{\lambda^2} + \frac{\pi^2}{h^2} \simeq \frac{\pi^2}{h^2},$$

or

$$T = \frac{2\pi}{|\omega|} \simeq \frac{2h^2}{\pi\eta_0},$$

where $T$ is the wave's period. To reproduce the observed 22-years period $\eta_0$ should be $\simeq 4 \times 10^{11}\ \mathrm{cm^2\ s^{-1}}$. Since in the solar convection zone the molecular diffusivity, $\eta$, is of the order $10^5\ \mathrm{cm^2\ s^{-1}}$, we conclude that the dominant term in $\eta_0$ must be the turbulent diffusivity $\tilde{\eta}$. The next question is whether $\tilde{\eta}$ is compatible with the required value of $\approx 10^{12}\ \mathrm{cm^2\ s^{-1}}$. Reasonable estimates based on the properties of the solar turbulent convective eddies give values close to those quoted above. It seems therefore that even a simplified dynamo model is able to reproduce the basic features of solar activity. However, more realistic models might give a completely different picture and the preceding discussion should be taken cautiously.

## Cylindrical Geometry

This geometry is the relevant one to discuss magnetic field generation in the Galaxy, where the field is mostly confined in the galactic disc, whose radius $R \gtrsim 10\,kpc$ is much greater than its thickness $h \simeq 300\,pc$, so that $h/R \lesssim 10^{-2}$. This suggests that the galactic disc can be treated as a thin disc, which implies that in cylindrical coordinates the vertical gradients are much larger than the radial ones. As a consequence, to lowest order, $r$-derivatives are dropped and only $z$-derivatives are kept. An order of magnitude estimate of the condition $\boldsymbol{\nabla} \cdot \boldsymbol{B} = 0$ implies that $B_z/B_r \simeq h/R \ll 1$, which allows us to eliminate $B_z$. The dynamo equations therefore reduce to

$$\frac{\partial B_r}{\partial t} = -\frac{\partial}{\partial z}(\alpha B_\theta) + \eta_0 \frac{\partial^2 B_r}{\partial z^2}, \tag{11.22}$$

$$\frac{\partial B_\theta}{\partial t} = B_r\left(r\frac{\partial \Omega}{\partial r}\right) + \frac{\partial}{\partial z}(\alpha B_\theta) + \eta_0 \frac{\partial^2 B_\theta}{\partial z^2}. \tag{11.23}$$

The total magnetic diffusivity, $\eta_0$, has been taken to be constant, while $\alpha = \alpha(z)$, is an unspecified odd function of $z$. Introducing the vector potential $\boldsymbol{A} = A(z)\boldsymbol{e}_\theta$, so that

$$B_r = -\frac{\partial A}{\partial z},$$

and noticing that in the disc $\boldsymbol{U} = U\boldsymbol{e}_\theta$, with $U = r\,\Omega$ almost constant, we find that $(r\partial\Omega/\partial r) = -\Omega$. Replace now Eq. (11.22) with

$$\frac{\partial A}{\partial t} = (\alpha B) + \eta_0 \frac{\partial^2 A}{\partial z^2},$$

and Eq. (11.23) with

$$\frac{\partial B}{\partial t} = \Omega\frac{\partial A}{\partial z} + \frac{\partial}{\partial z}(\alpha B) + \eta_0 \frac{\partial^2 B}{\partial z^2},$$

where, to simplify the notations, we have written $B_\theta \equiv B$. Assuming $B\,, A \propto \mathrm{e}^{pt}$, where, as before, $p$ is in general a complex quantity, the following eigenvalue system results:

$$pA = \alpha\, B + \eta_0 \frac{\partial^2 A}{\partial z^2}, \tag{11.24}$$

$$pB = \Omega\frac{\partial A}{\partial z} + \frac{\partial}{\partial z}(\alpha B) + \eta_0 \frac{\partial^2 B}{\partial z^2}. \tag{11.25}$$

It is worth noticing that the thin-disc approximation, that leads to the preceding system of equations and controls the $z$–dependence of the field, does not eliminate completely the $r$–dependence, since $\Omega$, $h$ and possibly $\alpha$, still depend on the radial variable. As a result, the eigenvalues, $p$, also depend on $r$: i.e. the growth rates change with radius.[5]

If $\alpha$ is an antisymmetric function of $z$, the solutions of the above system of equations can be classified according to their symmetry properties for $z \to -z$, into *odd or dipole solutions* :

---

[5] A discussion of this point, which is beyond the aim of this introductory book, can be found in Zeldovich et al. (1983) [10].

$$B(z) = -B(-z), \quad A(z) = A(-z),$$

or *even or quadrupole solutions*:

$$B(z) = B(-z), \quad A(z) = -A(-z).$$

Obviously, we must supplement Eqs. (11.24) and (11.25) with the boundary conditions in $z = \pm h$. So-called vacuum boundary conditions that require

$$B_\theta(\pm h) = B(\pm h) = 0 \quad \text{and} \quad B_r(\pm h) = 0 \quad \text{or equivalently} \quad \left(\frac{\partial A}{\partial z}\right)_{z=\pm h} = 0$$

are usually chosen. Since the galactic field seems to be even, the quadrupole solution is the appropriate one.

In spite of the apparent simplicity, the reduced set of dynamo equations (11.24) and (11.25) is complicated to solve analytically even with the simplest forms of $\alpha(z)$. Examples of analytical solutions can be found e.g. in Parker (1979) [11] or in Moffat (1978) [12].

In the following we shall limit ourselves to a few elementary cases, whose only merit is to give some flavor of the more realistic numerical solutions. To simplify the notation, we first introduce, as in Zeldovich et al. (1983) [10], the following non-dimensional quantities:

$$\zeta = \frac{z}{h}, \qquad \tau = \frac{h^2}{\eta_0}t, \qquad \tilde{\alpha}(\zeta) = \frac{\alpha(\zeta)}{\alpha_0} \quad \text{with} \quad \alpha_0 \equiv \alpha_{max},$$

and

$$R_\alpha = \frac{\alpha_0 h}{\eta_0}, \qquad R_\Omega = \frac{\Omega h^2}{\eta_0}.$$

The non-dimensional form of the dynamo equations reads:

$$\frac{\partial A}{\partial \tau} = R_\alpha[\tilde{\alpha} h B)] + \frac{\partial^2 A}{\partial \zeta^2}, \tag{11.26}$$

$$\frac{\partial (hB)}{\partial \tau} = R_\Omega \frac{\partial A}{\partial \zeta} + R_\alpha \frac{\partial}{\partial \zeta}[\tilde{\alpha}(hB)] + \frac{\partial^2 (hB)}{\partial \zeta^2}, \tag{11.27}$$

with the boundary conditions:

$$B(\pm 1) = 0, \qquad \left(\frac{\partial A}{\partial \zeta}\right)_{\pm 1} = 0.$$

[The extra factor $h$ is due to the fact that $B_r = -\partial A/\partial z = -(1/h)(\partial A/\partial \zeta)$]

- $\alpha = \Omega = 0$: *decay modes.*

When no sources are present, the even solution of the system (11.26), (11.27) is easily found by writing $A(\zeta, \tau) = a(\zeta) f(\tau)$. Using this representation in (11.26), we get

$$\frac{\mathrm{d}f/\mathrm{d}\tau}{f} = \frac{\mathrm{d}^2a/\mathrm{d}\zeta^2}{a} = -\gamma,$$

or

$$A \propto \mathrm{e}^{-\gamma\tau} \sin(\gamma^{1/2}\zeta).$$

The boundary conditions in $\zeta = \pm 1$ give

$$\cos(\gamma^{1/2}) = 0 \quad \text{and therefore} \quad \gamma = (n + \tfrac{1}{2})^2\pi^2, \quad n = 0, 1, 2, \ldots$$

Clearly $A = 0$ is also a solution of Eq. (11.26). In the same way we find that

$$B = 0 \quad \text{or} \quad B \propto \mathrm{e}^{-\gamma\tau} \cos(\gamma^{1/2}\zeta), \qquad \gamma = (n + \tfrac{1}{2})^2\pi^2, \quad n = 0, 1, 2, \ldots$$

- $\alpha = 0$, $\Omega \neq 0$: *differential rotation only.*

The equation for $A$ is identical to that of the preceding example, therefore:

$$A = A_0\mathrm{e}^{-\gamma\tau} \sin(\gamma^{1/2}\zeta), \quad \gamma = (n + \tfrac{1}{2})^2\pi^2, \quad n = 0, 1, 2, \ldots.$$

The equation for $B$ becomes:

$$\frac{\partial B}{\partial \tau} = \frac{R_\Omega}{h}\frac{\partial A}{\partial \zeta} + \frac{\partial^2 B}{\partial \zeta^2} = \frac{\gamma^{1/2} A_0 R_\Omega}{h} \mathrm{e}^{-\gamma\tau} \cos(\gamma^{1/2}\zeta) + \frac{\partial^2 B}{\partial \zeta^2},$$

which suggests writing $B$ as $B(\zeta, \tau) = b(\tau)\cos(\gamma^{1/2}\zeta)$. Inserting this expression in the preceding equation we find

$$\frac{\mathrm{d}b(\tau)}{\mathrm{d}\tau} + \gamma b(\tau) = \gamma^{1/2} A_0 R_\Omega \mathrm{e}^{-\gamma\tau},$$

a linear differential equation whose general solution is:

$$b(\tau) = \mathrm{e}^{-\gamma\tau}[b(0) + (\gamma^{1/2} A_0 R_\Omega / h)\tau].$$

If we consider a situation in which initially the azimuthal field vanishes, $b(0) = 0$, then

$$B \equiv B_\theta = (\gamma^{1/2} A_0 R_\Omega / h)\tau\, \mathrm{e}^{-\gamma\tau} \cos(\gamma^{1/2}\zeta).$$

On the other hand at $\tau = 0$ the field is purely radial (neglecting the small $B_z$)

$$B_r(0, \zeta) = -(1/h)\frac{\partial A}{\partial \zeta} = -(\gamma^{1/2} A_0/h)\cos(\gamma^{1/2}\zeta),$$

so that the quantity $(\gamma^{1/2} A_0/h)$ is simply the magnitude of the initial field at the midplane of the disc, $B_0 = |\boldsymbol{B}(0, 0)|$. Finally,

$$B \equiv B_\theta = B_0 R_\Omega \tau\, \mathrm{e}^{-\gamma\tau} \cos(\gamma^{1/2}\zeta).$$

The azimuthal field increases linearly with time until the damping factor becomes dominant, bringing both $B_r$ and $B_\theta$ to zero. The maximum value of the azimuthal field, $(B_0 R_\Omega \mathrm{e}^{-\gamma})$, is reached at $\tau = 1$.

- $\alpha \neq 0$, $\Omega = 0$: $\alpha^2$-*dynamo.*

In this case it is possible to show (see Zeldovich et al 1983) that the problem can be reduced to the following equation:

$$\frac{\partial^2 B}{\partial \zeta^2} - \gamma B = \mathrm{i} R_\alpha \frac{\partial \tilde{\alpha} B}{\partial \zeta}.$$

Writing $B$ as

$$B = \psi \exp\left(\tfrac{1}{2}\mathrm{i} R_\alpha \int \alpha \mathrm{d}\zeta\right),$$

the preceding equation can be reduced to a Schrödinger equation with a generally complex potential:

$$\frac{\partial^2 \psi}{\partial \zeta^2} + (E - V)\psi = 0,$$

$$E = -\gamma, \qquad V = \tfrac{1}{4} R_\alpha^2 \tilde{\alpha}^2 + \tfrac{1}{2}\mathrm{i} R_\alpha \frac{\partial \tilde{\alpha}}{\partial \zeta}.$$

If $\tilde{\alpha}$ is chosen to be, as in Parker (1979), $\tilde{\alpha} = 1$ for $\zeta > 0$ and $\tilde{\alpha} = -1$ for $\zeta < 0$, the eigenvalues $\gamma$ turn out to be

$$\gamma = \tfrac{1}{4} R_\alpha^2 - k^2,$$

where $k$ is given by the solutions of the equation

$$k \cot k = \tfrac{1}{2} R_\alpha.$$

Leaving aside the problem of actually solving the dynamo equations in realistic geometries, it is important to estimate at least approximately the amount of field growth that can be expected in the conditions typical of our Galaxy. To do so, we shall completely neglect the boundary conditions and consider the system (11.24) and (11.25), where we drop the $\alpha$–term in the equation for B, since the galactic dynamo seems to be of the $\alpha\omega$ type. If we look for solutions of the form $A, B \propto \exp(pt + ikz)$ and assume that both $\alpha$ and $\Omega$ are constants and $k > 0$, the dispersion relation is given by

$$(p + \eta_0 k^2)^2 = ik\alpha\Omega,$$

namely

$$p = -\eta_0 k^2 + \frac{1 \pm i}{\sqrt{2}} \sqrt{k\alpha\Omega},$$

where the positive sign refers to positive values of $\alpha\Omega$. Introducing the dynamo number $N_D = |\alpha\Omega|/(2k^3\eta_0^2)$, we have

$$p = -\eta_0 k^2 (1 - \sqrt{N_D}) \pm i\eta_0 k^2 \sqrt{N_D},$$

which implies that growing solutions require $N_D > 1$. Taking into account the thickness of the galactic disc, $h$, which implies that $k \geq k_0 = \pi/2h$, we realize that the critical value for growth is:

$$\frac{4|\alpha\Omega|h^3}{\pi^3\eta_0^2} \geq N_D > 1.$$

Let's now evaluate $N_D$ for our Galaxy. The observational data are the following: $h = 300\ pc \simeq 9 \times 10^{20}$ cm, $\Omega = 9 \times 10^{-16}\,\mathrm{s}^{-1}$, the average turbulent velocity in the interstellar medium, $\bar{u} = \sqrt{\langle u^2 \rangle} \simeq 10^6\,\mathrm{cm\,s}^{-1}$. The typical collision time, $\tau_{corr}$, is of the order of $10^7\ y \simeq 3.16 \times 10^{14}$ s and the correlation length, $\lambda_{corr} = \bar{u}\,\tau_{corr} \simeq 3.16 \times 10^{20}$ cm. Using these data to estimate the values of $\alpha$ and $\tilde{\eta}$ we obtain (see Sect. 11.4):

$$\tilde{\eta} \simeq \frac{1}{3}\bar{u}^2\tau_{corr} \simeq 1.0 \times 10^{26}$$

$$\alpha \simeq -\frac{1}{3}\langle \mathfrak{h} \rangle \tau_{corr} \simeq \frac{1}{3}\left(\frac{\bar{u}^2}{\lambda_{corr}}\right)\tau_{corr} \simeq \frac{1}{3}\bar{u} \simeq 3.3 \times 10^5$$

Since in the interstellar gas the dominant diffusivity is the turbulent one, we put $\eta_0 \simeq \tilde{\eta}$ and we obtain:

$$\frac{4|\alpha\Omega|h^3}{\pi^3\tilde{\eta}^2} \simeq 2.6,$$

slightly above the critical value of 1. With these values the growth rate turns out to be:

$$p \simeq -\tilde{\eta}\left(\frac{\pi}{2h}\right)^2 (1 - \sqrt{N_D}) \simeq 1.9 \times 10^{-16}\,\mathrm{s}^{-1},$$

and the amplification factor is $e^{pt} \simeq e^{6t}$, with $t$ expressed in units of a billion years. During a period of 3 billion years, thus shorter than the lifetime of the Galaxy, that factor would be $e^{18} \simeq 6.6 \times 10^7$, which means that an initial field of the order of $2 \times 10^{-14}\,G$ would be amplified to the actual value of $\approx 10^{-6}\,G$.

## 11.5 The Creation of Magnetic Fields

Let's go back to the remark made in the first Section of this Chapter: we cannot be sure that primordial seed fields existed immediately after the Big Bang. If such a primordial field was not there, how is it possible to generate a field different from zero in subsequent epochs?

At first sight this seems to be an impossible task in view of the evolution equation for $\boldsymbol{B}$, Eq. (11.1):

$$\frac{\partial \boldsymbol{B}}{\partial t} = \nabla \times (\boldsymbol{U} \times \boldsymbol{B}) + \eta \nabla^2 \boldsymbol{B}.$$

If the diffusivity, $\eta$, is identically zero (ideal plasma), Alfvén's theorem implies the strict conservation of flux and a field that is identically zero at the beginning will remain so for all times. But, even if $\eta \neq 0$, $\boldsymbol{B} = 0$ is a solution to the evolution equation at all times compatible with the initial null condition and, since the solution is unique, no magnetic field can be generated, independently of the form of $\boldsymbol{U}$.

However, this particular form of the evolution equation is not the most general one. To obtain it, we used an approximate form of the generalized Ohm equation, Eq. (5.5),

$$E_i + \frac{1}{c}(\boldsymbol{U} \times \boldsymbol{B})_i - \frac{J_i}{\sigma} = \frac{m_e}{e^2 n_e}\left[\frac{\partial J_i}{\partial t} + \frac{\partial}{\partial x_k}\Big(J_i U_k + J_k U_i\Big)\right] + \frac{1}{e n_e c}\Big(\boldsymbol{J} \times \boldsymbol{B}\Big)_i - \frac{1}{e n_e}\frac{\partial \mathsf{P}^{(e)}_{ik}}{\partial x_k},$$

dropping all the rhs terms. The conditions under which this is allowed are specified in Eqs. (5.6)–(5.9). When $\boldsymbol{B} = 0$, $\omega_{cp} = 0$ and the condition expressed by (5.9) is certainly violated, so that the corresponding term in the generalized Ohm equation must be included. Thus, the approximate form of the Ohm equation to be used is

$$\boldsymbol{E} + \frac{1}{c}\boldsymbol{U} \times \boldsymbol{B} = -\frac{1}{e n_e}\nabla P_e,$$

where the usual resistive term has been dropped.

Taking the curl of the preceding equation and eliminating the electric field by using

$$\nabla \times \boldsymbol{E} = -\frac{1}{c}\frac{\partial \boldsymbol{B}}{\partial t},$$

we get:

$$\frac{\partial \boldsymbol{B}}{\partial t} = \nabla \times (\boldsymbol{U} \times \boldsymbol{B}) + \frac{c}{e}\nabla \times \left(\frac{\nabla P_e}{n_e}\right),$$

namely

$$\frac{\partial \boldsymbol{B}}{\partial t} = \nabla \times (\boldsymbol{U} \times \boldsymbol{B}) - \frac{c}{e}\left(\frac{\nabla n_e \times \nabla P_e}{n_e^2}\right). \tag{11.28}$$

The last term in this equation is known as the Biermann battery term. Since it does not depend on $B$, it can act as a source term, even when $\boldsymbol{B} = 0$ at $t = 0$. In spite of its promising look, however, it does not solve the problem, unless some additional conditions are imposed. In fact, $P_e$ is usually considered to be a unique function of $n_e$ (barotropic assumption), so that $\nabla P_e$ is parallel to $\nabla n_e$ and the Biermann term vanishes. To avoid this, some mechanism has to be found that breaks the barotropic assumption. This could be provided by the propagation of shocks of limited spatial extent, not an unlikely event in a turbulent medium like the one present in the primordial Universe or in a protogalaxy. The basic idea is the following. If a shock propagates in an essentially pressureless medium, behind the shock the density suddenly increases to relax to the the pre-shock value after a certain time. The pressure follows the same path, but when $n_e$ has returned to the initial value, $P_e \neq 0$ since the gas has been heated by the shock. However, in the regions unaffected by the shock the pressure is still zero, so that to the same value of density correspond different values of the pressure. $P_e$ therefore is not a single-valued function of $n_e$ and so the Biermann term does not vanish.

The growth of seed magnetic fields from an initial field-free situation based on Eq. (11.28) has been the object of numerical investigations (Kulsrud et al. 1997) [28] that prove that the Biermann battery could indeed create seed fields of the order of $10^{-21}$ $G$ prior to the formation of galaxies, when it can be assumed that the plasma is completely ionized.

Once the seed field has been created, it can be amplified by the Biermann battery, even if the plasma is only partially ionized. To see this, let's write $n_e = n_i = \chi(n_i + n_n)$, where $n_n$ is the density of neutrals. If the temperature has the same value for the three plasma components then

$$\frac{P_e}{P} = \frac{n_e}{n_e + n_i + n_n} = \frac{n_e}{(1+\chi)(n_i + n_n)} = \frac{n_e m_p}{(1+\chi)\rho}.$$

We now express $P_e$ in terms of $P$ by using the above equation, multiply Eq. (11.28) by $e/m_p c$, introduce the vector $\boldsymbol{\Omega}_c = e\boldsymbol{B}/m_p c$ and finally obtain

$$\frac{\partial \boldsymbol{\Omega}_c}{\partial t} = \nabla \times (\boldsymbol{U} \times \boldsymbol{\Omega}_c) - \frac{\nabla\rho \times \nabla P}{(1+\chi)\rho^2}, \tag{11.29}$$

where it has been assumed that $\chi$ is constant. If we compare this equation with the well-known equation for the fluid vorticity $\omega = \nabla \times \boldsymbol{U}$:

$$\frac{\partial \omega}{\partial t} = \nabla \times (\boldsymbol{U} \times \omega) + \frac{\nabla\rho \times \nabla P}{\rho^2},$$

we see the two equations are identical, apart from the factor $-1/(1+\chi)$. This shows that, for identical initial conditions,

$$\boldsymbol{\Omega}_c = -\frac{\omega}{1+\chi},$$

everywhere and for all times, a relationship confirmed by numerical simulations that also prove that, starting from an initial field $\approx 10^{-21}\,G$ the Biermann battery is capable of amplifying the field up to $\approx 10^{-18}\,G$. Once this stage is reached, other mechanisms, like those discussed earlier, can boost the field to the present value of $\approx 10^{-6}\,G$.

# References

The following short list includes some textbooks that may be useful to integrate our treatment. We have also indicated a selected list of original papers dealing with astrophysical application discussed in the text.


1. T. J. M. Boyd and Sanderson, J.J., 2003, *The Physics of Plasmas*, Cambridge University Press, Cambridge, UK.
2. Goedbloed, H., Poedts, S., 2004, *Principles of Magnetohydrodynamics*, Cambridge University Press, Cambridge, UK.
3. Gurnett, D. A., Battacharjee, A., 2005, *Introduction to Plasma Physics*, Cambridge University Press, Cambridge, UK.
4. Spitzer, L., 1962, *Physics of Fully Ionized Gases*, Interscience, New York, USA.
5. Kulsrud, R. M., 2004, *Plasma Physics for Astrophysics*, Princeton University Press, Princeton, USA
6. Celnikier, L. M., 1989, *Basics of Cosmic Structures*, Ed. Frontières, Gif-sur-Yvette, France.
7. Landau, L. D., Lifshitz, E. M., 1987, *A Course in Theoretical Physics, Vol. 6: Fluid Mechanics*, Pergamon Press, Oxford, UK.
8. Lifshitz, E. M., Pitaevskii, L.P., 2002, *A Course in Theoretical Physics, Vol.10: Physical Kinetics*, Pergamon Press, Oxford, UK.
9. Frisch, U., 1995, *Turbulence: The Legacy of A. N. Kolmogorov*, Cambridge University Press, Cambridge, UK.
10. Zeldovich, A.B., Ruzmaikin, A.A. and Sokoloff, D.D., 1983, *Magnetic Fields in Astrophysics*, Gordon and Breach, New York, USA
11. Parker, E.N., 1979, *Cosmical Magnetic Fields*, Glarendon Press, Oxford, UK
12. Moffat, H.K., 1978, *Magnetic field Generation in Electrically Conducting Fluids*, Cambridge University Press, Cambridge, UK.
13. Krause, F. and Rädler, K.-H., 1980, *Mean-field Electrodynamics and Dynamo Theory*, Pergamon Press, New York.
14. Jeans, J. H., 1915, On the theory of star-streaming and the structure of the universe, *Monthly Notices of the Royal Astronomical Society***76**, 70–84.
15. Chandrasekhar, S., 1942, Principles of Stellar Dynamics, University of Chicago Press, Chicago.




C. Chiuderi and M. Velli, *Basics of Plasma Astrophysics*,
UNITEXT for Physics, DOI 10.1007/978-88-470-5280-2

## Original Papers


16. Chew, G.F., Goldberger, M. L., Low, F. E., 1956, *The Boltzmann Equation and the One-Fluid Hydromagnetic Equations in the Absence of Particle Collsions*, Proceedings Royal Society London, **A236**, 112.
17. Parker, E. N., 1966, *The Dynamical State of the Interstellar Gas and Field*, Astrophysical Journal, **145**, 811, IoP Publishing Ltd., Philadelphia, USA.
18. Chiuderi, C., Giachetti, R., Van Hoven G., 1977, *The Structure of Coronal Magnetic Loops. I - Equilibrium Theory*, Solar Physics, **54**, 107, D. Reidel Publishing Co., Dordrecht, Holland.
19. Giachetti, R., Van Hoven G, Chiuderi, C., 1977, *The Structure of Coronal Magnetic Loops. II - MHF Stability Theory*, Solar Physics, **55**, 371, D. Reidel Publishing Co., Dordrecht, Holland.
20. Burgess, D., Scholer, M., 2013, *Microphysics of Quasi-parallel Shocks in Collisionless Plasmas*, Space Sci Rev **178**, 513–533, Springer Science, Dordrecht, Holland.
21. Krasnoselskikh, V., Balikhin, M. et al., 2013, *The Dynamic Quasiperpendicular Shock: Cluster Discoveries*, Space Sci Rev **178**, 535–598, Springer Science, Dordrecht, Holland.
22. Furth, H. P., Killeen, J., Rosenbluth, M. N., 1963, *Finite-Resistivity Instabilities of a Sheet Pinch*, Physics of Fluids, **6**, 459, American Institute of Physics, Melville, USA.
23. Balbus, S. A., Hawley, J. F., 1991, *A Powerful Local Shear Instability in Weakly Magnetized Disks*, Astrophysical Journal, **376**, 214, IoP Publishing Ltd., Philadelphia, USA.
24. Loureiro, N.F., Schekochihin, A.A., Cowley, S.C., 2007, *Instability of Current Sheets and Formation of Plasmoid Chains*, Physics of Plasmas 14, 100703, American Institute of Physics, USA.
25. Pucci, F. and Velli, M., 2013, *Reconnection of quasi-singular current sheets: the "ideal" tearing mode*, Astrophysical Journal, **780**, L19, IoP Publishing Ltd., Philadelphia, USA.
26. Parker, E.N., 1972, *Topological Dissipation and the Small-Scale Fields in Turbulent Gases*, Astrophysical Journal, **174**, 499, IoP Publishing Ltd., Philadelphia, USA.
27. Cowling, T.G., 1934, *The stability of Gaseous stars*, Montly Notices of Royal Astronomical Society, **94**, 39, Wiley & Sons Inc.
28. Bullard, E.C., 1949, *The Magnetic Field within the Earth*, Proceedings Royal Society London, **A199**, 433.
29. Parker, E.N., 1955, *Hydromagnetic Dynamo Models,* Astrophysical Journal, **121**, 491, IoP Publishing Ltd., USA.
30. Kulsrud, R.M and Anderson, S.W., 1997, *The spectrum of random magnetic fields in the mean field dynamo. Theory of the Galactic magnetic field*, Astrophysical Journal, **396**, 606, IoP Publishing Ltd., USA.
31. Vlaosv, A.A., 1938, *On Vibration Properties of Electron Gas*, J Exp Theor Phys **8**(3), 291. (in Russian).

# Index




C. Chiuderi and M. Velli, *Basics of Plasma Astrophysics*,
UNITEXT for Physics, DOI 10.1007/978-88-470-5280-2

Printed in the United States
By Bookmasters